American Women
of Science since 1900

American Women of Science since 1900

VOLUME 2

I–Z

Tiffany K. Wayne

 ABC-CLIO

Santa Barbara, California • Denver, Colorado • Oxford, England

Library of Congress Cataloging-in-Publication Data

Wayne, Tiffany K., 1968-
  American women of science since 1900 / Tiffany K. Wayne.
    p. cm.
  Includes bibliographical references and index.
  ISBN 978–1–59884–158–9 (hard copy : alk. paper) – ISBN 978–1–59884–159–6 (ebook)
1. Women scientists—United States—Biography—Dictionaries. 2. Women scientists—United States—History—20th century—Dictionaries. 3. Women in science—United States—Biography—Dictionaries. 4. Women in science—United States—History—20th century—Dictionaries. I. Title.
Q141.W42   2011
509.2′273—dc22   2010026838

ISBN: 978–1–59884–158–9
EISBN: 978–1–59884–159–6

15 14 13 12 11   1 2 3 4 5

This book is also available on the World Wide Web as an eBook.
Visit www.abc-clio.com for details.

ABC-CLIO, LLC
130 Cremona Drive, P.O. Box 1911
Santa Barbara, California 93116-1911

This book is printed on acid-free paper ∞

Manufactured in the United States of America

Contents

List of Essays and Entries

A–Z Entries, 179

VOLUME I

Volume II

Acknowledgments

I want to thank my editors at ABC-CLIO—Steven Danver for presenting this project to me; James Sherman for keeping me going on it; and Kim Kennedy-White for seeing it through to completion.

My most sincere thanks go to Martha J. Bailey, the author of the two earlier ABC-CLIO volumes on American women scientists, whose research provided the foundation for my work here and who is, technically, the co-author of many of the biographical entries. Her work on women scientists began more than 15 years ago, and I hope she is proud to have her name still attached to the project.

Thanks to Kally Kedinger and Michelle Delgado for tracking down scientists and citations, and handling my last-minute and often confusing research requests with grace and professionalism. Appreciation goes also to all of the scientists who responded to our email queries and provided up-to-date information on their work.

My family has shown unwavering enthusiasm and patience for my often slow-going work. David Wayne has encouraged me in all of my pursuits and has always given me the freedom and the space (mental and physical) to do my work. So many of the women scientists profiled here trace their own passions and interests back to childhood, and I hope that Miles and Lillian—with their love of nature and animals, and their innate curiosity about the world—will read this book and be inspired to hold onto those passions and to continue to dream big.

Introduction

Nineteenth-century astronomer Maria Mitchell noted in an 1875 address to the Association for the Advancement of Women (of which she was the first president) both "how much women need exact science" and "how much science needs women" (Wyer 2001, 3). Her words could not be truer today. Science and technology are more important than ever to our society as we become a postindustrial high-tech "knowledge" society. It is important that science takes women into account, but women want and *need* to participate in the creation of that knowledge as well.

Women have reached the heights of achievement in the sciences and hold some of the most visible positions. Several women scientists now serve as presidents of major research universities, and in 2009, President Barack Obama appointed women scientists as directors of government research agencies such as the U.S. Geological Survey (USGS) and the National Oceanic and Atmospheric Administration (NOAA). Women have soared, so to speak, to the stars with the National Aeronautics and Space Administration (NASA) as astronauts, shuttle pilots, and members of the International Space Station. Women scientists serve on international advisory councils, affecting policy on everything from global warming to public health, and have founded and led biotechnology, pharmaceutical, and computer software companies. Women are professors, deans of medical and engineering schools, directors of research centers, inventors, and Nobel Prize laureates. A girl growing up in the twenty-first century will seem to have no lack of role models, no limits to her own interests and pursuits.

The history of women in science, however, is the history of not only individual achievement, but of social attitudes, institutional barriers, and legislative and policy initiatives. It is important to realize that access, beginning with the early education and recruitment of girls and young women into the sciences, is the first step to women's success in the sciences. Women have fought hard for that access and for opportunities for gainful employment, confronting the attitudes of employers, coworkers, family members, and society at large, many of whom have been resistant to change. From the belief that women are not as interested or as capable in science as men, to society's inability (or refusal) to create family-friendly workplaces, women have been engaged in a century-long struggle for access to education and careers in the sciences, technology, engineering, and mathematics (collectively known as "STEM").

While individual women have achieved the pinnacles of success, women as a group are still underrepresented in professional recognitions, such as with membership in the National Academy of Science and the National Academy of Engineering. In 2009, for example, only 11 women out of 72 new members were elected to the National Academy of Sciences. Likewise, although several high-profile women have won Nobel Prizes in the sciences—and 2009 was a particularly good year for women, with three American women scientists (and one Israeli woman scientist) named as winners—in the history of the Nobel Prize, only 35 women have won in any category out of a total of 789 prizes awarded; of these 35 women, only 15 have been in the sciences.

The past several decades have seen steady increases in the numbers of women earning science degrees and entering science and engineering professions, but there is still a small percentage of women at the highest levels. In 2006, women earned 38.4% of STEM doctorates. The numbers and percentages vary greatly by field, with women earning nearly 50% of Ph.D.s in the biological sciences, 56% in anthropology, and greater than 70% in psychology, the most popular field for women in the sciences. But women earn significantly lower percentages of doctorates in other disciplines, including only 20% of doctorates in combined engineering fields and only 16.6% of Ph.D.s in physics (NSF Table F-2).[1] There are also fewer women the higher up the academic career ladder one goes. In 2006, women made up 31% of all science and engineering faculty, but only 25% of tenured faculty and only 19% of full professors in science and engineering fields (NSF Table H-25).

In her 1988 AAAS presidential address (speaking more than 100 years after Maria Mitchell), **Sheila Widnall** outlined the problem of this "leaky pipeline" that has come to define women's representation in the sciences. The problem begins as early as high school, when boys and girls still have nearly equal interests and grades in subjects such as math. By the end of high school, boys will slightly

---

[1]Note that these numbers include only Ph.D.s, or research-based fields, and do not include professional degrees of M.D., D.D.S., D.Pharm., or Psy.D.; nor do they include second doctorates, so that persons changing or combining fields may only be counted in the first field. It is worth noting that inclusion of these other degrees and occupations might significantly alter the overall numbers of women in science-related fields. See notes on "Survey of Earned Doctorates" at http://www .nsf.gov/statistics/srvydoctorates/.

outnumber girls in completion of higher-level mathematics courses, but the first major split occurs when choosing a college major, with nearly three times as many boys selecting science and engineering paths. A high percentage of women who do major in STEM subjects will complete their degrees and go on for a master's degree, but another drop or split occurs between men and women who continue on for the doctorate. Widnall created a hypothetical scenario (based on the current statistics) in which, out of 2,000 ninth graders (1,000 boys and 1,000 girls) taking comparable high-level mathematics courses, 140 men and 44 women will go on to major in STEM in college; of these, 46 men and 20 women will receive bachelor's degrees, but only 5 men and 1 woman will receive the science or engineering Ph.D. (Widnall 1988). The particular struggles facing women at each of these various stages along the educational and career path are addressed separately in the "Issues" section that follows.

The gender wage gap, glass ceiling, and work/life balance are not just issues for women in the sciences, but affect women across the professions. Women's access, opportunity, and success in the professions is intertwined with other issues in twentieth-century American history, such as social and economic changes, government needs and policy, and the rise of feminism. The larger question of "women in science" is actually twofold; it is the question of women's presence and representation in scientific disciplines and employment, as well as the question of what effect women's presence has on science itself. There are, therefore, both quantitative and qualitative questions to consider when talking about women in the sciences. The statistics show that the numbers of women in STEM disciplines and careers has steadily increased over the course of the twentieth century, but we must also consider how women (and feminism) have changed science itself in terms of the questions asked, the methodologies used, and the new knowledge discovered.

The present volume addresses both aspects of this history and the status of American women in science since 1900 by looking across the century at the work done by more than 500 individual women, and their innovations and contributions, as well as the challenges they faced in pursuing that work. The book includes the following sections: "Issues" (ten essays on specific topics related to American women in science, such as education, employment sectors, minority women, etc.), "Disciplines" (entries on the presence and impact of women in 29 different scientific fields, such as biology, chemistry, physics, etc.), and the biographical entries from A to Z.

This work is an update, revision, and expansion of Martha Bailey's original two-volume biographical dictionaries, *American Women in Science: Volume I* (1994) and *American Women in Science: 1950 to the Present* (1998). The present volume focuses only on those scientists who lived and had significant career activity after 1900, updating or revising many of Bailey's original entries and adding new entries on significant early women scientists not included in Bailey's original volumes. Carrying the story of American women in science forward, this volume also updates the career information and accomplishments of many scientists still working since Bailey's report of 1998, and adds entries on a new generation of scientists emerging in the late twentieth and early twenty-first centuries.

Reaching back to the nineteenth century, Bailey included many more women who were not necessarily professionally trained as scientists or did not conduct scientific research, but who *supported* scientific work through writing, indexing, cataloging, or popularizing scientific information. In preparing the present volume, however, I eliminated many entries on women who were certainly pioneers in their fields, but who did not hold regular positions as researchers or teachers of science; these were usually in fields opened to and heavily dominated by women in the late nineteenth and early twentieth centuries, such as nutrition, botany, ornithology, and nature writing and illustrating. Women in the earlier part of the century often had more eclectic careers than their later counterparts—they combined research and travel (often self-funded) with writing, illustrating, and teaching, not only in colleges or universities, but in public schools as well. Many (although not all) women of that crucial turn-of-the-century era did gain access to higher education, but still were not always able (or chose not) to secure permanent, formal, or regular employment. In some cases, I eliminated entries for which there simply was not enough specific career information available. And although I cut out much of the personal information Bailey had collected for the original volumes, some of that information on individual women's experiences of combining work and family life, advice to young women scientists, and specific instances of discrimination or other bias over the course of their careers has made its way into my summaries in the "Issues" and "Disciplines" sections of this book. I refer readers to Bailey's *Volume I* for a more thorough overview of women's roles in the scientific disciplines in the nineteenth and early twentieth centuries and, although our projects and purposes were originally quite different, I am indebted to her recovery of and preliminary research on many of those early scientists.

The focus in this volume is primarily on women who made significant impacts in their fields and who received professional recognition for their work, whether through career positions and advancement, membership in professional societies, or scientific awards and honors. However, my criteria for inclusion itself was an inexact science, and I also maintained a sampling of early women who worked in less represented fields, even if they did not have significant research contributions (for example, including a few representative women as early astronomers, entomologists, botanists, mathematicians, paleontologists, and geologists). I also sought to emphasize women who accomplished "firsts" in their careers, to emphasize to readers the relatively recent history in which women scientists have begun to break down the barriers in specific disciplines. I included many women who were the first presidents of professional scientific societies, the first to receive doctorates in specific disciplines, the first faculty members in specific institutions, or women scientists who worked at high levels of government or academia, on presidential councils, cabinets, or as university presidents. In deciding which disciplines to include, I looked to the sections of the National Academy of Sciences; the emphasis, therefore, is on the physical and natural sciences, although some social scientists are included. The resulting list is certainly not inclusive, and undoubtedly there will be names or accomplishments or disciplines I have missed. As with any reference work, the hope is that readers and students will be inspired to further research these and other women in the history of American science.

Although the women scientists profiled here lived and worked within the specific social and political contexts of twentieth-century America, it is worth noting that, in terms of research commitments, career paths and affiliations, and scientific advances, a somewhat false line is drawn between the work of U.S. and non-U.S. scientists. Indeed, many women profiled here participated in projects and professional networks that were international in scope. Non-U.S. women came to the United States for education or jobs, and American women pursued fellowships or visiting appointments abroad. Some of the greatest achievements by individual women scientists of the twentieth century belong to European researchers, such as physicist Marie Curie of France and her daughter, Iréne Joliet-Curie, both of whom won Nobel Prizes; English crystallographer Rosalind Franklin; German physicist Lise Meitner; or British primatologist Jane Goodall. While these figures are not included in the present volumes, other foreign-born women who spent the majority of their careers or achieved their highest successes employed in American institutions are included. Of course, women scientists around the world continue to work together through collaboration and through professional organizations that recognize the broader challenges to women's education and advancement in the sciences, regardless of national origin. Many other fine volumes exist that take a broader view of women's scientific contributions and work, either across regional boundaries or with a longer chronological view.

References and Further Reading

Abir-Am, Pnina G. and Dorinda Outram. *Uneasy Careers and Intimate Lives: Women in Science, 1789–1979*. New Brunswick, NJ: Rutgers University Press, 1987.

Ambrose, Susan A. 1997. *Journeys of Women in Science and Engineering: No Universal Constants*. Philadelphia, PA: Temple University Press.

Bart, Jody, ed. 2000. *Women Succeeding in the Sciences: Theories and Practices across Disciplines*. West Lafayette, IN: Purdue University Press.

Hanson, Sandra L. 1996. *Lost Talent: Women in the Sciences*. Philadelphia, PA: Temple University Press.

Herzenberg, Caroline L. 1986. *Women Scientists from Antiquity to the Present*. West Cornwall, CT: Locust Hill Press.

Kass-Simon, G. and Patricia Farnes, eds. 1990. *Women of Science: Righting the Record*. Bloomington: Indiana University Press, 1990.

Kirkup, Gill and Laurie Smith Keller, eds. 1992. *Inventing Women: Science, Technology, and Gender*. Cambridge, MA: B. Blackwell.

Morse, Mary. 1995. *Women Changing Science: Voices from a Field in Transition*. New York: Insight Books.

National Science Foundation. "Table F-2. S&E doctoral degrees awarded to women, by field: 1999–2006." *Women, Minorities, and Persons with Disabilities in Science and Engineering*. National Science Foundation, Division of Science Resources Statistics, Survey of Earned Doctorates, 1999–2006. http://www.nsf.gov/statistics/wmpd/pdf/tabf-2.pdf.

National Science Foundation. 2006. "Table H-25. S&E doctorate holders employed in universities and 4-year colleges, by broad occupation, sex, race/ethnicity, and faculty rank: 2006." *Women, Minorities, and Persons with Disabilities in Science and Engineering*. National Science Foundation, Division of Science Resources Statistics, Survey of Doctorate Recipients, 2006. http://www.nsf.gov/statistics/wmpd/pdf/tabh-25.pdf.

Rosser, Sue V. 1997. *Re-engineering Female Friendly Science*. New York: Teachers College Press.

Rosser, Sue V. 2004. *The Science Glass Ceiling: Academic Women Scientists and the Struggle to Succeed*. New York: Routledge.

Rosser, Sue v. 2008. *Women, Science, and Myth: Gender Beliefs from Antiquity to the Present*. Santa Barbara, CA: ABC-CLIO.

Rossiter, Margaret. 1982. *Women Scientists in America: Struggles and Strategies to 1940*. Baltimore, MD: Johns Hopkins University Press.

Bossiter Margaret. 1995. *Women Scientists in America: Before Affirmative Action, 1940–1972*. Baltimore, MD: Johns Hopkins University Press.

Ruddick, Sara and Pamela Daniels, eds. 1977. *Working It Out: 23 Women Writers, Artists, Scientists, and Scholars Talk about Their Lives and Work*. New York: Pantheon Books.

Schiebinger, Londa L. 1989. *The Mind Has No Sex?: Women in the Origins of Modern Science*. Cambridge, MA: Harvard University Press.

Stanley, Autumn. 1995. *Mothers and Daughters of Invention: Notes for a Revised History of Technology*. Metuchen, NJ: Scarecrow Press.

Tang, Joyce. 2006. *Scientific Pioneers: Women Succeeding in Science*. Lanham, MD: University Press of America.

Vare, Ethlie Ann and Greg Ptacek. 1988. *Mothers of Invention: From the Bra to the Bomb; Forgotten Women & Their Unforgettable Ideas*. New York: Morrow.

Warren, Wini. 1999. *Black Women Scientists in the United States*. Bloomington: Indiana University Press.

Wasserman, Elga. 2002. *The Door in the Dream: Conversations with Eminent Women in Science*. Washington D.C.: Joseph Henry Press, 2002.

Whaley, Leigh Ann. 2003. *Women's History as Scientists: A Guide to the Debates*. Santa Barbara, CA: ABC-CLIO.

Widnall, Sheila E. 1988. "AAAS Presidential Lecture: Voices from the Pipeline." *Science* 241: 1740–1745. (September 30, 1988).

Wyer, Mary et al. 2009. *Women, Science, and Technology: A Reader in Feminist Science Studies*. 2nd edition. New York: Routledge.

I

Intriligator, Devrie (Shapiro)

b. 1941
Astrophysicist

Education: B.S., physics, Massachusetts Institute of Technology, 1962, M.S., 1964; Ph.D., planetary and space physics, University of California, Los Angeles, 1967

Professional Experience: assistant research geophysicist, Institute of Geophysics and Planetary Physics, University of California, Los Angeles (UCLA), 1967; research associate, Space Science Division, Ames Research Center, National Aeronautics and Space Administration (NASA), 1967–1969; research fellow, physics, California Institute of Technology, 1969–1972, assistant professor, 1972–1980, member, Space Science Center, 1978–1983; staff member, Stauffer Hall of Science, University of Southern California, 1974–1977, assistant professor, physics, 1977–1979; senior research physicist, Carmel Research Center, 1979–, director, Space Plasma Laboratory, 1980–

Devrie Intriligator is renowned for her research in space physics and astrophysics, and for her expertise in designing measurement instruments for interplanetary spacecraft. Among the projects in which she has participated are the Pioneer 10 and 11 missions to the outer planets, the Pioneer-Venus Orbiter, and the Pioneer 6, 7, 8, and 9 heliocentric missions. Her research includes high-energy nuclear physics, plasma physics, and astrophysics. She began doing physics experiments as a high school sophomore and won a national prize in a Future Scientist of America contest in her senior year. She received financial aid to enroll in college, but the dean of women at the first school she attended would not permit her to enroll in physics, and she had to give up the financial aid when she transferred to the Massachusetts Institute of Technology (MIT) the following year. She was unable to secure any funding at MIT due to prejudices against women due to the belief that women would not put their education to use and find work as scientists. Instead, Intriligator held a number of jobs in college to support herself. She was a research assistant in the cosmic ray group at MIT in 1960, and prior to her senior year, was a consulting physicist for the Institute of Physics, University of Milan, where she consulted on cosmic-ray balloon experiments. She continued as a graduate student at MIT and worked as a physicist in the cosmic-ray branch of the Air Force's Cambridge Research Laboratory from 1962 to 1963. When her husband received an appointment to teach at UCLA, she transferred to that school to complete her doctorate.

Since UCLA would not accept her credits from MIT, she had to repeat a number of courses, but in the course of the three years she spent studying at UCLA, she became interested in solar wind plasma physics and decided to add it as a specialty. The solar wind plasma is a stream of particles—electrons, protons, and other ions—that continually flow from the sun and that is responsible for many features of the solar system and the Earth's environment. After graduation, she won a prestigious National Academy of Sciences Resident Research Associateship for use at NASA's Ames Research Center, where she was the principal investigator of the positive-ion probe on the UCLA Small Scientific Satellite. She also was a co-investigator of the Ames solar wind plasma probes on several Pioneer spacecraft in orbit around the sun.

At the California Institute of Technology, where she began working in 1969, she analyzed data sent back from instruments aboard the Pioneer spacecraft in orbit around the sun. She was co-investigator of the Ames solar wind plasma probe for the Pioneer 10 and 11 missions to Jupiter, and she was also a member of the plasma measurement team for the outer planet missions to Jupiter, Saturn, Uranus, Neptune, and Pluto. In her current position as director of the Space Plasma Laboratory, she is continuing her research on cosmic rays and solar winds.

Intriligator is co-editor of the book *Exploration of the Outer Solar System* (1976) and has written numerous scientific papers. She has received three achievement awards from NASA and is a member of the American Geophysical Union, American Physical Society, and American Association for the Advancement of Science.

Irwin, Mary Jane

b. 1949
Computer Scientist

Education: B.S., mathematics, Memphis State University, 1971; M.S., computer science, University of Illinois, Urbana-Champaign, 1975, Ph.D., computer science, 1977

Professional Experience: associate to assistant professor, computer science, Pennsylvania State University, 1977–1989, professor, computer science and engineering, 1989–1999, Distinguished Professor, 1999–2003, A. Robert Noll Chair in Engineering, 2003–, Evan Pugh Professor, computer science and engineering, 2006–

Concurrent Positions: research staff, Supercomputer Research Center, Institute for Defense Analysis, Maryland, 1986

Mary Jane Irwin is a computer sciences engineer whose research focuses on computer architecture, computer arithmetic, embedded and mobile computing systems design, energy and reliability aware systems design, and emerging technologies in computing systems. She received a Ph.D. in computer science from the University of Illinois and has been a faculty member in computer sciences and engineering at Pennsylvania State University since 1977. She is co-director of the Microsystems Design Lab at Pennsylvania State University, a project funded collaboratively by both government and corporate research interests, including the National Science Foundation, Gigascale Systems Research Center, Semiconductor Research Corporation, Pennsylvania Technology Collaborative, Intel, Microsoft, Honda, and Toyota. She has been an invited lecturer and speaker at conferences and universities worldwide and has served on numerous government, corporate, and academic research councils and advisory committees, including Microsoft's External Research Advisory Board. She was a founding editor of the Association for Computing Machinery's *Journal on Emerging Technologies in Computing Systems*.

Irwin was elected to the National Academy of Engineering in 2003. She is a fellow of the American Academy of Arts and Sciences, Association for Computing Machinery (ACM), and Institute of Electrical and Electronics Engineers (IEEE), and a member of the International Federation for Information Processing. Her awards and honors include an honorary doctorate from Chalmers University, Sweden (1997), Pennsylvania State University Engineering Society's Premier Research Award (2001), IEEE/CAS Best Paper Award (2003), DAC Marie R. Pistilli Women in EDA Award (2004), ACM/SIGDA Distinguished Service Award (2005 and 2007), ACM Distinguished Service Award (2005), Computing Research Association's (CRA) Distinguished Service Award (2006), IEEE/ICPADS Best Paper Award (2006), and Anita Borg Technical Leadership Award (2007).

Further Resources

Pennsylvania State University. Faculty website. http://www.cse.psu.edu/research/mdl/mji/.

J

Jackson, Jacquelyne Mary (Johnson)

b. 1932
Sociologist

Education: B.S., University of Wisconsin, Madison, 1953, M.S., 1955; Ph.D., sociology, Ohio State University, 1960

Professional Experience: assistant to associate professor, Southern University, 1959–1962; professor and department chair, sociology, Jackson State College, 1962–1964; assistant professor, Howard University, 1964–1966; instructor, medical sociology, Duke University Medical Center, 1967–1968, assistant to associate professor, 1968–1998, emerita

Concurrent Positions: visiting professor, sociology, St. Augustine's College, 1969–

Jacquelyne Jackson is known for her research on minority aging and for her participation in the civil rights movement. She also had a number of "firsts" in her career. She was the first black woman to receive a doctorate in sociology from Ohio State University, the first full-time black faculty member to be hired at the Duke University Medical Center, and the first black tenured faculty member at the medical school. After receiving her doctorate, she did postdoctoral study at the University of Colorado before becoming a faculty member at Southern University, a professor of sociology at Jackson State College, an assistant professor at Howard University, and then joining the faculty of the Duke University Medical Center as an assistant professor of medical sociology. Her work has always been connected to real people and real issues. Her interest in minority aging grew out of the experience of elderly friends who had to sell their houses to pay for medical care. Later, one friend was in a racially segregated ward in New Orleans's Charity Hospital, and Jackson organized her students to donate "black" blood for the woman because blood was segregated at the time. In 1974, she and colleague Frank Cantor made a short documentary film called *Old, Black and Alive*, which investigated the living conditions and needs of elderly African Americans in one Alabama county. She helped found the *Journal of Minority Aging*, and in 1980, she published *Minorities in Aging*, which has become a classic in the field.

Jackson became involved in the civil rights movement while teaching at Jackson State College. When a group of civil rights advocates was forbidden to hold a meeting at Jackson State for fear of creating racial unrest, she secured the support of Charles Evers, brother of Medgar Evers, to schedule the meeting at another site in the city. She took part in the 1963 march in Washington, D.C., and in 1962, she published *These Rights They Seek*, a study of the Tuskegee Civic Association, the Montgomery Improvement Association, and the Alabama Christian Movement for Human Rights.

Jackson was elected a fellow of the National Science Foundation in 1961. In addition to her teaching and research, she has also served as a consultant to the National Center for Health Statistics and to the U.S. Senate's Special Committee on Aging. She was a member of numerous professional and civic organizations, including but not limited to the Association of Social and Behavioral Scientists, National Council on Family Relations, American Sociological Association, Caucus of Black Sociologists, National Caucus on the Black Aged, Gerontological Society, and Carver Research Foundation of Tuskegee Institute.

Jackson, Shirley Ann

b. 1946
Physicist

Education: B.S., Massachusetts Institute of Technology, 1968, Ph.D., physics, 1973

Professional Experience: research associate, theoretical physics, Fermi National Accelerator Laboratory, 1973–1974; visiting science associate, European Organization for Nuclear Research (CERN), Geneva, 1974–1975; research associate, Stanford Linear Accelerator Center, 1975–1976; technical staff, theoretical physics, AT&T Bell Laboratories, 1976–1991; professor, physics, Rutgers University, 1991–1995; chair, Nuclear Regulatory Commission (NRC), 1995–1999; president, Rensselaer Polytechnic Institute, 1999–

Concurrent Positions: chair, International Nuclear Regulators Association (INRA), 1997–1999.

Shirley Ann Jackson is a theoretical physicist whose research has focused on particle physics and condensed matter physics. Theoretical physics uses theories and mathematics to predict the existence of subatomic particles and the forces that bind them together. One method for this research uses a particle accelerator, a

Physicist Shirley Ann Jackson, President of Rensselaer Polytechnic Institute, with Senator Hillary Rodham Clinton of New York, 2005. (AP/Wide World Photos)

device in which nuclei are accelerated to high speeds and then forced to collide with a target to separate them into subatomic particles. Another method detects their movements using certain types of nonconducting solids. Jackson has conducted research using both methods at a number of prestigious physics laboratories in both the United States and Europe, such as the Fermi National Accelerator Laboratory in Illinois, the European Organization for Nuclear Research in Switzerland, and the Stanford Linear Accelerator Center in California.

Jackson was the first African American woman to receive a doctorate in any field from the Massachusetts Institute of Technology, and she was the first woman and the first African American to serve as chair of the Nuclear Regulatory Commission (NRC), the federal agency that regulates the uses of nuclear materials and technology throughout the United States to ensure the protection of public health, safety, and the environment. At the NRC, she oversaw the process for renewing the licenses of existing nuclear power plants, ensuring public safety as electric utilities were deregulated, and ensuring safety in the disposal of spent reactor fuel. In 1997, the International Nuclear Regulators Association was formed with Jackson elected as its first chair. She became president of Rensselaer in 1999, the first black woman to lead a major technology institute. Jackson brings to this position her commitment to the presence of more women and minorities in science and technology careers.

In 2001, Jackson was the first African American woman to be elected to the National Academy of Engineering. She has received numerous honorary degrees and awards, including the Thomas Alva Edison Science Award (1993), the New Jersey Governor's Award in Science (1993), the Golden Touch Award for Lifetime Achievement from the National Society of Black Engineers (2000), the Black Engineer of the Year Award from *US Black Engineer & Information Technology* magazine (2001), and the Vannevar Bush Award from the National Science Board (2007). She has been inducted into the National Women's Hall of Fame (1998) and the Women in Technology International (WITI) Hall of Fame (2000), and was a fellow of the Association of Women in Science (2004). She is a member of the American Physical Society, New York Academy of Sciences, National Society of Black Physicists (president, 1980), American Academy of Arts and Sciences, and American Association for the Advancement of Science (president, 2004). In 2002, she was named one of "The 50 Most Important Women in Science" by *Discover* magazine. In 2009, Jackson was appointed to President Obama's Council of Advisors on Science and Technology.

Further Resources

Rensselaer Polytechnic Institute. "Shirley Ann Jackson, Ph.D." http://www.rpi.edu/president/profile.html.

Williams, Clarence G. 2003. *Technology and the Dream: Reflections on the Black Experience at MIT, 1941–1999*. Cambridge, MA: MIT Press.

Jameson, Dorothea A.

1920–1998
Psychologist

Education: B.A., Wellesley College, 1942

Professional Experience: research assistant, Harvard University, 1941–1947; research psychologist, color control department, Eastman Kodak Company, 1947–1957; research scientist, psychology, New York University (NYU),

1957–1962; research associate, psychology, University of Pennsylvania, 1962–1968, research professor, 1968–1972, professor, psychology and visual science, 1972–

Concurrent Positions: visiting professor, University of Rochester, 1974–1975; visiting professor, Columbia University, 1974–1976

Dorothea Jameson was an expert in the new field of color vision, and she combined her work in psychology with work in optics, visual mechanisms, and human perception. While still an undergraduate at Wellesley, she worked as a research assistant at Harvard where, during World War II, she worked on improving the accuracy of visual rangefinders. It was at Harvard that she met her future husband, psychologist Leo Hurvich, the beginning of a lifelong professional collaboration. The couple (who married in 1948) worked together as researchers at Eastman Kodak in Rochester, New York, spent five years at NYU, and then moved to the department of psychology and Institute of Neurological Sciences at the University of Pennsylvania. Even without an advanced degree, Jameson was hired based on her experience as a researcher. A bigger problem for her was that, at that time, most universities were opposed to hiring husband-and-wife faculty teams, so Jameson was not appointed a regular faculty position until the rules were loosened in 1968; she was promoted to full professor at the University of Pennsylvania in 1972. Jameson and Hurvich published dozens of scientific papers together and were renowned for their innovations in color vision research. Much of Jameson's early research at NYU and the University of Pennsylvania was supported by grants from the National Institutes of Health (NIH) and the National Science Foundation. She was a fellow of the Center for Advanced Study in Behavioral Sciences in 1981 and 1982, and served on the national advisory eye council for the NIH starting in 1985.

Jameson was elected to membership in the National Academy of Sciences in 1975. Between 1983 and 1986, she was chair of the National Academy of Sciences psychology section. She received honorary degrees from the University of Pennsylvania (1972) and the State University of New York (1989). She was a fellow of the Society of Experimental Psychologists and the American Academy of Arts and Sciences, and a member of the Optical Society of America. Her numerous awards and honors included the Warren Medal of the Society of Experimental Psychologists (1971), the Distinguished Science Contribution Award of the American Psychological Association (1972), the Inter-Society Color Council's Godlove Award for Research in Color Vision (1973), the Wellesley College Alumnae Achievement Award for Scientific Research (1974), the Tillyer Medal of the Optical Society of America (1982), the Judd Award of the Association Internationale de Couleur (1985), and the Helmholtz Award from the Cognitive Neuroscience Association (1987).

Further Resources

University of Pennsylvania. 1998. "Dorothea Jameson, Pioneer in Color Perception." *Almanac.* 44(30). (21 April 1998). http://www.upenn.edu/almanac/v44/n30/deaths.html.

Jan, Lily

Neurobiologist

Education: B.Sc., physics, National Taiwan University, 1968; M.Sc., physics, California Institute of Technology, Ph.D., physics and biophysics

Professional Experience: postdoctoral research fellow, California Institute of Technology (CalTech) and Harvard University; Lange Professor of Physiology and Biophysics, University of California, San Francisco (UCSF); Howard Hughes Medical Investigator, 1984–

Lily Jan is a neurobiologist and biophysicist whose research focuses on the development and function of the nervous system and, in particular, how potassium (regulated through "potassium channels") affects the electrical impulses sent from the brain throughout the body. Born in China and raised in Taiwan, Jan chose physics to study in high school due to the inspiration of recent Nobel Prize winners in China. She graduated from National Taiwan University in 1968 and moved to the United States to attend CalTech for graduate work in theoretical physics. She was inspired again by another Nobel Prize winner, Max Delbrück, one of her professors, who encouraged her interest in biology, and she earned her doctorate in physics and biophysics from CalTech. Another Taiwanese student who had come to study at CalTech was Yuh Nung Jan. The two were lab partners and then postdoctoral research fellows together in neurobiology, married in 1971, and began collaborating on projects involving genetic explanations for certain behaviors in the fruit fly. They were the first to identify the DNA sequence responsible for potassium channels and mutations in the channels, linking it to behavioral changes in the fly. The Jans conducted some postdoctoral work in neurophysiology at Harvard and then returned to the West Coast as faculty members at UCSF. They raised two children together and regularly collaborate in the lab, but have also developed their own individual research interests and groups at UCSF and as Howard Hughes Medical Institute investigators.

Jan was elected a member of the National Academy of Sciences in 1995, and is a fellow of the American Academy of Arts and Sciences and Academia Sinica (Taiwan). Her awards and honors include a Javits Neuroscience Investigator

Award, W. Alden Spencer Award from Columbia University, K. S. Cole Award, Distinguished Alumni Award from California Institute of Technology, and Presidential Award of the Society of Chinese Bioscientists in America. She was named Harvard Foundation's 2005 Scientist of the Year.

Further Resources

University of California, San Francisco. "Jan Laboratory." http://physio.ucsf.edu/jan/index.html.

Howard Hughes Medical Institute. "Lily Y. Jan, Ph.D." http://www.hhmi.org/research/investigators/janly_bio.html.

"Biophysicists in Profile: Lily Jan." Biophysical Society Newsletter. (September/October 2002). http://www.biophysics.org/Portals/1/PDFs/Career%20Center/Profiles/jan.pdf.

Jeanes, Allene Rosalind

1906–1995
Chemist

Education: A.B., Baylor University, 1928; A.M., University of California, Berkeley, 1929; Ph.D., organic chemistry, University of Illinois, 1938

Professional Experience: high school teacher, mathematics and physics, 1930; department head, science, Athens College, Alabama, 1930–1935; instructor, chemistry, University of Illinois, 1936–1937; research fellow, National Institutes of Health, 1938–1940; research chemist, Northern Regional Research Laboratory, U.S. Department of Agriculture (USDA), 1941–1976

Allene Jeanes was an organic chemist whose research group isolated and characterized over 100 different dextrans that have great value in research, especially in immunology and immunochemistry. She received one of the first Corn Industries Research Foundation fellowships at the National Institutes of Health, where she co-developed a new technique of periodate oxidation of starches. She joined the staff at Northern Regional Research Laboratory in Peoria, Illinois, a regional laboratory of the USDA, in 1941, three months after it opened. Initially she studied the nature and structural role of the branch points in starch and developed xanthan gum, a thickening substance used in numerous food and cosmetic products. During the Korean War, there was a need for a blood-plasma substitute, and she and her group were able to find a chemical, dextran, that was used successfully to expand plasma volume. Her technique was used for isolating and characterizing dextrans, and she held several patents for her work.

In 1953, Jeanes was the first woman in the Chemistry Bureau to receive the USDA Distinguished Chemist Award. She was also the recipient of a Garvan Medal of the American Chemical Society (1956) and a Federal Woman's Award of the U.S. Civil Service Commission (1962). In 1999, she was posthumously inducted in the USDA's Agricultural Research Service (ARS) Science Hall of Fame. She was a member of the American Chemical Society.

Jemison, Mae Carol

b. 1956
Physician, Astronaut

Education: B.S., chemical engineering, B.A., African and Afro-American Studies, Stanford University, 1977; M.D., Cornell University Medical School, 1981

Professional Experience: intern, University of Southern California Medical Center, 1981–1982; physician, INA–Ross Loos Medical Group, Los Angeles, 1982; medical officer, Peace Corps, 1983–1985; physician, Cigna Health Plan of California, 1985–1987; astronaut, National Aeronautics and Space Administration (NASA), 1987–1993, mission specialist, *Endeavour*, 1992; founder and director, Jemison Group, 1993–; founder, BioSentient, 1999–

Concurrent Positions: professor-at-large, Cornell University; teaching fellow, environmental studies, Dartmouth College, 1995–2002

Mae Jemison is a physician and astronaut who was the first black woman to travel in space. She began her career as a Peace Corps medical officer in Africa and then as a physician and biomedical researcher investigating hepatitis B vaccine, schistosomiasis, and rabies. She entered astronaut training in 1987 and was assigned to the space shuttle *Endeavour* mission that flew September 12–20, 1992. Aboard the *Endeavour*, she conducted experiments concerning weightlessness, tissue growth, and the development of semiconductor materials. One of the experiments was to test whether motion sickness in space could be alleviated by the use of biofeedback techniques. She also investigated the loss of calcium in human bones in space and the effects of weightlessness on the fertilization and embryologic development of frogs.

While in medical school, Jemison traveled to a Thai refugee camp and received a grant to conduct health studies in Kenya. She joined the Peace Corps and traveled to Sierra Leone and Liberia, where she managed healthcare for volunteers, developed and taught health classes for volunteers, and implemented public health and safety guidelines for the program. When NASA announced in 1986 that it was seeking candidates for the space shuttle program, she applied and was one of 15 chosen from a field of some 2,000 applicants. After five years in the astronaut program, Jemison left for a teaching and science advocacy career. She has since founded two companies and worked on projects such as establishing a space-based telecommunication system to facilitate healthcare delivery in countries of the developing world, and marketing mobile medical technologies. She also directed the Jemison Institute for Advancing Technology in Developing Countries at Dartmouth College.

Jemison was elected to the Institute of Medicine of the National Academy of Science in 2001. She is a popular public figure and role model committed to inspiring young people, women, and minorities in the sciences. She has appeared in television shows and documentaries, and in 1994, she founded a science camp program for children aged 12 to 16 called "The Earth We Share" (TEWS). She is also the national science spokesperson for the Bayer pharmaceutical and medical research company. Her autobiography, *Find Where the Wind Goes: Moments from My Life*, was published in 2001.

Further Resources

Kevles, Bettyann H. 2003. *Almost Heaven: The Story of Women in Space*. New York: Basic Books.

National Aeronautics and Space Administration "Mae C. Jemison (M.D.)." http://www .jsc.nasa.gov/Bios/htmlbios/jemison-mc.html.

"Meet Our National Spokesperson, Mae C. Jemison M.D." Making Science Make Sense. Bayer US. http://www.bayerus.com/msms/MSMS_About/NationalSpokesperson/ Spokesperson.aspx.

Johnson, Barbara Crawford

b. 1925
Aerospace Engineer

Education: B.S., general engineering, University of Illinois, 1946

Professional Experience: engineer, Rockwell International Space Division, 1950s, project leader and supervisor, Entry Performance Analysis, 1961–1968, system engineer and manager, Apollo program, 1968–1972, manager, Mission Requirements and Integration, Rockwell Space Systems Group, 1973–1983

Barbara Johnson is one of the many women scientists and engineers who have played significant supplementary roles in the National Aeronautics and Space

Administration (NASA) space program. She spent her career at the Rockwell International Space Division in support of the manned space flight program. Rockwell was one of the primary contractors for NASA, and one of Johnson's major contributions was to create the Entry Monitor System (EMS), the backup entry guidance system designed for the Apollo space missions. The EMS is a graphic display for the astronauts to use in the case of a primary guidance failure, and similar graphic displays are now a part of the instrument panels of virtually all spacecraft and aircraft, and are even currently available in many automobiles. She was supervisor of the Entry Performance Analysis team, which determined the trajectories that enabled the Apollo aircraft to reenter the Earth's atmosphere safely; if it entered on too shallow a trajectory, there was a danger of overheating; if too deep, the astronauts would experience unbearable gravitational forces. Before the 1960s, a spacecraft had never reentered the Earth's atmosphere from hypervelocity, which is a speed greater than that of the Earth's rotation. As system engineering manager for the Apollo program, she supervised system analysis in support of a lunar landing and exploration. In 1973, she was named manager of Mission Requirements and Integration for Rockwell, which meant she directed the mission, flight performance, and trajectory design analysis of the space shuttle and orbiter projects.

Johnson received a medallion from NASA for her role in the first Apollo landing on the moon, and she has also received the Achievement Award of the Society of Women Engineers (1974), the Distinguished Alumni Merit Award from the University of Illinois (1975), and the Outstanding Engineer Merit Award of the Institute for the Advancement of Engineers (1976). She is a member of the American Institute of Aeronautics and Astronautics, and a fellow of the Institute for the Advancement of Engineers.

Further Resources

Society of Women Engineers. "Barbara Crawford Johnson." http://societyofwomenengineers .swe.org/index.php?option=com_content&task=view&id=46&Itemid=68.

Johnson (Masters), Virginia (Eshelman)

b. 1925
Psychologist, Sex Therapist

Education: student, Drury College, 1940–1942, University of Missouri, 1944–1947; student, Washington University, St. Louis

Professional Experience: research staff, Division of Reproductive Biology, School of Medicine, Washington University, St. Louis, 1957–1960, research assistant and instructor, 1960–1964; research associate, Reproductive Biology Research Foundation, St. Louis, 1964–1969, assistant director to co-director, 1969–1973; co-director, Masters and Johnson Institute, 1973–1994; director, Virginia Johnson Masters Learning Center, St. Louis, 1994–

Virginia Johnson is renowned for her pioneer studies with William H. Masters and her unique contribution to our knowledge of human sexuality.

Psychologist and sex therapist Virginia Johnson and physician William H. Masters were known for their pioneering studies of human sexuality. (AP/Wide World Photos)

At the Reproductive Biology Research Foundation in St. Louis, and later at the Masters and Johnson Institute, she counseled clients and taught sex therapy to practitioners. By the late 1950s, William Masters was a respected professor of obstetrics and gynecology who hired Virginia Johnson to interview volunteers for his research project on reproductive biology. Soon she was promoted to research assistant, instructor, and eventually co-director of the project. Gathering scientific data by electroencephalography (EEG), electrocardiography, and the use of color monitors, the two measured and analyzed 694 volunteers. They gathered data allowing them to identify the four stages of sexual arousal, the efficacy of contraceptives, and the observation that sexual enjoyment need not decrease with age. They created the nonprofit Reproductive Biology Research Foundation in 1964, began training couples to combat their sexual problems, and wrote a scientific text, *Human Sexual Response* (1966), describing their research. Although the book was advertised only in scientific journals, within a few months, it had become a bestseller.

Masters and Johnson married in 1971, founded the Masters and Johnson Institute in 1973, and went on to publish several books for a general audience, always inciting controversial reactions to their findings on sensitive topics. In *Human Sexual Inadequacy* (1970), they discussed the possibility that sex problems are more cultural than physiological or psychological. *The Pleasure Bond: A New Look at Sexuality and Commitment* (1975) advised total commitment and fidelity to the partner as the basis for an enduring sexual bond. In *Homosexuality in Perspective* (1981), they came to the controversial conclusion that homosexuality is a "learned" behavior and that homosexuals can be "converted." In *Crisis: Heterosexual Behavior in the Age of AIDS* (1988; co-authored with Dr. Robert Kolodny), they accurately predicted a large-scale outbreak of the virus in the heterosexual community. However, due to exaggerated and erroneous claims about how AIDS could be transmitted, many in the medical community, including the surgeon general C. Everett Koop, criticized the study, and the negative publicity hurt the couple's reputation. They divorced in 1992, and the Institute was closed in 1994 with Masters's retirement. Johnson, however, retained the Institute's records and went on to found the Virginia Johnson Masters Learning Center in St. Louis, which produces instructional material for couples with sexual problems. William H. Masters died in 2001.

Further Resources

Maier, Thomas. 2009. *Masters of Sex: The Life and Times of William Masters and Virginia Johnson, the Couple Who Taught America How to Love.* New York: Basic Books.

Johnston, Mary Helen

b. 1945
Metallurgical Engineer

Education: B.S., engineering science, Florida State University, 1966, M.S., 1969; Ph.D., metallurgical engineering, University of Florida, 1973

Professional Experience: metallurgical staff, University of Alabama, Huntsville, 1969–; materials engineer, George Marshall Space Flight Center, National Aeronautics and Space Administration (NASA), 1969–1983, payload specialist, astronaut program, 1983–1985

Mary Johnston is known for her expertise in failure analysis while working at the George Marshall Space Flight Center of NASA. As a metallurgist, she was concerned with the stability of the metal and materials parts of which the spacecraft was composed. There always is a possibility that a part might malfunction or break when exposed to the extremes of soaring heat or frigid cold in space, and the failure could occur in any part of a spacecraft, including bolts and screws. Although she worked for NASA for a number of years, she was never part of the astronaut program. In the 1970s, she started planning for the time when women would be accepted into the space program, and she was among the women employees who taught themselves how to function in a weightless environment. When she chose her major of metallurgical engineering, materials processing in space did not exist as a specialty. However, in 1974, she participated in an all-woman crew of experimenters in a five-day simulation of a Spacelab mission set up by NASA at Marshall because NASA needed to know how difficult it would be to handle materials-processing experiments in space. These experiments required a lot of power and put out a lot of heat, and Johnston predicted that nuclear radiation detector material would be a good material for a Spacelab experiment. One advantage of metallurgical research in space is that the zero-gravity environment in space allows for more control; on Earth, it is more difficult to study the processes involved in metals when the metals are cooled. Later, she was assigned to be the backup payload specialist on Spacelab 3, but she did not go into space.

Johnston was the first woman to graduate from Florida State University in engineering, and went on to graduate school at the University of Florida, one of the few female engineering students who completed that program. She is a member of American Society for Metals and National Society of Professional Engineers.

Jones, Anita Katherine

b. 1942
Computer Scientist

Education: B.A., mathematics, Rice University, 1964; M.A., English, University of Texas, Austin, 1966; Ph.D., computer science, Carnegie Mellon University, 1973

Professional Experience: programmer, International Business Machines Corporation (IBM), 1966–1968; assistant to associate professor, computer science, Carnegie Mellon University, 1973–1981; vice president and founder, Tartan Laboratories, Pittsburgh, 1981–1987; freelance consultant, 1987–1988; professor and department chair, computer science, University of Virginia, 1988–1993; director, Defense Research and Engineering, U.S. Department of Defense, 1993–1997; professor, computer science, University of Virginia, 1997–

Anita Jones is renowned for her work in the area of computer software and systems. Her research includes design and implementation of programmed systems on computers, including enforcement of security policies on computers, operating systems, and scientific databases. She was director of Defense Research and Engineering for the U.S. Department of Defense (DOD) (the highest-level defense job ever held by a woman), the department's senior official for research and technology matters. Her responsibilities included management of DOD science and technology programs; all in-house laboratories and research, development, and engineering centers; university research initiatives; and overseeing the

Advanced Research Projects Agency, which was responsible for the development of ARPAnet, the predecessor of the Internet. The DOD engineering research would eventually serve as a basis for both commercial and military information technology. Early on, Jones predicted that virtual reality (VR) simulations would be used extensively in education and job training, pointing out that the military invented high-fidelity simulations for flight training, and it still bankrolls the most cutting-edge applications.

Jones is married to fellow University of Virginia computer science professor William Wulf, who served a 10-year term as president of the National Academy of Engineering. Together they have formed a formidable power couple of engineering. In 1981, they launched a software firm, Tartan Laboratories, which specialized in research for optimizing compilers. Six years later, they sold the company to Texas Instruments and accepted faculty positions at the University of Virginia. Jones took a leave from her academic position to work for the DOD, but returned to academia in 1997. She has edited two books—*Foundations of Secure Computation* (1971) and *Perspectives in Computer Science* (1977)—in addition to writing numerous scientific papers. She has been a consultant to or member of the National Research Council, the Defense Science Board (1985–1993), and the U.S. Air Force Science Advisory Board (1980–1985), and served as vice chair of the National Science Foundation (2000–2006).

Jones was elected to membership in the National Academy of Engineering in 1994. She is a fellow of the Association for Computing Machinery and the Institute of Electrical and Electronics Engineers. She is the recipient of an Air Force Meritorious Civilian Service Award (1985) and a Distinguished Service Award of the Computing Research Association (1997).

Further Resources

Schrof, Joannie M. "Keeping Up with Anita Jones." http://www.cs.virginia.edu/misc/news-jones-keeping_up.html.

University of Virginia. Faculty website. http://www.cs.virginia.edu/brochure/profs/jones.html.

Jones, Mary Ellen

1922–1996
Biochemist

Education: B.S., biochemistry, University of Chicago, 1944; Ph.D., biochemistry, Yale University, 1951

Professional Experience: research chemist, Armour and Company, 1942–1948; U.S. Public Health Service Fellow, physiological chemistry, Yale University, 1950–1951; postdoctoral fellow, biochemistry research laboratory, Massachusetts General Hospital, 1951–1957; assistant to associate professor, biochemistry, Brandeis University, 1957–1966; associate professor to professor, biochemistry and zoology, University of North Carolina, Chapel Hill, 1966–1971; professor, biochemistry, University of Southern California, 1971–1978; professor, biochemistry and nutrition, University of North Carolina, Chapel Hill, 1978–1995

Mary Ellen Jones was a distinguished biochemist who contributed to early cancer research through her studies of DNA and RNA. Her research interests included biosynthetic and transfer reactions, metabolic regulation of enzymes, multifunctional proteins, and pyrimidine and amino acid biosynthesis. Her studies of metabolic pathways increased understanding of how cells, including cancer cells, divide and differentiate. This laid the groundwork for later, continued cancer research studies. She worked as a research chemist for Armour and Company while obtaining her undergraduate degree from the University of Chicago. She continued her education at Yale, where her husband Paul Munson was a faculty member in pharmacology, and received her doctorate in 1951 under a prestigious U.S. Public Health Service fellowship. She held a postdoctoral fellowship at Massachusetts General Hospital under Fritz Lipmann, who went on to win the Nobel Prize in Physiology or Medicine in 1953. She worked at Brandeis University as a biochemist until 1966, when the couple both moved to the University of North Carolina, Chapel Hill (UNCCH). Jones became a full professor at UNCCH in 1968, but moved to the University of Southern California in 1971. She returned to her position at UNCCH in 1978, where she was the first woman to head a medical school department at that institution. Jones was associate editor of the *Canadian Journal of Biochemistry* from 1969 to 1974. She was co-editor of a book, *Purine and Pyrimidine Nucleotide Metabolism* (1978), which is volume 51 in the *Methods in Enzymology* series.

Jones was elected to the Institute of Medicine in 1981 and the National Academy of Sciences in 1984, and held several distinguished appointments, such as member of the grants committee of the American Cancer Society (1971–1973), member of the metabolic biology study section of the National Science Foundation (1978–1981), and member of the science advisory board for the National Heart, Lung, and Blood Institute, National Institutes of Health (1980–1984). Her extensive professional service included terms as president of the Association of Medical School Departments of Biochemistry (1985), American Society for Biochemistry and Molecular Biology (1986), American Society of Biological Chemists (1986), and American Association of University Professors (1988).

Among her numerous awards were the Wilbur L. Cross Medal from Yale University (1982), a Distinguished Chemist award of the North Carolina American Chemical Society (1986), the Thomas Jefferson Award from the University of North Carolina (1990), and an Award in Science from the state of North Carolina (1991). A major research building is named after her at the University of North Carolina, Chapel Hill medical school. She was elected a fellow of the American Association for the Advancement of Science, and was also a member of the American Chemical Society and the American Philosophical Society.

Further Resources

Traut, Thomas W. "Mary Ellen Jones, December 25, 1922–August 23, 1996." http://www.nap.edu/html/biomems/mjones.html.

K

Kalnay, Eugenia

b. 1942
Meteorologist

Education: license, meteorology, University of Buenos Aires, Argentina, 1965; Ph.D., meteorology, Massachusetts Institute of Technology, 1971

Professional Experience: assistant professor, University of Montevideo, Uruguay, 1971–1973; assistant to associate professor, Massachusetts Institute of Technology (MIT), 1973–1979; senior research meteorologist, Global Modeling and Simulation, National Aeronautics and Space Administration (NASA)/ Goddard Space Flight Center, 1979–1984, director, 1984–1986; director, Environmental Modeling Center (EMC), National Centers for Environmental Prediction (NCEP), Maryland, 1987–1997, senior scientist, 1998–2000; Robert E. Lowry Chair in Meteorology, University of Oklahoma, 1997–1999; professor, meteorology, University of Maryland, 1999–2001, Distinguished University Professor, 2001–, Eugenia Brin Professor in Data Assimilation, 2008–

Eugenia Kalnay is a meteorologist who studies global weather forecasting and atmospheric weather dynamics. She uses computer modeling for numerical weather predictions. She was the first woman to get a Ph.D. in meteorology from MIT. She has been an outspoken critic of those, including other scientists, who deny humankind's role in global climate change. She is the author of a popular textbook, *Atmospheric Modeling, Data Assimilation and Predictability* (2002).

Kalnay was born in Argentina, the seventh of eight children, to Hungarian and Swiss parents. Her father died when she was a teenager, but her mother encouraged and supported her education. Kalnay enrolled at the University of Buenos Aires intending to study physics, but her mother chose meteorology as her major due to the availability of scholarships and job opportunities. She received her degree in 1965 and relocated to Massachusetts to study at MIT. She married and had a child while in graduate school, and was the only female student in the meteorology program at MIT. She received her doctorate in 1971 and then returned to South America to teach in Uruguay for two years. She returned to MIT in 1973 as a research associate and then faculty member. She was then offered a position with NASA in the Global Modeling and Simulation laboratory. The NASA job required

Meterologist, Eugenia Kalnay. (Courtesy of University Publications, University of Maryland)

that she finally secure U.S. citizenship, which she did in 1978. She then joined the EMC/NCEP, where she oversaw the work of a team of scientists compiling computer modeling information on atmospheric and ocean climates for the National Weather Service. After 10 years, however, she stepped down as director of the EMC and returned to academic research as professor of meteorology at the University of Oklahoma and then the University of Maryland.

Kalnay was elected to the National Academy of Engineering in 1996 and has been named a foreign member of the Academia Europaea (2000) and a corresponding member of the Argentine National Academy of Physical Sciences (2003). She is also a fellow of American Geophysical Union, American Meteorological Society, and the American Academy of Arts and Sciences. She has received Gold and Silver Medals from the U.S. Department of Commerce, a NASA Medal for Exceptional Scientific Achievement (1981), the Jule G. Charney Award of AMS (1995), the Presidential Rank Award for Meritorious Achievement (1997), the Kirwan Award of the University of Maryland (2006), and the IMO Prize of the World Meteorological Organization (2009). In 2008, she received an honorary doctorate from the University of Buenos Aires.

Further Resources

University of Maryland. Faculty website. http://www.atmos.umd.edu/~ekalnay/.

Kanter, Rosabeth (Moss)

b. 1943
Sociologist, Management Consultant

Education: student, University of Chicago, 1962–1963; B.A., Bryn Mawr College, 1964; M.A., University of Michigan, 1965, Ph.D., sociology, 1967

Professional Experience: instructor, sociology, University of Michigan, 1967; assistant professor, sociology, Brandeis University, 1967–1973; associate professor, administration, Harvard University, 1973–1974; associate professor, sociology, Brandeis, 1974–1977; associate professor, sociology, Yale University, 1977–1978, professor, 1978–1986; professor, Harvard University Business School, 1986–2000, Ernest L. Arbunkle professor of business administration, 2000–

Concurrent Positions: visiting scholar, Newberry Library, 1973; visiting scholar, Harvard University, 1975; faculty member, Young President's Organization of International University, Hong Kong, 1976; founding partner, Goodmeasure, Inc., 1977– ; scholar-in-residence, Miami University, Ohio, 1978; visiting professor, Organizational Psychology and Management, Sloan School of Management, Massachusetts Institute of Technology, 1979–1980; director, American Center for Quality of Work Life, 1978–1982; director, Educational Fund for Individual Rights, 1979–1984; director, Legal Defense and Education Fund, National Organization for Women, 1979–1986 and 1993–1995; visiting scholar, Norwegian Research Council on Science and Humanities, 1980; editor, *Harvard Business Review*, 1989–1992

Rosabeth Moss Kanter brought a multidisciplinary perspective to the study of organizations and revolutionized management by introducing humanism into the workplace. In her landmark book, *Men and Women of the Corporation* (1977), she debunked the notion that the right personality is the key ingredient for success. Her research indicated that the structure of a company and a person's position within it determines her or his behavior and chances of promotion. Her statements that people can be products of their jobs, not the reverse, was particularly important for women, who usually are told they do not have the personality to be managers, when they have never been able to develop leadership skills in low-level, powerless jobs.

Kanter's earlier research was on the sociology of communal living. She moved from the study of communes to corporations and, in 1977, she and her husband, Barry Stein, established their own management consulting firm, Goodmeasure, Inc. They co-authored *A Tale of "O"* (1980), which described in a whimsical manner how "x's" and "o's" are treated differently and revealed the insidious effect of discrimination in organizations. In *The Change Masters: Innovation for Productivity in the American Corporation* (1983), she advised companies on the idea of "intrapreneurship," or how to stimulate entrepreneurial efforts from employees within an organization. Kanter has kept up with changes affecting American corporations, and in *World Class: Thriving Locally in the Global Economy* (1995), she emphasized the alternatives to job insecurity and economic chaos that have been brought on by the increasing globalization of industry. Many of her published articles on topics of strategy, innovation, and leadership were collected in the book *Rosabeth Moss Kanter on the Frontiers of Management* (1997; 2nd ed., 2003). Her other books include *Innovation: Breakthrough Thinking at Du Pont, GE, Pfizer, and Rubbermaid* (1997), *Evolve!: Succeeding in the Digital Culture of Tomorrow* (2001), and *Confidence: How Winning Streaks and Losing Streaks Begin and End* (2004).

Kanter has been a board member, trustee, or consultant to numerous businesses, organizations, and government entities. She has also been an advisor to political campaigns, working closely with Governor Michael Dukakis of Massachusetts in his campaign for the presidency in 1988. She and Dukakis wrote *Creating the Future: The Massachusetts Comeback and Its Promise for America* (1988). In 1994, Massachusetts governor William Weld appointed her to his Council on Economic Growth and Technology and named her co-chair of his International Trade Task Force. She is also a member of the American Sociological Association, American Association for Higher Education, and Society for the Study of Social Problems.

Further Resources

Harvard Business School. Faculty website. http://drfd.hbs.edu/fit/public/facultyInfo.do ?facInfo=ovr&facId=6486.

Kanwisher, Nancy

Psychologist

Education: B.S., biology, Massachusetts Institute of Technology, 1980, Ph.D., cognitive psychology, 1986

Professional Experience: visiting scholar, Institute for War and Peace Studies, Columbia University, 1986–1987; postdoctoral fellow, psychology, Harvard University, 1987–1988; assistant research psychologist, psychology, University of California, Berkeley, 1988–1990; assistant and associate professor, psychology, University of California, Los Angeles, 1990–1994; assistant professor and John L. Loeb Associate Professor of the Social Sciences, psychology, Harvard University, 1994–1997; associate professor, Brain and Cognitive Sciences, Massachusetts Institute of Technology (MIT), 1997–2001, professor, 2001–, Ellen Swallow Richards Professor, 2004–2009, Walter A. Rosenblith Professor, 2009–

Concurrent Positions: assistant in neuroscience, Department of Radiology, Massachusetts General Hospital, 2000–; investigator, McGovern Institute for Brain Research, MIT, 2000–

Nancy Kanwisher is a psychologist who studies visual perception, including object recognition, attention, number recognition, and social cognition. Her work combines cognitive and neurological research methods and tools. One of her main contributions to the field of cognitive neuroscience has been the identification of an area of the brain she terms FFA (Fusiform Face Area), which is dedicated to processing facial recognition. Using magnetic resonance imaging, or an MRI, to track brain activity, Kanwisher has found that, even when vision is not impaired, neurological injury or problems with this specific area can impact a patient's ability to recognize faces. Her research has also revealed other dedicated areas of the brain that process specific imagery related to other body parts, such as feet or elbows. These findings have unlimited implications for uncovering the previously unknown function of other brain regions, and for further research into the role of genetics, evolutionary biology, and environmental or social conditioning on the development of specific areas of the brain.

Kanwisher received her doctorate in cognitive psychology from MIT in 1986 and taught at several universities on the East Coast and California before returning to MIT as a faculty member in the Department of Brain and Cognitive Sciences in 1997. Since 2000, she has also been an Investigator at MIT's McGovern Institute for Brain Research. The McGovern Institute brings together researchers on brain function specifically for the purpose of understanding physical and cognitive brain disorders, diseases, and injuries. She has been an invited lecturer, committee member, and advisor for numerous schools and institutions throughout the United States and abroad. She has published widely, including collaborations with Harvard psychologist **Elizabeth Spelke**, and served on the editorial boards of professional journals such as *Current Opinion in Neurobiology, Cognition, Journal of Neuroscience, Journal of Experimental Psychology*, and several others.

Kanwisher was elected to the National Academy of Sciences in 2005. She is a fellow of the American Academy of Arts and Sciences and the Society of Experimental Psychologists, and was named a MacVicar Faculty Fellow at MIT (2002). She is also the recipient of a MacArthur Foundation Fellowship in Peace and International Security (1986–1988), a National Institute of Mental Health FIRST Award (1988–1992), a National Academy of Sciences Troland Research Award (1999), and a Golden Brain Award of the Minerva Foundation (2007).

Further Resources

McGovern Institute. "Nancy Kanwisher." http://web.mit.edu/mcgovern/html/Principal _Investigators/kanwisher.shtml.

Massachusetts Institute of Technology. "Kanwisher Lab." http://web.mit.edu/bcs/nklab/.

Karle, Isabella Helen Lugoski

b. 1921
Crystallographer

Education: B.S., University of Michigan, M.S., physical chemistry, 1942, Ph.D., physical chemistry, 1944

Professional Experience: associate chemist, University of Chicago, 1944; instructor, chemistry, University of Michigan, Ann Arbor, 1944–1946; physicist, Naval Research Laboratory (NRL), 1946–1959, scientist, x-ray defraction and structural chemistry, 1959–2009

Isabella Karle is a chemist and physicist who, along with her husband Jerome Karle and others, developed a new mathematical technique called "direct methods" in crystallography, or the study of the atomic structure and composition of crystals. Her research interests have included application of electron and x-ray diffraction to structure problems, phase determination in crystallography, elucidation of molecular formulae, peptides, and configurations and conformations of natural products and biologically active materials. Jerome (who was a co-recipient of the Nobel Prize in Chemistry in 1985) was the theorist and Isabella the experimentalist. He and others developed the theory of direct method, but she applied their theories by designing a machine that could diffract and photograph images of crystals to determine their atomic structures, speeding up the process and thus revolutionizing the field. Karle's study of frog venom and other biological materials allowed advances in creating synthetic chemicals for everything from insect repellents to medicines. She published more than 200 scientific papers and has received several honorary degrees. She has been a consultant or advisor to government agencies, including the National Committee on Crystallography, the National Research Council, and the Atomic Energy Commission.

Isabella Karle completed her undergraduate degree on a four-year fellowship and received her doctorate at the age of 22, but she was unable to secure a graduate teaching assistantship in chemistry at Michigan because women had never held such a position. She was granted a fellowship by the American Association of University Women to start her graduate studies. After she received her doctorate, she and Jerome worked at the University of Chicago on the Manhattan Project for six months, and then Isabella returned to the University of Michigan for a short time. The couple was unable to obtain suitable employment together in a university due to anti-nepotism rules, but the U.S. Naval Research Laboratory (NRL) offered them an opportunity to work together beginning in 1946, and they were affiliated with the NRL until their joint retirement in July 2009.

Isabella Karle was elected to membership in the National Academy of Sciences in 1978 and received a National Medal of Science from President Clinton in 1995. Her awards and honors over the course of a long career are numerous, but they include eight honorary doctorates as well as a Superior Civilian Service Award of the Navy Department (1965), Annual Achievement Award of the Society of Women Engineers (1968), Hildebrand Award from the American Chemical Society (1969), Federal Woman's Award (1973), Garvan Medal of the American Chemical Society (1976), Pioneer Award from the American Institute of Chemists (1984), Women in Science and Engineering Lifetime Achievement Award (1986), Gregori Aminoff Prize of the Swedish Academy of Sciences (1988), Bijvoet Medal from the University of Utrecht, Netherlands (1990), Bower Award and Prize for Achievement in Science from the Franklin Institute (1993), U.S. Department of Defense Distinguished Civilian Service Award (1995), and Merrifield Award of the American Peptide Society (2007). She is a member of the American Crystallographic Association (president, 1976), the American Physical Society, and the American Chemical Society. Jerome and Isabella Karle had three daughters, all of whom pursued degrees in the sciences.

Further Resources

Naval Research Laboratory. "Jerome and Isabella Karle Retire from NRL Following Six Decades of Scientific Exploration." Press release. (21 July 2009). http://www.nrl.navy .mil/pao/pressRelease.php?Y=2009&R=58-09r.

Wasserman, Elga. 2002. *The Door in the Dream: Conversations with Eminent Women in Science*. Washington, D.C.: Joseph Henry Press.

Karp, Carol Ruth (Vander Velde)

1926–1972
Mathematician

Education: B.A., Manchester College, 1948; M.A., Michigan State University, 1950; Ph.D., mathematics, University of Southern California, 1959

Professional Experience: instructor, mathematics, New Mexico Agricultural and Mechanical College (now New Mexico State University), 1953–1954; instructor, mathematics, University of Maryland, 1958–1960, assistant to associate professor, 1960–1966, professor, 1966–1972

Carol Karp was renowned for her research on logic, particularly infinitary logic in mathematics. Logic is the science that investigates the principles governing correct or reliable inference, and her book *Languages with Expressions of Infinite Length*

(1964), based on her doctoral thesis, was the first systematic explanation of the theory of infinitary logic. In infinitary logic, a modification of calculus, the formulas are formed from symbols representing variables, constants, functions, and relations. Karp introduced four new symbols representing conjunction of infinite sets. Her work was internationally recognized, and she was able to recruit other faculty and a steady stream of graduate students to the University of Maryland. She was instrumental in bringing several important participants to the colloquia that she sponsored, and she and her husband even had a home with an extra apartment in which visiting logicians were frequently housed. Karp's intellectual standards were extremely high, and she was unfailingly honest in appraising the mathematical contributions and research promise of her students, refusing to let anyone graduate until their results met her own high standards for publishability.

Karp developed breast cancer in 1969, but she continued her schedule of teaching and research until 1971, when she was too ill to work. At the time of her death in 1972, she was working on a second book, but it was still too incomplete to publish. Colleagues and friends prepared a memorial volume, *Infinitary Logic: In Memoriam Carol Karp* (1975), which incorporates many of her ideas and notes. She was a member of the American Mathematical Society, Mathematical Association of America, and Association for Symbolic Logic.

Further Resources

Agnes Scott College. "Carol Karp." Biographies of Women Mathematicians. http://www.agnesscott.edu/lriddle/women/karp.htm.

Kaufman, Joyce (Jacobson)

b. 1929
Chemist, Pharmacologist

Education: B.S., Johns Hopkins University, 1949, M.A., 1959, Ph.D., chemistry and chemical physics, 1960; DES, theoretical physics, Sorbonne, 1963

Professional Experience: research chemist, U.S. Army Chemical Center, Maryland, 1949–1952; research assistant, Johns Hopkins University, 1952–1960; staff scientist, Martin Company Research Institute for Advanced Studies, 1960–1962, head, quantum, chemistry group, 1962–1969; associate professor of anesthesiology, School of Medicine, and principal research scientist in chemistry, Johns Hopkins University, 1969–1977, associate professor, Department of Surgery, 1977–

Joyce Kaufman has gained a distinguished national and international reputation in a wide variety of fields—chemistry, physics, biomedicine, and supercomputers—on both the experimental and the theoretical levels. Her specialties include theoretical quantum chemistry, experimental physical chemistry, and chemical physics of energetic compounds; the last includes explosives, rocket fuels, oxidizers, and energetic polymers. She has examined the application of those techniques and experimental animal studies to biomedical research, including pharmacology, drug design, and toxicology. She is also knowledgeable in nuclear chemistry and radiochemistry, and has been successful in using experimental chemical techniques in determining the guidelines for effective drug action in a number of different areas. She published a landmark paper in 1980 in which she introduced a new theoretical method for coding and retrieving certain carcinogenic polycyclic aromatic hydrocarbons. Since that time, at least 30 papers have been written by other researchers using and expanding upon her concept. At Johns Hopkins School of Medicine, she works with interns and residents studying the effect of drugs, such as narcotics, tranquilizers, psychotropic drugs, general anesthetics, and spinal anesthetics, on the central nervous system.

Kaufman completed high school in two years and, after receiving her undergraduate degree, worked as a librarian at the Army Chemical Center, where she set up a scientific indexing system for their technical reports. During the 1950s, it was common practice for companies and agencies to hire women scientists as librarians specializing in scientific literature rather than to employ them in the laboratories. However, Kaufman was able to transfer to a position as a research chemist after one year. A chemistry professor at Johns Hopkins invited her to work with him on a research contract, and he later convinced her to obtain a doctorate. She later joined Martin Company's Research Institute for Advanced Studies to do theoretical research on the application of quantum mechanics to problems in chemistry, but returned to Johns Hopkins in 1969 as a professor and research scientist.

Kaufman has received numerous awards, including the Gold Medal of the Martin Company each year for three years (1964–1966), the Dame Chevalier of the Centre National de la Recherche Scientifique, France (1969), the Garvan Medal of the American Chemical Society (1974), and a Woman of Achievement Award from the Jewish National Fund (1974). She is a fellow of the American Physical Society, American Institute of Chemists, and the European Academy of Science, Arts, and Letters.

Keller, Evelyn Fox

b. 1936
Physicist, Mathematical Biologist, Molecular Biologist

Education: student, Queens College, 1953; B.A., Brandeis University, 1957; M.A., Radcliffe College, 1959; Ph.D., physics, Harvard University, 1963

Physicist, biologist, and feminist scholar, Evelyn Fox Keller. (Photograph by Marleen Wynants)

Professional Experience: instructor, New York University, 1962–1963, assistant research scientist, 1963–1966; assistant professor, Graduate School of Medical Science, Cornell University, 1966–1969; associate professor of molecular biology, New York University, 1970–1972; associate professor, Division of Natural Science, State University of New York (SUNY) at Purchase, 1972–1982; professor of humanities and mathematics, Northeastern University, 1982; senior fellow, Cornell University, 1986–1987; professor of rhetoric, Women's Studies and History of Science, University of California, Berkeley, 1989–1992; professor, history and philosophy of science, Massachusetts Institute of Technology, 1992–

Concurrent Positions: visiting fellow, Massachusetts Institute of Technology, 1979–1980, visiting scholar, 1980–1984, visiting professor, 1985–1986

Evelyn Fox Keller is known for her work in the fields of theoretical physics, molecular biology, and mathematical biology, as well as her feminist critique of scientific methods and beliefs. She was drawn to physics as a means for deep inquiry into nature, and received a National Science Foundation fellowship to attend Harvard. She did not enjoy the competitive and discriminatory atmosphere at Harvard and was ready to quit school after two years. A summer at the Cold Spring Harbor Laboratory, however, inspired her finish her thesis on molecular biology, and she received her doctorate in physics in 1963.

While teaching a women's studies course in New York in 1974, she began to question the treatment of women in the sciences. An article on geneticist **Barbara McClintock** turned into a full biography, *A Feeling for the Organism: The Life of Barbara McClintock*, which Keller published in 1983. McClintock had worked for years in relative obscurity at Cold Spring Harbor on the genetics of maize. She had discovered that some genes move from one area on the chromosome to another, but her work was ignored for many years. McClintock received the Nobel Prize for this discovery more than 30 years after publishing her first findings. Keller generated controversy with her next book, *Reflections on Gender and Science* (1985), in which she emphasized the importance of intuition in science and speculated on what a truly gender-free science might look like. Her recent works include *The Century of the Gene* (2000), and *Making Sense of Life: Explaining Biological Development with Models, Metaphors and Machines* (2002).

Fox Keller has received numerous awards and recognitions, including the Blaise Pascal Research Chair by the Préfecture de la Région D'Ile-de-France (2005–2007). She is a member of the American Philosophical Society and the American Academy of Arts and Science.

Further Resources

Massachusetts Institute of Technology. Faculty website. http://web.mit.edu/sts/people/keller.html.

Kempf, Martine

b. 1958
Computer Scientist

Education: student, astronomy, Friedrich Wilhelm University, Bonn, 1981–1983

Professional Experience: founder and CEO, Kempf USA, 1985–, CEO, Kempf SAS, 2002–

Martine Kempf is known for her research on voice commands for computer programs. In 1985, she invented a breakthrough voice recognition microcomputer dubbed Katalavox, a name derived from the Greek word *katal*, "to understand," and *vox*, which is Latin for "voice." While she was a student in Bonn, she saw many German teenagers who had been born without arms because their mothers had taken thalidomide during pregnancy, and reasoned that a voice recognition system would enable them to drive cars. Learning to program on an Apple computer, she succeeded in directly transforming the human voice's analog signals into the computer's digital signals. Further refinements enable Katalavox to respond to a spoken command in a fraction of a second, compared with one or two seconds for competing systems. She was unable to secure financing to start a company in France, so she moved to Sunnyvale, California, to create and market the voice-recognition device not only for drivers, but also for people confined to wheelchairs or who suffer from cerebral palsy or strokes, and for doctors to use surgical tools and microscopes hands-free.

Kempf was an astronomy student who does not hold a higher degree, but taught herself electronics and computers. She not only designed the software for her device but designed and built the hardware, designing the board and soldering the circuits herself. She also invented the Comeldir Multiplex Handicapped Driving Systems for people who must operate cars with their feet rather than their hands. Kempf's own father was a polio survivor who designed a car he could drive with his hands and created a business customizing more than 1,000 cars per year for others with disabilities. Martine's company, Kempf USA, is still headquartered in California, and she became CEO of Kempf SAS in Europe (her father's business) after his death in 2002.

Further Resources

Kempf USA. "Who is Martine KEMPF?" http://www.kempf-usa.com/Kempf _Martine.html.

Kenyon, Cynthia J.

b. 1955
Molecular Biologist

Education: B.S., chemistry and biochemistry, University of Georgia, 1976; Ph.D., biology, Massachusetts Institute of Technology, 1981

Professional Experience: postdoctoral fellow, Medical Research Council Laboratory of Molecular Biology, Cambridge, England, 1982–1986; assistant professor, biochemistry and biophysics, University of California, San Francisco (UCSF), 1986–1992, associate professor, 1992–1994, professor, 1994–

Concurrent Positions: director, UCSF Hillblom Center for the Biology of Aging, 2002–

Cynthia Kenyon is a molecular biologist known for her studies of the genes of a microscopic roundworm or nematode called *Caenorhabditis elegans*, or *C. elegans*. Her findings that gene mutations were responsible for determining the life span of *C. elegans* led to further research on the genetic role in aging and age-related diseases (such as heart disease, cancer, diabetes, or Huntington's disease) in other organisms, such as mice or humans. By altering the genes of the *C. elegans* hormonally and environmentally, Kenyon found that she could increase the worm's life span by as much as 50% compared to normal, from 21 days to 45 days in some cases. While the implications for the human aging process are still being researched, she has reported that her findings have at least prompted her to think about her own aging; for example, finding that too much sugar shortened the worm's life span, she made dietary changes to limit the amount of high-glycemic index carbohydrates she eats, avoiding white flour and sugar. These dietary changes also promote weight loss and regulate insulin production, which can also ward off disease. In 1999, she co-founded a company, Elixir Pharmaceuticals, to research the development of medications that could slow down the aging process and treat metabolic disorders.

Kenyon earned her undergraduate degree in chemistry from the University of Georgia in 1976 and went on to receive a doctorate from the Massachusetts Institute of Technology in 1981. Her thesis was on DNA damage in *E. coli*. She then went to Cambridge, England as a postdoctoral molecular biology researcher in the laboratory of Nobel Laureate Sydney Brenner, where she began studying *C. elegans*. She joined the faculty at UCSF in 1986, where she was Herbert Boyer Distinguished

Molecular biologist Cynthia Kenyon researches the genetic role in aging and age-related diseases. (AP/Wide World Photos)

Professor of Biochemistry and Biophysics (1997–2004), and in 2005 was named an American Cancer Research Society Professor. She is also the founding director of the Hillblom Center for the Biology of Aging at UCSF, established in 2002.

Kenyon was elected to both the National Academy of Sciences and the Institute of Medicine in 2003. She is a fellow of the American Academy of Arts and Sciences and the Genetics Society of America (president, 2003). Her most recent awards and honors include the King Faisal International Prize for Medicine (2000), Life Extension Prize (2002), Discover Prize for Basic Research (2004), American Association of Medical Colleges Award for Distinguished Research in Biomedical Sciences (2004), Ilse & Helmut Wachter Award for Exceptional Scientific Achievement in the Field of Medicine (2005), and La Fondation IPSEN Prize (2006).

Further Resources

University of California, San Francisco. "Kenyon Lab." http://kenyonlab.ucsf.edu/.

Kidwell, Margaret Gale

b. 1933
Geneticist, Evolutionary Biologist

Education: B.Sc., Nottingham University, 1953; M.S., animal breeding, Iowa State University, 1962; Ph.D., genetics, Brown University, 1973

Professional Experience: officer, Ministry of Agriculture, London, 1955–1960; research fellow, Brown University, 1973–1974, research associate, 1974–1975, investigator, 1975–1977, assistant to associate professor, 1977–1984, professor, 1984–1985; professor, ecology and evolutionary biology, University of Arizona, Tucson, 1985–

Margaret Kidwell is renowned for her research on *Drosophila*, the common fruit fly. Her research interests include *Drosophila* genetics and evolution, recombination transposable elements, and speciation. In the 1990s, her team discovered that sometime around 1950, genes of one fruit fly jumped to another species. Since that time, "the jumping genes" have spread like wildfire, so that today, essentially all fruit fly populations, except those maintained in isolation in laboratories, carry the same elements. The theory is that a tiny parasitic mite lives in association with both species. Although there have been reports of other possible gene transfers between species, principally by viruses, this discovery was the first indication that a mite or anything like it can transfer genetic material. The transfer of genetic material between species has a major impact on our understanding of evolution, as the "transposons" cause mutations if they happen to land in a gene. However, if lateral transfers of genetic material between species occur frequently, that could complicate the work of researchers who are attempting to study the evolutionary relationships among species. Kidwell, a pioneer in this research, was the one who zeroed in on the mite. Since the two species of fruit flies cannot breed, the team recognized that the material had to have been transferred by some agent.

Born in England, Kidwell came to the United States in 1960 with a fellowship to pursue graduate study, receiving a master's degree from Iowa State and her doctorate from Brown University in 1973 at the age of 40. She had originally planned to return to England, but discovered she wanted to pursue a research career in the United States, and had married an American and started a family as well. While still in graduate school, she accepted a position at Brown as a research scientist, then moved into the academic ranks as an assistant professor in 1977. She became a full professor at Brown in 1985 but was recruited to the University of Arizona as professor of ecology and evolutionary biology and in affiliation with the Interdisciplinary Program in Genetics.

Kidwell was elected to membership in the National Academy of Sciences in 1996. She is a fellow of the American Association for the Advancement of Science and of the American Academy of Arts and Letters, and a member of the American Genetics Association (president, 1992), Society for Molecular Biology and Evolution (1996), American Society of Naturalists (vice president, 1984), Genetics Society of America, and Society for the Study of Evolution. Kidwell had two daughters, both of whom pursued advanced degrees in the biological sciences.

Further Resources

Wasserman, Elga. 2002. *The Door in the Dream: Conversations with Eminent Women in Science*. Washington, D.C.: Joseph Henry Press.

University of Arizona. Faculty website. http://www.eebweb.arizona.edu/Faculty/Bios/kidwell.html.

Kieffer, Susan Werner

b. 1942
Geologist, Volcanologist, Mineral Physicist

Education: B.S., physics and mathematics, Allegheny College, 1964; student, astrogeophysics (solar physics), University of Colorado, Boulder, 1964–1965; M.Sc., geological sciences, California Institute of Technology, 1967, Ph.D., planetary science, 1971

Professional Experience: postdoctoral research fellow, University of California, Los Angeles, 1971–1973, assistant to associate professor, geology, 1973–1979; geologist, U.S. Geological Survey (USGS), Flagstaff, Arizona, 1979–1990, scientist emeritus; professor, geology, Arizona State University, 1990–1993; professor and department head, geological science, University of British Columbia, 1993–1995; co-founder and head, Kieffer & Woo, Inc., Ontario, 1996–2000; professor, geology and physics, Center for Advanced Study, and affiliated faculty member, civil and environmental engineering, University of Illinois, Urbana-Champaign, 2000–

Concurrent Positions: visiting professor, geology, California Institute of Technology, 1982; research professor, geology, Arizona State University, 1989; co-founder and president, Kieffer Institute for Development of Science-Based Education, Arizona, 1999; founder, S. W. Kieffer Science Consulting Inc., 2000–

Susan Kieffer is renowned as an expert on volcanoes both on Earth and on Io, Venus, Mars, and other planets. Her research includes geological physics, high-pressure geophysics and impact processes, shock metamorphism of natural materials, thermodynamic properties of minerals, mechanisms of geyser and volcano eruptions, and river hydraulics. Her expertise on the hydraulics of lava flow also transfers to her studies of the hydraulics, sediment transfer, rapids, and waves in rivers. She also participated in the studies of asteroid impact on Earth at the Chicxulub crater in Mexico. She has studied geysers, volcanoes, and the volcanic environment on Earth as well as on other planets and has found that simulated volcanic eruptions on Earth, Venus, and Mars produce plumes with different fluid dynamic regimes. A major portion of the differences are caused by differing atmospheric pressures and ratios of volcanic vent pressure to atmospheric pressure. She did extensive studies of the hydraulics of lava flow and erosion furrows after the eruption of Mount St. Helens in Washington State in 1980, research which earned her team a USGS Group Achievement Award. She has also studied the hydraulics of river flow in areas such as the Colorado River, and explored Old Faithful in Yellowstone National Park by lowering a robot down its vent.

Although it is commonplace for asteroids to strike other planets, it is a comparatively rare occurrence when they strike Earth. Kieffer collaborated with Walter Alvarez in his study of the crater left by an asteroid striking the Earth at Chicxulub in Mexico, a study that supported Alvarez's theory that the dust cloud from this impact blotted out the sun while circling the Earth, thus killing the vegetation that was the food supply for the dinosaurs. Kieffer and Alvarez co-authored a paper about the study, and the research is also discussed in Alvarez's book *T-Rex and the Crater of Doom* (1997).

She has worked for government agencies as well as for several universities. After receiving her doctorate, she secured a position as a research geophysicist at the University of California, Los Angeles, and then transferred to a tenure-track academic position. She worked for the USGS for more than a decade, then returned to academia as a professor at Arizona State University before moving to the University of British Columbia in 1993. In 1996, she co-founded Kieffer & Woo, a consulting firm in Canada, and was chair of the Canadian Geoscience Council (CGC) committee on Geologic Disposal of High-Level Nuclear Fuel Waste. Most recently, she has consulted on volcanic intrusions into waste repositories for the Nuclear Waste Technical Review Board. In 2000 she became a professor of geology and physics at the University of Illinois, Urbana-Champaign.

Kieffer was elected to membership in the National Academy of Sciences in 1986. She received a prestigious five-year MacArthur fellowship (1995–2000), and has also received the Mineralogical Society of America Award (1980), Distinguished Alumnus Award of California Institute of Technology (1982), Meritorious Service Award of the Department of Interior (1987), Spendiarov Prize of the Soviet Academy of Sciences (1989), and Day Medal of the Geological Society of America (1992). She is the co-editor of *Microscopic to Macroscopic Atomic Environments to Mineral Thermodynamics* (1985). She is a fellow of the American Geophysical Union and the American Association for the Advancement of Science (chair of Geology/Geography Section, 2002–2005), and a member of the Meteoritical Society, Geological Society of America, Geological Association of Canada, Society of Canadian Women in Science and Technology, and American Academy of Arts and Sciences.

Further Resources

University of Illinois. "The Geological Fluid Dynamics Group." http://www .geology.uiuc.edu/~skieffer/.

Kimble, Judith

b. 1949
Geneticist

Education: B.A., University of California, Berkeley, 1971; Ph.D., biology, University of Colorado, Boulder, 1978

Professional Experience: postdoctoral fellow, Laboratory of Molecular Biology, Cambridge, England, 1978–1982; assistant to associate professor, molecular biology and biochemistry, University of Wisconsin, Madison, 1983–1992, professor, 1992–

Concurrent Positions: investigator, Howard Hughes Medical Institute, 1994–

Judith Kimble is renowned for her research on elegans, a type of nematode—unsegmented worms of the phylum Nematoda that have elongated, cylindrical bodies. Her research concerns understanding animal development at the molecular and cellular levels. She became interested in this field after studying human embryology as an undergraduate and realized that understanding stem cells and organ development of the simplest animals would have implications for understanding all animals, including humans. After completing her undergraduate degree at Berkeley, Kimble spent two years at the University of Copenhagen Medical School as an assistant before she received a National Science Foundation predoctoral fellowship at the University of Colorado, Boulder. After receiving her doctorate, she was a postdoctoral fellow at the Laboratory of Molecular Biology in Cambridge, England, and went on to receive prestigious postdoctoral fellowships, including one from the National Institutes of Health (1980–1982). Kimble joined the faculty in molecular biology, biochemistry, and medical genetics at the

University of Wisconsin, Madison. She has published widely in scientific journals such as *Developmental Biology; Genetics; Cell; Developmental Genetics;* and *Proceedings of the National Academy of Sciences.*

Kimble was elected to membership in the National Academy of Sciences (NAS) in 1995 and was recently elected to a term on the Council of the NAS (2008–). She received a National Institutes of Health Research Career Development Award (1984–1989), and has served on several prestigious committees, such as the Damon Runyon–Walter Winchell Cancer Research Fund Scientific Advisory Board (1992–1996) and the Searle Scientific Advisory Board (1997–). She has been active in professional organizations, such as the Society of Developmental Biology (secretary, 1987–1990; president, 2004–2005), Genetics Society of America (president, 2000), and American Society for Cell Biology (council member, 1994–). She is also a member of the American Academy of Arts and Sciences and the American Philosophical Society.

Further Resources

University of Wisconsin. Faculty website. http://www.biochem.wisc.edu/faculty/kimble/.

King, Helen Dean

1869–1955
Geneticist

Education: B.A., Vassar College, 1892; Ph.D., morphology, Bryn Mawr College, 1899

Professional Experience: fellow, biology, Bryn Mawr College, 1896–1897, assistant, 1899–1904; instructor, science, Baldwin School, Pennsylvania, 1899–1907; fellow, biology, University of Pennsylvania, 1906–1908; assistant, anatomy, Wistar Institute of Anatomy and Biology, 1909–1912, assistant professor, anatomy, 1912–1927, professor, embryology, 1927–1949

Helen King's outstanding contribution to science was her success in breeding pure strains of laboratory animals, including 150 generations of rats and the "Wistar" rat (named after the Institute where she worked for 40 years) that became widely used as a lab animal. In addition to discovering new types of rats, her research shed light on inquiries into heredity, sex determination, fertility, and longevity. Through careful inbreeding experiments with brother and sister rats, a practice unpopular at the time, she demonstrated the capacity to improve the strain, knowledge that has been applied to other animals, such as racehorses. Her research was reported in the newspapers in the 1910s and 1920s at a time when there was great interest in eugenics. Her work on rats sparked controversy and outrage over whether she was advocating "human inbreeding" or incest. Her other research interests included sex determination in amphibians and mammals, germ cells in amphibians and mammals, parthenogenesis, growth and reproduction of the white rat, and modification of the sex ratio.

After she received her doctorate from Bryn Mawr, she remained at the school as an assistant in biology and also taught science at the Baldwin School before accepting a research fellow position at the University of Pennsylvania. She moved to the Wistar Institute of Anatomy and Biology in Philadelphia in 1909, where she spent the remainder of her career. She served on the institute's advisory board for 24 years, was editor of its bibliographic service for 13 years, and was editor of the *Journal of Morphology and Physiology* for 3 years.

King received many honors and awards for her work, including the Ellen Richards Prize of the Association to Aid Scientific Research for Women in 1932. She was elected a fellow of the New York Academy of Sciences. Her other memberships included the American Association for the Advancement of Science, the American Society of Naturalists, the American Society of Zoologists, and the American Association of Anatomists.

King, Mary-Claire

b. 1946
Geneticist, Epidemiologist

Education: B.A., mathematics, Carleton College, 1966; Ph.D., genetics, University of California, Berkeley, 1973

Professional Experience: visiting professor, University of Chile, Santiago, 1973; assistant professor, epidemiology, School of Public Health, University of California, Berkeley, 1974–1980, associate professor, epidemiology, 1980–1984, professor, epidemiology, 1984–1996, professor, genetics and molecular biology, 1989–1996, American Cancer Society Professor, genetics and epidemiology, 1994–1996; professor, Genome Sciences and Medicine, University of Washington, 1996–

Mary-Claire King is renowned for her research on breast cancer. In 1990, she predicted the existence of the gene BRCA1 that, if damaged, can predispose women to breast and ovarian cancer. The next year, she and other researchers discovered the chromosomal location of a gene that causes a form of inherited deafness, and

Geneticist Mary-Claire King. (Courtesy of University of Washington/UnivPhoto)

another of her discoveries consists of the genetic clues to the reason some men infected with HIV-1 develop AIDS more rapidly than others. She began her research on breast cancer in the 1970s, but made very little headway until the early 1980s, when breakthroughs in molecular biology led to the mapping of more genetic markers. Her team was very close to finding the gene BRCA1 when it was located by a team at the University of Utah in 1994. Although disappointed, she continued her work on the location of other genes and on gene mutations. As a scientist, but also as a person with a family history of breast and ovarian cancer, she was convinced that there was a hereditary link to breast cancer and continues to call for the development of new tests to detect the gene.

King has worked with the Human Genome Diversity Project to examine why some early humans became ill when exposed to viruses or bacteria, while others did not. This research has been applied to her study of possible genetic reasons why some homosexual men who have been exposed to HIV develop AIDS, while others do not. She is the recipient of the Genetics Prize of the Peter Gruber Foundation (2004), the Heineken Prize for Medicine from the Royal Netherlands Academy of Arts and Sciences (2006), and the Weizmann Women & Science Award (2006). She has served on various committees of the National Cancer Institute, Institute of Medicine, Special Commission on Breast Cancer of the President's Cancer Panel, and National Institutes of Health (NIH). She was considered for the directorship of the NIH in 1991, but she declined this administrative position in order to focus on her research. She is a member of the American Society of Human Genetics and Society for Epidemiologic Research. She was elected to the National Academy of Sciences in 2005.

Further Resources

University of Washington. Faculty website. http://www.gs.washington.edu/faculty/king.htm.

Davies, Kevin and Michael White. 1996. *Breakthrough: The Race to Find the Breast Cancer Gene.* NY: John Wiley & Sons.

Klinman, Judith (Pollock)

b. 1941
Biochemist, Physical Organic Chemist

Education: B.A., chemistry, University of Pennsylvania, 1962, Ph.D., organic chemistry, 1966

Professional Experience: postdoctoral fellow, Isotope Department, Weizmann Institute of Science, Israel, 1966–1967; affiliate, chemistry, University College, London, 1967–1968; postdoctoral research associate, Institute for Cancer Research, Philadelphia, 1968–1972, assistant to associate member, 1972–1978; associate professor, chemistry, University of California, Berkeley, 1978–1982, professor, 1982–, professor, molecular and cell biology, 1993–

Concurrent Positions: assistant professor, medical biophysics, University of Pennsylvania, 1974–1978

Judith Klinman is renowned for bringing the principles and tools of physical organic chemistry to bear on biological processes. Her research has led to important breakthroughs in our understanding of protein function and structure, including the discovery of new cofactors (or vitamins) and the effect of oxygen on proteins. She also has been a leading figure in the use of isotope effects to probe enzymatic-reaction mechanisms and transition states.

After she received her doctorate, she was a postdoctoral fellow at the Weizmann Institute of Science in Israel and then was affiliated with the Department of Chemistry at the University College, London. First a postdoctoral associate at the Institute for Cancer Research in Philadelphia, she was promoted to research associate and later became an associate member there. Concurrently, she was assistant professor of biophysics at the University of Pennsylvania, where she taught and supervised graduate students. She joined the University of California, Berkeley in 1978 as an associate professor of chemistry and the first female faculty member in the physical sciences. She was promoted to full professor in 1982, and has held a joint appointment as professor of molecular and cell biology since 1993. Klinman is very active professionally, serving on the editorial board of scientific journals and giving lectures at universities and scientific organizations internationally.

In 1994, she was elected to the National Academy of Sciences and awarded the Repligen Award of the Division of Biological Chemistry of the American Chemical Society (ACS). She also received the Remsen Award of the ACS, a Merit Award from the National Institutes of Health (1992), and honorary doctorates from the University of Uppsala, Sweden (2000) and the University of Pennsylvania (2006). She has been named a fellow of the American Academy of Arts and Sciences, the Japanese Ministry of Science, and the American Philosophical Society. She is a member of the American Chemical Society, American Society of Biochemists and Molecular Biologists (president, 1998–1999), and Protein Society.

Further Resources

Wasserman, Elga. 2002. *The Door in the Dream: Conversations with Eminent Women in Science*. Washington, D.C.: Joseph Henry Press.

University of California, Berkeley. "Klinman Research Group." http://www.cchem .berkeley.edu/jukgrp/index.html.

Knopf, Eleanora Frances Bliss

1883–1974
Geologist

Education: A.B., A.M., Bryn Mawr College, 1904; Ph.D., petrology, University of California, Berkeley, 1912; Johns Hopkins University, 1917–1918

Professional Experience: assistant curator, geology museum, Bryn Mawr College, 1904–1905, 1908–1909, demonstrator, geology laboratory, 1905–1906; aide, U.S. Geological Survey (USGS), 1912–1917, assistant geologist, 1917–1918, associate geologist, 1918–1928, geologist, 1928–1970

Concurrent Positions: geologist, Maryland Geological Survey, 1917–1920; research associate, department of earth sciences, Stanford University, 1951–1966

Eleanora Knopf introduced rock fabric analysis and structural petrology in the United States. She gained recognition for her work after publishing her methods in the book *Structural Petrology* (1938). In 1913, Knopf announced her discovery of the mineral glaucophane in Pennsylvania; this was the first sighting of the substance east of the Pacific coast. One of her most important projects was at Stissing Mountain, a region on the New York–Connecticut border. In the course of her studies for this project, she decided to use structural petrology in analyzing her data, which led to the publication of her book on the topic.

After she received her undergraduate degree at Bryn Mawr, studying with **Florence Bascom**, she continued working at the school in the geology museum and the geology laboratory. She received her doctorate from the University of California, Berkeley in 1912, and began working with the USGS, where she spent her entire career. The USGS hired women geologists on a contract basis, and Knopf accepted other work at times. She worked for the Maryland Geological Survey before her marriage, and she had a long-term association with the department of earth sciences at Stanford from 1951 to 1966, when her husband, Adolph Knopf, was a faculty member there. While at Stanford, she made studies of several locations in the Rocky Mountains in Montana and the Spanish peaks in Colorado. When she lived in New Haven, she was a visiting lecturer at Harvard and Yale, but she did not have formal appointments.

Knopf wrote a chapter on "The Geologist" for a 1920 guide to *Careers for Women* (edited by Catherine Filene). She was elected a fellow of the Geological Society of America, and was a member of the American Geophysical Union.

Kopell, Nancy J.

b. 1942
Applied Mathematician

Education: B.S., mathematics, Cornell University, 1963; Ph.D., mathematics, University of California, Berkeley, 1967

Professional Experience: instructor, mathematics, Massachusetts Institute of Technology (MIT), 1967–1969; faculty, Northeastern University, 1969–1986; professor, mathematics, Boston University, 1986–2000, William Goodwin Aurelio Professor of Mathematics and Science, 2000–

Concurrent Positions: visiting faculty, Centre National de la Recherche Scientifique, France 1970, MIT, 1975, 1976–1977, California Institute of Technology, 1976

Nancy Kopell is one of the few mathematicians working in the field of applied biomathematics. Her research uses mathematical models to analyze biological and neurophysiological features of neurons, networks of cells that are responsible for physical and cognitive functions such as motor skills, behaviors, perception, learning, and sensory processing. She studied math as an undergraduate at Cornell University at a time when there were few women in the discipline and received her doctorate in mathematics from the University of California, Berkeley in 1967. She taught at

Mathematician, Nancy Kopell. (Courtesy of BU Photo Services)

MIT and Northeastern before joining the faculty at Boston University in 1986, where she is also co-director of the Center for BioDynamics.

Kopell was elected to the National Academy of Sciences in 1996 for her work in systems neuroscience. She has held several prestigious grants and fellowships, including a Guggenheim fellowship and a five-year MacArthur "genius grant" (1990–1995). She has been an invited speaker and lecturer at numerous professional organizations and universities, and was named the John von Neumann Lecturer by the Society for Industrial and Applied Mathematics in 2007.

Further Resources

Boston University. Faculty website. http://cbd.bu.edu/members/nkopell .html.

Case, Bettye Anne and Anne Leggett, eds. 2005. *Complexities: Women in Mathematics*. Princeton, NJ: Princeton University Press.

Wasserman, Elga. 2002. *The Door in the Dream: Conversations with Eminent Women in Science*. Washington, D.C.: Joseph Henry Press.

Koshland, Marian Elliott

1921–1997
Immunologist

Education: B.A., Vassar College, 1942; M.S., bacteriology, University of Chicago, 1943, Ph.D., immunology, 1949

Professional Experience: assistant, cholera project, Office of Scientific Research and Development, University of Chicago, 1943–1945; assistant, Commission on Air Borne Diseases, University of Colorado, 1943–1944; junior chemist, atomic bomb project, Manhattan district, Tennessee, 1945–1946; associate bacteriologist, Brookhaven National Laboratory, 1953–1962, bacteriologist, 1962–1965; associate research immunologist, University of California, Berkeley, 1965–1969, research immunologist, 1969–1970, professor, bacteriology and immunology, 1970–1997

Concurrent Positions: fellow, bacteriology and immunology, Harvard University, 1949–1951; visiting professor, Cancer Center, Massachusetts Institute of Technology, 1980, 1985–1986

Marian Koshland was an immunologist who made important contributions to the study of disease. She became interested in science as a child after watching a younger brother suffer from typhoid fever. Koshland (known as "Bunny" to her family and colleagues) worked her own way through Vassar College and went on to earn a master's degree from the University of Chicago in 1943. She had intended to go to medical school, but decided to pursue research while working at Chicago on a project to develop a cholera vaccine. This project was funded by the government in an effort to develop vaccines for soldiers serving overseas. She was also engaged in research to prevent respiratory infections and spent time at the University of Colorado on a World War II–era government project for the Commission on Air Borne Diseases. She had a contract assignment as a junior chemist with the Manhattan Project studying the effects of radiation before finally receiving her doctorate in immunology in 1949. She spent two years as a postdoctoral fellow at Harvard before returning to government work with the Brookhaven National Laboratory, where her husband, Daniel Koshland, was employed as a research scientist. When her husband accepted a position in California, she moved to the University of California, Berkeley as a researcher and then full-time professor of bacteriology and immunology. She made some of her most important scientific contributions early in her career while working only part-time and raising five children. She eventually published more than 200 articles.

Koshland was elected to the National Academy of Sciences in 1981. She was a member of the governing board of the National Science Foundation from 1976 to 1982 and was selected the R. E. Dyer Lecturer by the National Institutes of Health in 1988. She served on numerous boards and committees, including the National Institutes of Health, the National Science Board, and the National Council of the National Institute of Allergy and Infectious Diseases. She was a member of the American Association of Immunologists (president, 1982–1983), the Institute of Medicine, the American Society of Biological Chemists, and the American Academy of Microbiologists. She was a longtime member of the Board of Trustees for Haverford College in Connecticut, which has named its science research complex—the Marian E. Koshland Integrated Natural Sciences Center—in her honor. Haverford awarded her an honorary doctorate in 1995. In 2004, the Marian Koshland Science Museum of the National Academy of Science was established in Washington, D.C., with a gift from her husband.

Further Resources

Wasserman, Elga. 2002. *The Door in the Dream: Conversations with Eminent Women in Science*. Washington, D.C.: Joseph Henry Press.

Guyer, Ruth Levy. 2007. *Marian Elliot Koshland, 1921–1997: A Biographical Memoir*. Washington, D.C.: National Academy of Sciences.

Marian Koshland Science Museum. http://www.koshland-science-museum.org.

Kreps, Juanita (Morris)

b. 1921
Economist

Education: B.A., Berea College, 1942; M.A., Duke University, 1944, Ph.D., economics, 1948

Professional Experience: instructor, economics, Denison University, 1945–1946, assistant professor, 1947–1950; lecturer, Hofstra College, 1952–1954, and Queens College, 1954–1955; visiting assistant professor, economics, Duke University, 1955–1958, assistant to associate professor, 1958–1967, professor, 1967–1977, director, undergraduate economics studies and Dean of Women's College, 1969–1972, vice president, Duke University, 1973–1977; Secretary of Commerce, U.S. Department of Commerce, 1977–1979

Juanita Kreps is an economist focused on women's employment and was the first woman Secretary of Commerce of the U.S. Department of Commerce. She was also the first professional economist to hold that cabinet post. Prior to that time, the Secretaries had supported the interests of business, but she stated that she would support the interests of the public, including consumers, as well as those of business. The late 1970s was a period of high unemployment owing to the restructuring of industries, and corporations were experiencing increased competition from abroad in industries such as steel and automobiles, traditionally the strong sectors of U.S. industry. While she was working to revitalize industry, she was also working to increase social consciousness among businesspeople.

Kreps specialized in labor demographics with particular emphasis on the employment of women and older workers. In *Sex in the Marketplace: American Women at Work* (1971), she explored such questions as why women enter into the same types of occupations, why their proportion of advanced degrees remains so low, and why so many exchange the monotony of housework for equally dull and low-paying office and factory jobs. Another book, *Sex, Age, and Work: The Changing Composition of the Labor Force* (1975), explored the effect of women's increased presence in the workplace. In 1975, she organized a conference called "Women and the American Economy" that produced a statement endorsing the Equal Rights Amendment, recommending stronger affirmative action programs at universities, and urging public education for preschool children.

Kreps was born in a coal-mining region of Kentucky and grew up during the Great Depression. She worked her way through college on a work-study program. She decided to major in economics after her first class in the subject because it seemed especially relevant to her situation. As a highly respected economist, Kreps attracted attention from leading corporations that were under pressure in the early 1970s to add women to their boards of directors. She was named to the board of the New York Stock Exchange

Economist Juanita Kreps served as Secretary of Commerce under President Jimmy Carter from 1977 to 1979. (Department of Commerce)

plus the boards of several companies such as Western Electric and Eastman Kodak. After she completed her term as Secretary of the Department of Commerce, she continued to write and lecture, and she served on many committees and commissions. She is a fellow of the Gerontological Society of America and of the American Academy of Arts and Sciences, and a member of the American Economic Association.

Krim, Mathilde (Galland)

b. 1926
Geneticist, Virologist

Education: B.S., genetics, University of Geneva, Switzerland, 1948, Ph.D., cytogenetics, 1953

Professional Experience: junior scientist and research associate, Weizmann Institute, Israel, 1953–1959; research associate, virology, Division of Virus Research, Cornell University Medical College, 1959–1962; associate, Sloan-Kettering Institute of Cancer Research, 1962–1975, associate and member, 1975–1986; associate research scientist, St. Luke's Roosevelt Hospital Center and College of Physicians and Surgeons, New York City, 1986–

Concurrent Positions: adjunct professor, public health and management, Columbia University; founder and co-chair, American Foundation for AIDS Research (amfAR), 1985, chair, 1990–2004

Mathilde Krim is a distinguished geneticist and virologist who, since the 1980s, has devoted her time to raising funds for AIDS research. While working at the Sloan-Kettering Institute for Cancer Research on cancer viruses, she became intrigued with the possibility that the protein interferon, which is produced naturally by almost all animal species and even some plants, might inhibit tumors and modify some properties of the immune system in animals. She felt this would be a significant area of research, particularly in 1974, when a Swedish physician announced some success with interferon's stopping the recurrence of highly malignant bone cancer in a number of patients. She pressed Sloan-Kettering to establish an interferon laboratory and also sought funding from the National Institutes of Health and the National Cancer Institute. When the results of the Swedish tests were discredited, there was great controversy over the efficacy of interferon, which was very expensive as a natural substance. After a researcher cloned the interferon gene, it was possible to produce interferon in large quantities, and Krim was then appointed as head of Sloan-Kettering's interferon evaluation program; the Institute won Food and Drug Administration (FDA) approval to use interferon to treat certain types of leukemia.

Krim initially became involved in AIDS research through studies of the effectiveness of interferon in treating Kaposi's sarcoma, a cancer that afflicts many AIDS patients. She realized that the funding for AIDS research was inadequate, and in 1983, she founded the AIDS Medical Foundation, later merged with another

Mathilde Krim at a benefit for amfAR, the American Foundation for AIDS Research, in New York, 2007. (AP/Wide World Photos)

group to form the American Foundation for AIDS Research; amfAR is now the largest nonprofit AIDS research organization, funding all areas including gene therapy, prevention, drug treatments, and public-policy initiatives. Krim was able to unite the scientific and entertainment communities through her husband, the founder of Orion Pictures, and celebrity spokespersons such as actress Elizabeth Taylor helped raise public awareness about AIDS and the efforts of amfAR. In recent years, Krim has also focused on another health threat, multidrug-resistant tuberculosis (MDRTB), which is a serious problem among people who are HIV-positive as well as the homeless, people in prison, and the poor.

Krim has been a member of the Committee of 100 for National Health Insurance and president of the Commission to Study Ethical Problems in Medical, Biomedical, and Behavioral Research. She is a member of the American Association for the Advancement of Science, American Cancer Society, and American Association on Mental Deficiency. She was awarded the Presidential Medal of Freedom in 2000 by President Clinton in recognition of her "extraordinary compassion and commitment."

Further Resources

amfAR Aids Research. http://www.amfar.org.

Krueger, Anne (Osborn)

b. 1934
Economist

Education: B.A., economics, Oberlin College, 1953; M.S., University of Wisconsin, 1956, Ph.D., economics, 1958

Professional Experience: instructor, economics, University of Wisconsin, Madison, 1958–1959; assistant to associate professor, economics, University of Minnesota, 1959–1966, professor, 1966–1982; vice president, economics and research, World Bank, Washington, D.C., 1982–1986; professor, economics, Duke University, 1987–1993; professor, economics, Stanford University, 1993–2001; First Deputy Managing Director, International Monetary Fund, 2001–2006; professor, international economics, Johns Hopkins University, 2007–

Anne Krueger is an economist with expertise in international trade and economic development. She has been involved in developing international economic policy as a vice president and consultant for the World Bank and director of the International Monetary Fund from 2001 to 2006. She has been a longtime member or consultant on a variety of government and academic councils, including the

National Bureau of Economic Research, Institute for Global Economics, and Center for Policy Studies. While a professor of economics at Stanford, she was also director of the Center for Research on Economic Development and Policy Reform. She has authored, co-authored, or edited more than 15 books on trade, development, economic change, developing countries, exchange rates, and economic aid. The National Bureau of Economic Research sponsored her three-volume *Trade and Employment in Developing Countries* (1983; 2nd ed., 1988), and her other books include *Political Economy of Policy Reform in Developing Countries* (1994), *American Trade Policy: A Tragedy in the Making* (1995), *The World Trade Organization as an International Organization* (2000), *Transforming India's Economic, Financial and Fiscal Policies* (co-author, 2003), *Latin American Macroeconomic Reform: The Second Stage* (co-author, 2003), and *Economic Policy Reform and the Indian Economy* (2003).

Krueger was elected to membership in the National Academy of Sciences in 1995. Due to her international reputation, she has received honorary doctorates from Hacettepe University in Turkey (1990), Georgetown University (1993), Monash University in Melbourne, Australia (1996), and Chinese University of Hong Kong (2003). Among the other honors she has received are the Robertson Prize of the National Academy of Sciences (1984), Bernhard-Harms Prize of the Kiel Institute of World Economics (1990), Kenan Enterprise Award of the Kenan Charitable Trust (1990), and Frank E. Seidman Distinguished Award in Political Economy (1993). She is a fellow of the American Academy of Arts and Sciences and the Econometric Society, and a member of the American Economic Association (vice president, 1977–1978; president, 1996) and the Royal Economic Society.

Further Resources

Johns Hopkins University. Faculty website. http://www.sais-jhu.edu/faculty/krueger/index.htm.

Kübler-Ross, Elisabeth

1926–2004
Psychiatrist

Education: M.D., University of Zurich, 1957

Professional Experience: intern, Community Hospital, Glen Cove, New York, 1958–1959; research fellow, Manhattan State Hospital, 1959–1962; fellow, psychiatry, Psychopathic Hospital, University of Colorado Medical School, 1962–1963, instructor, psychiatry, Colorado General Hospital, 1962–1965; assistant professor, psychiatry, Billings Hospital, University of Chicago, 1965–1970, assistant director, Psychiatric Consultation and Liaison Service, 1965–1969, acting chief, Psychiatric Inpatient Service, 1965–1966, associate chief, Psychiatric Inpatient Service, 1966–1967; medical director, Family Service and Mental Health Center, South Cook County, Illinois, 1970–1973; president, Ross Medical Associates, 1973–1976; president and chair of the board, Shanti Nilaya Growth and Health Center, Escondido, California, 1977–1995; founder and president, Elisabeth Kübler-Ross Center, Virginia, 1990–1995

Concurrent Positions: resident, Montefiora Hospital, 1961–1962; staff member, LaRabida Children's Hospital and Research Center, Chicago, 1965–1970; clinical professor, Behavioral Medicine and Psychiatry, University of Virginia, 1985

Elisabeth Kübler-Ross was a psychiatrist who challenged the taboos surrounding death in our culture. She pioneered a new field of healthcare, "thanatology," the study of the effects of death and dying, especially the investigation of ways to lessen the suffering and address the needs of the terminally ill and their survivors. While teaching psychiatry courses in the 1960s, Kübler-Ross ran a series of conversations with the terminally ill in order to assess their feelings about the process of dying. She pointed out that treatment of the dying had changed over time, from taking place at home in the comforting presence of family and friends to occurring in impersonal institutional settings where death is seen as a failure of the technological expertise of physicians, who wish to prolong life. In her landmark book, *On Death and Dying* (1969), she identified the five stages that dying patients experience— denial, anger, bargaining, depression, and acceptance—and her work paved the way for the more humane treatment of the terminally ill by medical personnel. Hospice care was established as an alternative to dying in hospitals, and more emphasis was put on the emotional needs of patients and their families. The rights of terminally ill patients, however, have been the topic of much debate in recent years around the issue of assisted suicide.

After *Life* magazine published an article about her work, Kübler-Ross gained public attention and began receiving invitations to speak at seminars throughout the United States and Canada. She also continued to see patients and their families in her regular practice. Her book *AIDS: The Ultimate Challenge* (1987) was written for those suffering from the disease and focused on the medical, moral, and social implications of AIDS. She was committed to working with the patients directly and, in 1977, she created a center for the terminally ill and their families called "Shanti Nilaya" (Home of Peace) in Escondido, California, which continues her work to this day. She also operated the Elisabeth Kübler-Ross Center in Virginia to train those working professionally with the terminally ill, and co-founded the American Holistic Medical Association.

She received numerous honorary degrees and was a member of the American Association for the Advancement of Science, American Holistic Medical Association, American Medical Women's Association, American Psychiatric Association, and American Psychosomatic Society. She published numerous books, including her autobiography, *The Wheel of Life: A Memoir of Living and Dying* (1997), in which she claimed she had out-of-body experiences, meetings with spirit guides, and visions of fairies. An earlier biography was published by Derek Gill entitled *Quest: The Life of Elisabeth Kübler-Ross* (1980).

Further Resources

Elisabeth Kübler-Ross. http://www.elisabethkublerross.com/.

Kuhlmann-Wilsdorf, Doris

b. 1922
Physicist, Metallurgist

Education: B.S., materials science, University of Göttingen, 1944, M.S., physics, 1946, Ph.D., materials science, 1947

Professional Experience: fellow, materials science, University of Göttingen, 1947–1948; fellow, physics, Bristol University, England, 1949–1950; lecturer, physics, University of Witwatersrand, Johannesburg, 1950–1956; associate professor, metallurgy, University of Pennsylvania, 1957–1961, professor, 1962–1963; professor, engineering physics, University of Virginia, 1963–1966, professor, physics and metallurgical science, 1966–2005

Concurrent Positions: visiting professor, physics, Pretoria University, 1982–1983

Doris Kuhlmann-Wilsdorf is a metallurgist and materials scientist renowned for her design for electrical metalfiber brushes to be used as sliding electrical contacts. She holds patents on six inventions related to the electrical brushes. The brushes have application in electric motors that could replace the heavier and less efficient diesel engines. Her area of expertise is called *tribology*, which is the study of the effects of friction on moving machine parts and of methods of lubrication. Another of her contributions is the development of a model for surface deformation, which takes into account erosion as well as friction and wear. She has also investigated the behavior and properties of various metals, such as studying why rolled aluminum sheets crinkle under pressure, while other sheet metals break.

Prior to entering college in Germany, she served as an apprentice metallographer and materials tester for two years. After receiving her doctorate from Göttingen, she continued her research and studied under Nobel Laureate Nevill F. Mott. She and her husband, Heinz G. F. Wilsdorf, came to the United States and eventually both received appointments at the University of Virginia as professors in the Physics and Materials Science Departments. In 1994, the Wilsdorfs funded a professorship in their name, and in 2001, a gift from one of Doris Kuhlmann-Wilsdorf's former students established a memorial building on campus in their name.

Kuhlmann-Wilsdorf has published over 250 scientific papers and has served as a consultant to corporations such as General Motors Technical Center, Chemstrand Research Laboratories, and General Dynamics Corporation, as well as for the National Institute for Standards and Technology. She was elected to membership in the National Academy of Engineering in 1994. She has received numerous honors and awards, including the Society of Women Engineers Achievement Award (1989), Ragnar Helm Scientific Achievement Award of the Institute of Electrical and Electronics Engineers (1991), Medal for Excellence in Research of the American Society of Engineering Education (1965 and 1966), and Heyn Medal of the German Society of Materials Science (1988), and was named Christopher J. Henderson Inventor of the Year by the University of Virginia Patent Foundation in 2001 and 2006 Fellow of TMS-AIME (the Minerals, Metals, and Materials Society and the American Institute of Mining, Metallurgical, and Petroleum Engineers). She is also a fellow of the American Society for Metals and the American Physical Society, and a member of the American Society of Mechanical Engineers and Society of Women Engineers.

Kurtzig, Sandra L. (Brody)

b. 1946
Computer Scientist, Aeronautical Engineer

Education: B.S., chemistry and mathematics, University of California, Los Angeles, 1968; M.S., aerospace engineering, Stanford University, 1968

Professional Experience: mathematical analyst, TRW Systems, 1967–1968; marketing representative, General Electric Corporation, 1969–1972; president and CEO, ASK Computer Systems, 1972–1993; co-founder, eBenefits, 1996–

Sandra Kurtzig is a computer pioneer who founded ASK Computer Systems in 1972, the largest public company founded by a woman and one of the biggest success stories of the 1970s minicomputer boom. The company's integrated software products for manufacturers, primarily its MANMAN Information System, are industry standards and are available as turnkey solutions for minicomputers, particularly those manufactured by Digital Equipment and Hewlett-Packard. In the 1990s, Kurtzig expanded the product line by developing portable applications software to run on multiple computer platforms and adapted the software to specific niche markets, such as the automotive industry. Kurtzig was a young mother when she started ASK out of her apartment with only a $2,000 investment. The name of the company was derived from her and her husband's initials—Arie and Sandra Kurtzig. She started developing innovative programs for businesses, such as one for a newspaper company to monitor its carriers, and later created minicomputer programs and information systems to help manufacturers optimize inventory, improve product quality, reduce operating expenses, and improve customer service. She had the foresight to design software to run on minicomputers when they were just starting to become popular.

In 1994, her company was purchased by Computer Associates International for $310 million. She has since formed an online business software consulting firm with her son. Kurtzig's autobiography, *CEO: Building a $400 Million Company from the Ground Up* (1991), describes her experiences starting and running a successful business in a male-dominated field. She was one of only a few women studying math and aeronautical engineering in college, and she has rarely encountered other women in manufacturing companies or in upper management.

Kwolek, Stephanie Louise

b. 1923
Polymer Chemist

Education: B.S., chemistry, Carnegie Institute of Technology, 1946

Professional Experience: chemist, Fibers Department, Experimental Station, E. I. Du Pont de Nemours and Company, 1946–1959, research chemist, 1959–1967, senior research chemist, 1967–1974, research associate, 1974–1986; consultant, E. I. Du Pont de Nemours and Company, 1986–

Stephanie Kwolek invented a polymer that is manufactured by Du Pont under the trade name Kevlar. After graduating from college, she took what was supposed to be a temporary job at Du Pont while saving money to attend medical school. Her work was so interesting, however, that she stayed on with the company and became involved in the research that led to the discovery of low-temperature polymerization. She gained national attention in 1960 for her work creating long molecule chains at low temperatures and her discovery of the method to spin synthetic, petroleum-derived fibers in a liquid crystalline solution. The compound had such high tensile strength that she ran the tests again and again to make sure she had not made an error before reporting her discovery to the laboratory director. The resulting product, Kevlar, eventually led to a multimillion-dollar industry with more than 200 commercial applications, including use in radial tire cords, composites, rope, thermal insulating clothing, and bulletproof vests. At the time of her retirement, Kwolek owned 17 U.S. patents.

The use of Kevlar in bulletproof vests has earned Kwolek many fans and accolades. More than 2,000 police officers whose lives were saved due to wearing Kevlar vests formed a Survivors Club, a joint venture between Du Pont and the International Chiefs of Police Association. Kwolek is regularly contacted by individuals thanking her and even asking for an autograph. Even after her retirement, Kwolek continues to consult with Du Pont, as well as to give public and school lectures about careers in science. In 1996, she was featured along with other Du Pont employees in a series of print and television ads describing the company's research.

Kwolek was elected to the National Academy of Engineering in 2001 for "the discovery, development, and liquid-crystal processing of high-performance aramid fibers." She won an early publication prize from the American Chemical Society (1959), and has also received the Creative Invention Award of the ACS (1980) and the Perkin Medal of the Society of Chemical Industry (1997), only the second woman to receive that prize. She was inducted into the National Inventors Hall of Fame (1995) and the Women in Technology International (WITI) Hall of Fame (1996), and has received the nation's highest technology honor, the National Medal of Technology (1996). She has received honorary doctorates from Worcester Polytechnic Institute (1981) and from the University of Delaware (2008). She is a member of the American Chemical Society and American Institute of Chemists.

L

LaBastille, Anne

b. 1938
Ecologist

Education: B.S., conservation, Cornell University, 1955; M.S., wildlife management, Colorado State University, 1958; Ph.D., wildlife ecology, Cornell University, 1969

Professional Experience: wildlife tour leader, National Audubon Society, Palm Beach, Florida, 1955–1956; organizer and co-leader, Caribbean Wildlife Tours, Miami, Florida, 1956–1963; owner, co-manager, and naturalist, Covewood Lodge, Big Moose, New York, 1956–1964; ranger-naturalist, Everglades National Park, Florida, 1964; assistant professor, Department of Natural Resources, Cornell University, 1969–1971, research associate, Laboratory of Ornithology, 1971–1973; freelance wildlife ecologist, consultant, writer, and photographer, 1971–

Concurrent Positions: commissioner, Adirondack Parks Agency, 1976–1993; visiting lecturer and Basler Chair of Excellence for the Integration of the Arts, Rhetoric and Science, East Tennessee State University, 2001; owner, West of the Wind Publications, Eagle Bay, New York

Anne LaBastille is an ecologist who has fulfilled a variety of roles in support of wilderness conservation. She has done extensive work on preserving the wildlife habitat of several species of birds, including a project with a flightless bird known as the giant pied-billed grebe that was found at only one large lake in Guatemala. There was little known about this water bird until she began the first systematic study of its characteristics, and no photographs or drawings of it had ever been made. She established a sanctuary for the birds and monitored the population, obtaining grants from the World Wildlife Fund and the Smithsonian Institution to support her work. She persuaded the Guatemalan government to designate the grebe's habitat as the country's first wildlife refuge, but even so, the population dwindled. She published two early books on the folklore of birds, *Birds of the Mayas* (1964) and *Bird Kingdom of the Mayas* (1967). The local people called her "Mama Poc," based on the Indian name for the grebe, and she recorded her experiences in her book *Mama Poc: Story of the Extinction of a Species* (1990).

Ecologist Anne LaBastille has written numerous books and articles on wildlife and wilderness conservation. (AP/Wide World Photos)

After receiving her undergraduate degree from Cornell University, LaBastille spent her summers conducting wildlife tours in Florida and winters operating a lodge in upstate New York while working on her master's degree at Colorado State University. She returned to Cornell to obtain her doctorate and worked as a research associate in the internationally known Laboratory of Ornithology at Cornell while she started working freelance as a wildlife ecologist, writer, and photographer for organizations such as *National Geographic*. She has lived alone in two cabins she built in the upstate forestland of New York and is best known for her four-part autobiographical series describing her life among the plant and animal life in the Adirondacks: *Woodswoman* (1976), *Woodswoman II* (1987), *Woodswoman III* (1997), and *Woodswoman IV* (2003). Her book, *The Wilderness World of Anne LaBastille* (1993), consists of selections from never-before-published poems and short stories as well as color photographs. In *Women and Wilderness* (1980), she examines the historical role of other women living and studying in wilderness, including scientists such as **Eugenie Clark**, Jane Goodall, and others employed as park rangers, marine and wildlife biologists, professional environmentalists, or naturalists.

LaBastille's writings and activism have earned her the devotion of fans, both local and international. She has been a wilderness guide as well as an invited lecturer at universities and conservation groups. Her many awards and honors include a World Wildlife Fund Gold Medal (1974), Literature Award of the New York State Outdoor Education Association (1977), Citation of Merit from the Explorers Club (1987), Chevron Conservation Award (1988), Jade Chief's Award of the Outdoors Writer Association of America, and research grants from the International Union for Conservation of Nature and Natural Resources, Caribbean Research Institute, World Wildlife Fund, Smithsonian Institution, and other agencies. She has been a member of the Society of Women Geographers, American Women in

Science, Association for Tropical Biology, Wildlife Society, Outdoor Women Writers of America, and Explorer's Club.

Further Resources

Holmes, Madelyn. 2004. *American Women Conservationists: Twelve Profiles*. Jefferson, NC: McFarland.

Ladd-Franklin, Christine

1847–1930
Psychologist

Education: A.B., mathematics, Vassar College, 1869; Ph.D., mathematics, Johns Hopkins University, 1926; University of Göttingen, 1891–1892; University of Berlin, 1892, 1894, 1901

Professional Experience: high school teacher, 1869–1878; lecturer, psychology and logic, Johns Hopkins University, 1904–1909; lecturer, psychology and logic, Columbia University, 1914–1927

Christine Ladd-Franklin was one of the foremost women psychologists of the early twentieth century. Her research interests included color vision, deductive reasoning, the doctrine of histurgy, the one-time one-place theory of judgment, and proof that a nerve when stimulated emits physical light. Although she published papers on symbolic logic, her primary contribution to the history of psychology is her emphasis on the evolutionary development of increased differentiation in color vision, known as the Ladd-Franklin color theory. She published a compilation of her papers in *Colour and Colour Theories* (1929) and was invited to contribute an appendix to the English translation of Hermann von Helmholtz's classic *Handbook of Physiological Optics* (1924).

Ladd-Franklin studied mathematics at Vassar because there were no laboratory facilities available for study in physics. After she graduated in 1869, she taught high school science for 10 years, during which time she published articles on mathematics in the British journal *Educational Times* and the American journal *Analyst*. Originally denied admission to graduate study due to her sex, she was eventually admitted to Johns Hopkins University on a fellowship due to the recommendation of a mathematics professor who had read her papers. Although she fulfilled the requirements for a Ph.D. by 1882 with a thesis on "The Algebra of Logic," the trustees refused to grant the degree to a woman, as was the custom at that time; she finally received the degree in 1926. Still, even without the formal

Psychologist and mathematician Christine Ladd-Franklin. (National Library of Medicine)

degree, she held a lectureship in logic and psychology at Johns Hopkins from 1904 to 1909. In 1882, she married Fabian Franklin, a member of the mathematics department, and the couple moved to New York City in 1910 when he was appointed to an associate newspaper editor position. She spent the remainder of her career lecturing on logic and psychology at Columbia University.

Ladd-Franklin was a strong supporter of higher education, and she was instrumental in establishing research fellowships and even giving her money directly to women scientists who needed funds for research or travel. She published newspaper articles and editorials on women's education and status. Vassar College awarded her an honorary degree in 1887. She was a member of the American Association for the Advancement of Science, the American Society of Naturalists, the American Psychological Association, the Optical Society of America, and the American Philosophical Association.

Further Resources

Scarborough, Elizabeth and Laurel Furumoto. 1987. *Untold Lives: The First Generation of American Women Psychologists*. New York: Columbia University Press.

Laird, Elizabeth Rebecca

1874–1969
Physicist

Education: B.A., University of Toronto, 1896; Ph.D., physics and mathematics, Bryn Mawr, 1901

Professional Experience: instructor, mathematics, Ontario Ladies' College, 1896–1897; assistant, physics, Mount Holyoke College, 1901–1902, instructor,

1902–1903, acting head, 1903–1904, professor, 1904–1940; physicist, radar development, University of Western Ontario, 1941–1945, honorary professor of physics, 1945–1953

Concurrent Positions: Cavendish Laboratory, Cambridge University, 1909; physics laboratory, University of Chicago, 1919; honorary research fellow, Yale University, 1925

Elizabeth Laird was regarded as a notable physicist who spent most of her career teaching at Mount Holyoke College. Her research interests included spectroscopy, thermal conductivity, spark radiation, soft x-rays, the Raman effect, and electrical properties of biological material in the microwave region.

A native of Canada, she received her undergraduate degree at the University of Toronto, where she was awarded honors and fellowships in mathematics and physics. Denied a scholarship reserved for men to continue their graduate studies, she instead taught math for two years at an Ontario women's college before applying for admission at Bryn Mawr in Pennsylvania. For her research, she received a physics fellowship to work at the University of Berlin and received the doctorate in physics and mathematics from Bryn Mawr in 1901. She immediately joined the faculty of Mount Holyoke College, advancing very quickly in three years from instructor to professor. She stayed at Mount Holyoke for 40 years, training an entire generation of young women in the sciences. During World War II, she returned to Canada, where she spent four years as a physicist in radar development at the University of Western Ontario and also taught radio techniques for the Royal Canadian Air Force. Even after she officially retired, she continued her research on microwave radiation until at least 1953 as an honorary professor. She received honorary degrees from the University of Toronto (1927) and the University of Western Ontario (1954).

Laird was elected a fellow of the American Physical Society and was a member of the American Association for the Advancement of Science, the Optical Society of America, the American Association of Physics Teachers, the Canadian Association of Physicists, and the History of Science Society. She received several awards and honors. She received the American Association of University Women Sarah Berliner Research fellowship for study at the University of Wurzburg (1913–1914).

La Monte, Francesca Raimond

1895–1982
Ichthyologist

Education: B.A. and certificate of music, Wellesley College, 1918

Professional Experience: secretary, Department of Ichthyology, American Museum of Natural History, 1919–1923, 1925–1928, staff assistant, Department of Fishes and Aquatic Biology, 1928–1929, assistant curator, 1929–1935, associate curator, 1935–1962

Francesca La Monte was recognized for her work as an ichthyologist at the American Museum of Natural History. Her primary interests were marlin and swordfish (she participated in big-game fishing as a hobby), and she developed exhibits at the museum on these and other species. Soon after receiving her undergraduate degree from Wellesley, she joined the museum and she rose through the ranks to become associate, curator, retiring in 1962. It was not unusual for a woman to be appointed curator of a museum. In the nineteenth century, many women worked with fathers, husbands, or brothers as underpaid or unpaid staff in museums, arboreta, and herbaria; these positions sometimes evolved into paid professional jobs. La Monte was a specialist in taxonomic ichthyology and a valued member of the staff at the American Museum of Natural History. She was a member of the museum's Lerner-Cape Breton expeditions of 1936 and 1938, the Lerner-Bimini expedition of 1937, and the Chile-Peru expedition of 1940. She was a member of the fisheries committee for the 1939–1940 World's Fair in New York City. At the museum, she worked on the *Bibliography of Fishes* and, having grown up in Russia and England and also having spent time in France, Italy, and Germany as a child, she was able to translate numerous documents for the American bibliography as well as articles for an English-speaking audience. She was the museum's delegate to the International Zoological Congress in Padua, Italy in 1930, one of only five representatives from U.S. institutions.

La Monte was co-editor of *Field Book of Fresh Water Fishes of North America* (1938), *Game Fish of the World* (1949), and *The Fisherman's Encyclopedia* (1950). She was co-author of *Vanishing Wilderness* (1934) and author of *North American Game Fishes* (1945), *Marine Game Fishes of the World* (1952), and *Giant Fishes of the Ocean* (1966). She was elected a fellow of the New York Academy of Sciences. Her other professional memberships included the American Association for the Advancement of Science, the American Society of Ichthyologists and Herpetologists, and the Society of Systematic Zoology.

Further Resources

Brown, Patricia Stocking. 1994. "Early Women Ichthyologists." *Environmental Biology of Fishes*. 41:9–30. http://swfsc.noaa.gov/uploadedFiles/Education/Women%20in%20 Ichthyology.pdf.

Lancaster, Cleo

b. 1948
Physiologist

Education: B.S., Elizabeth City State University, 1971; M.S., biomedical science, Western Michigan University, 1979

Professional Experience: research assistant, Brookhaven National Laboratory, 1971; research associate, Upjohn Company, 1971–1989, senior research associate, pharmacology, 1989–

Cleo Lancaster is a pioneer in biological research leading to new ulcer therapies and an expert in the field of prostaglandin cytoprotection, which is the cellular protection of the gastric lining by the use of hormonelike fatty acids. At Upjohn Company (later acquired by Pfizer), she has developed experimental models of such gastrointestinal diseases as ulcers, diarrhea, pancreatitis, and colitis in order to discover natural or synthetic chemicals to treat such conditions. In the early 1970s, she studied the ulcer-causing effects of nicotine and linked smoking to duodenal ulcers in humans. She has examined the effects of ibuprofen, aspirin, and alcohol as irritants to the gastrointestinal tract, and she has also examined a steroid used in organ transplant patients that causes ulcers. Her research revealed that fatty acids known as prostaglandins can be used to inhibit gastric acid secretion by stimulating mucus/bicarbonate production and increasing the cell resistance of the stomach lining, thus preventing ulcers. She holds two patents: one for a treatment of pancreatitis and the second for treating ulcers with oxalate derivatives. She also contributed to developing surgical techniques for the research of gastric secretion.

Lancaster grew up on a farm learning about anatomy and veterinary science from working with animals. She originally planned to be a biology teacher, but in her third year of college, she decided on a career in research because she wanted the challenge of discovery. She worked as a research assistant in radiation genetics the summer after she received her undergraduate degree, and she joined the Upjohn Company in the fall as a research associate in gastrointestinal, or ulcer, research. She received her master's degree in biomedical science from Western Michigan University while working for Upjohn. She received the Laboratory Special Recognition Award of the Upjohn Company and the Mary McLeod Bethune Award for Science and Technology. She is a member of the American Association for the Advancement of Science and the New York Academy of Sciences.

Lancefield, Rebecca Craighill

1895–1981
Bacteriologist

Education: A.B., Wellesley College, 1916; A.M., Columbia University, 1918, Ph.D., immunology and bacteriology, 1925

Professional Experience: high school teacher, 1917; technical assistant, Rockefeller Institute, 1918–1919; department of genetics, Carnegie Institution, 1919–1921; instructor, bacteriology, University of Oregon, 1921–1922; technical assistant, Rockefeller Institute, 1922–1929, associate, 1929–1958, professor, microbiology, 1958–1965

Rebecca Lancefield was recognized among microbiologists as the outstanding authority on streptococci. Her research was in immunochemical studies of strepto-cocci, and the chemical composition and antigenic structure of hemolytic strepto-cocci. Both national and international organizations devoted to streptococcal problems have renamed their groups the Lancefield Society in her honor. While she was attending Wellesley College, she became interested in the biology course her roommate was taking, and she switched her major from French and English to biology. She was able to receive a scholarship offered specifically for daughters of Army and Navy officers to attend Columbia University. Lancefield obtained a position as a technical assistant at the Rockefeller Institute for Medical Research, working on streptococci. The group identified four distinct serological types that served to classify 70% of the 125 strains studied; her name was included as a co-author of the paper reporting this work, a distinct honor so early in her career.

After teaching for a year at the University of Oregon, she and her husband returned to Rockefeller, where Lancefield remained the rest of her career. She worked with rheumatic fever research and received her doctorate in 1925. She returned to her studies of hemolytic streptococci, in which she provided a basis for understanding the clinical and epidemiological patterns of disease caused by these organisms. The research at that time was concentrated on puerperal fever, wound infections, and pneumonia that followed measles or influenza. Later research involved scarlet fever and rheumatic fever. In the mid-1920s, she suc-ceeded in obtaining two antigens in soluble form from hemolytic streptococci, one that was type-specific and one that was species-specific. She continued her research on streptococci until a few months before her death in 1981.

Lancefield was elected a member of the National Academy of Sciences in 1970. Among the other honors and awards she received were the Jones Memorial Award of the Helen Hay Whitney Foundation (1960), the Research Achievement Award of the American Heart Association (1964), and the Medal of the New York

Academy of Medicine (1973). As further recognition within the field, she was elected president of the American Society for Microbiology in 1943, the second woman to be elected president of the organization, and served as the first woman president of the American Association of Immunologists in 1961 and 1962. She also was a member of the American Association for the Advancement of Science and the Harvey Society. She received honorary degrees from the Rockefeller Institute (1973), and from Wellesley College (1976) on the sixtieth anniversary of her graduation.

Leacock, Eleanor (Burke)

1922–1987
Cultural Anthropologist

Education: student, Radcliffe College, 1939–1942; B.A., Barnard College, 1944; M.A., Columbia University, 1946, Ph.D., anthropology, 1952

Professional Experience: research assistant, psychiatry, Cornell University Medical College, 1952–1955; lecturer, anthropology and sociology, Queens College, 1955–1956; special consultant, U.S. Department of Health, Education, and Welfare, 1957–1958; co-director of research, suburban interracial housing, Teaneck, New Jersey, 1958–1960; senior research associate, schools and mental health project, Bank Street College of Education, 1958–1965; lecturer, history and economics, Polytechnic Institute of Brooklyn, 1962–1963, associate professor to professor, anthropology, 1963–1972; professor, anthropology, City College of New York, 1972–1987

Concurrent Positions: lecturer, City College of New York, 1956–1960, 1966–1967, and Washington Square College, 1960–1961

Eleanor Leacock was a prominent cultural anthropologist known for her studies of the changing social and gender relations among the natives of Labrador, her reevaluations of the work of the Marxist Friedrich Engels, her contributions to feminist theory, and her analyses of racism in American education. When she accompanied her first husband, a filmmaker, to Europe in 1948, she began archival research on changes in the social organization of an Indian people in Labrador, the Montagnais-Naskapi (Innu), following the introduction of the fur trade. The next year, she started her field research in Labrador, and her research changed the prevailing interpretation of private property in hunter-gatherer societies. She found that although the rights to trap in given places were privatized, the rights to gather, fish, or hunt for food were still communal. It had been thought that these societies were patriarchal, but she found that there was flexibility in the relations between women and men. She recorded stories that the residents told her, typed them, and presented them to the tribes.

Leacock was exposed to radical social theories early in life, for her father was a literary critic and social philosopher whose social circle included artists, political radicals, and writers in Greenwich Village. As a college student, she was active in student radical groups, and when she applied for a job in Washington, D.C., in 1944, the Federal Bureau of Investigation denied her clearance. She held various research and teaching positions before becoming a professor of anthropology and achieved recognition for her work on anthropology and education, on class and culture in urban schools, and on reevaluating the work of early Marxists. She published more than 70 papers and books before dying unexpectedly in Honolulu in 1987 after suffering a stroke in Western Samoa, where she was conducting fieldwork. She was a fellow of the American Anthropological Association and the Society for Applied Anthropology, and a member of the American Ethnological Society.

Further Resources

Gacs, Ute et al. 1988. *Women Anthropologists: Selected Biographies*. Westport, CT: Greenwood Press.

Leavitt, Henrietta Swan

1868–1921
Astronomer

Education: Oberlin College, 1885–1886; A.B., Radcliffe College, 1892

Professional Experience: volunteer research assistant, Harvard College Observatory, 1895–1900, staff member, 1902–1921

Henrietta Leavitt discovered the period-luminosity law, that is, the relation between a star's magnitude and its period of luminosity. This work involved determining the magnitude (brightness) of a star from a photographic image. At the turn of the century, visual photometry was superseded by photographic methods because the photographic plate is more sensitive to light of certain wavelengths than is the human eye. Another of her contributions to astronomy was the discovery of 2,400 variable stars, about half of the total known at the time. Her most important scientific contribution resulted from her study of the Cepheid variable stars in the Magellanic Clouds. She also studied color indices, which is the difference in magnitude of a star depending on the color-sensitivity of photographic plates.

Born in Massachusetts, she began her college studies in music at Oberlin College in Ohio. She moved to Radcliffe College in 1888, where she earned a bachelor's degree in 1892. She took a course in astronomy during her senior year and another after graduation. She returned to Harvard College Observatory as a volunteer research assistant in 1895 and was appointed to the permanent staff in 1902. She was assigned by the director of the Harvard College Observatory to develop photographic measurements that were eventually accepted among the astronomers of the world and became known as the Harvard Revised Magnitude Scale, or Harvard Standard. Leavitt soon became head of the department of photographic stellar photometry at the Observatory, although much of her work was published in reports under the name of the Observatory Director at the time, Edward C. Pickering. Although she never received the recognition as some other female astronomers of her generation, such as **Annie Jump Cannon**, Leavitt certainly deserved it. Her research revealed what are now known to be satellite galaxies of the Milky Way, and her methods helped later astronomers to determine the distances from the Earth of similar stars within our own galaxy and in distant galaxies. Interestingly, in a discipline that relied so heavily on sight and analysis of detailed imagery, both Cannon and Leavitt were partially deaf.

Leavitt was a member of the American Association of University Women, the American Association for the Advancement of Science, and the American Astronomical and Astrophysical Society. She was elected an honorary member of the American Association of Variable Star Observers. Both an asteroid and a moon crater are named after her.

Further Resources

Byers, Nina and Gary A. Williams. 2006. *Out of the Shadows: Contributions of Twentieth-Century Women to Physics*. New York: Cambridge University Press.

Johnson, George. 2005. *Miss Leavitt's Stars: The Untold Story of the Woman Who Discovered How to Measure the Universe*. New York: W.W. Norton.

Ledley, Tamara (Shapiro)

b. 1954
Climatologist

Education: B.S., University of Maryland, 1976; Ph.D., meteorology, Massachusetts Institute of Technology, 1983

Professional Experience: research associate, Space Physics and Astronomy and Earth Systems and The Energy and Environment Systems Institute, Rice University, 1983–1985, assistant research scientist, 1985–1990, senior faculty fellow, 1990–1998; senior scientist, TERC, Cambridge, Massachusetts, 1997–, interim director, Center for Science Teaching and Learning, 2009–

Concurrent Positions: consultant, Houston Museum of Natural Science, 1989–1990; director, teacher training program, George Observatory, Rice University, 1990–1992; visiting lecturer, geology and geophysics, Rice University, 1993; assistant director, Summer Solar Institute, Rice–Houston Museum of Natural Science, 1993; associate editor, *Journal of Geophysical Research—Atmosphere*, 1993; associate research scientist, Texas A&M University, 1995–1996; lecturer, mathematics and sciences, Babson College, Massachusetts, spring 1997; visiting scientist, Department of Earth, Atmospheric and Planetary Sciences, Massachusetts Institute of Technology, 1997–1998; adjunct professor, University of Massachusetts, Dartmouth, 2008–2009

Tamara Ledley is known for her research on the role of the polar regions in shaping climate and has examined how the interaction of atmosphere and sea with ice and oceans influences climate change. She has conducted research in both Alaska and Antarctica, and been active in presenting information on climatology to elementary school children as well as to university students. There is sometimes confusion about the difference between meteorology and climatology. Meteorology is the science dealing with the atmosphere and its phenomena, including weather, while climatology is the science that deals with the phenomenon of climate or climatic conditions. Ledley has consulted on numerous private and government projects related to climate change. She was a member of the working team at the Alaska facility for the National Aeronautics and Space Administration (NASA) (1988) and a member of the McMurdo Sound working team (1990). She also was a participant in the workshop on the Arctic initiative of the Office of Naval Research (1988), a participant in the U.S. Global Change Research program's climate modeling forum (1988), and a member of the committee on global and environmental change of the American Geophysical Union (1993).

In addition to several academic affiliations, Ledley has participated in many outreach programs bringing science to the public and the schools through science curriculum building and teacher training programs on climatology, and is a senior scientist at TERC, Inc., a producer of science and math education curriculum and programs. She has received grants from the National Science Foundation to fund various classroom and teacher professional development materials, including the Earth Exploration Toolbook and the Digital Library for Earth System Education.

Ledley is a member of the American Association for the Advancement of Science, American Meteor Society, and Ocean Society.

Further Resources

TERC. "Tamara Shapiro Ledley." Earth Exploration Toolbook, TERC, Carleton College. http://serc.carleton.edu/eet/people/ledley.html.

Leeman, Susan (Epstein)

b. 1930
Endocrinologist, Physiologist

Education: B.A., Goucher College, 1951; M.A., Radcliffe College, 1954, Ph.D., physiology, 1958

Professional Experience: instructor, physiology, Harvard Medical School, 1958–1959; fellow, neurochemistry, Brandeis University, 1959–1962, senior research associate, biochemistry, 1962–1966, adjunct assistant professor, 1966–1968, assistant research professor, 1968–1971; assistant to associate professor, physiology, Laboratory of Human Reproduction and Reproductive Biology, Harvard Medical School, 1972–1980; professor, physiology, Medical School, University of Massachusetts, 1980–1992, director, Interdepartmental Neuroscience Program, 1984–1992; professor, pharmacology, Boston University Medical School, 1992–

Endocrinologist and physiologist, Susan Leeman. (Courtesy of BU Photo Services)

Susan Leeman is considered one of the founders of the field of neuroendocrinology based on her research on peptides. She is renowned for her work with substance P and neurotensin, peptides that help govern the functioning of the nervous, endocrine, and immune systems. Neuroendocrinology is the study of the anatomical and physiological interactions between the nervous and endocrine systems. During the 1960s, she made a chance finding of a chemical that turned out to be substance P, a transmitter that is distributed throughout both the central and the peripheral nervous systems and the spinal cord, which had been discovered in the 1930s but had never been isolated. She and her colleagues isolated and characterized the peptide as well as discovering another one, neurotensin, which is involved in the relaxation and contraction of the blood vessels and may be involved in psychiatric disorders and, perhaps, regulation of the menstrual cycle. While in graduate school, Leeman began her work on corticotropin, a hormone used in the treatment of rheumatoid arthritis and rheumatic fever. It was while she was trying to purify corticotropin that she made the chance finding of substance P.

Leeman was elected to membership in the National Academy of Sciences in 1991 and received the Academy's Fred Conrad Koch Award in 1994. She has received numerous other awards, including the Excellence in Science Award of Eli Lilly and Company (1993). She is a member of the Endocrine Society, Society for Neuroscience, American Association for the Advancement of Science, and American Physiological Society.

Further Resources

Boston University Medical School. Faculty website. http://www.bumc.bu.edu/Dept/Content.aspx?DepartmentID=65&PageID=7764.

LeMone, Margaret Anne

Meteorologist

Education: A.B., mathematics, University of Missouri, 1967; Ph.D., atmospheric sciences, University of Washington, 1972

Professional Experience: postdoctoral fellow, Advanced Study Program, National Center for Atmospheric Research (NCAR), 1972–1973, acting project leader, GATE Group, 1974–1975, Ph.D. scientist, 1973–1978, staff scientist, Mesoscale Research Section, 1978–1980, staff scientist, GATE Group, Cloud Systems Division, 1980–1982, scientist, 1982–1992, senior scientist, 1992–

Concurrent Positions: affiliate professor, Colorado State University, 1984–1990; adjoint professor, University of Colorado, 1994–; affiliate professor, Colorado State University, 1996–; Advanced Study Program and National Oceanic and Atmospheric

Administration (NOAA) Aeronomy Lab, 1998–1999; Chief Scientist, Global Learning and Observations to Benefit the Environment (GLOBE), 2003–

Margaret "Peggy" LeMone is a meteorologist whose research focuses on storm and cloud systems. She is considered an observational meteorologist because she focuses on the lower area of the Earth's atmosphere known as the *planetary boundary layer*. She combines aircraft and radar observations with mathematical models to understand the relationship between atmospheric weather systems and the Earth's surface in terms of vegetation, soil properties, and terrain. She has conducted weather-watching fieldwork around the world, including in West Africa, Australia, the Solomon Islands, Mexico, and Taiwan. She has been affiliated with NCAR in a variety of staff scientist and researcher positions since 1972, and has been senior scientist there since 1992. Since 2003, she has served as Chief Scientist of GLOBE, an international earth sciences educational program supported in part by the National Science Foundation and National Aeronautics and Space Administration (NASA).

LeMone has written numerous scientific, articles, pamphlets, encyclopedia entries, and weather portions of elementary and high school textbooks. She has served and consulted on numerous government research boards, including National Research Council Committees on Road Weather, on Tools for Tracking Chemical/Biological/Nuclear Releases in the Atmosphere, on Improving the Effectiveness of U.S. Climate Modeling, and on Atmospheric Sciences and Climate. She has also consulted for the U.S. Department of Energy, National Science Foundation, and NOAA. LeMone has also been committed to science education from elementary through high school and college, and has been an invited speaker and mentor for organizations promoting women and minorities in the sciences, including as founding chair (1975–1978) of the American Meteorological Society (AMS) Board on Women and Minorities. She has also written several articles on women scientists working in meteorology.

LeMone was elected to the National Academy of Engineering in 1997. She is a fellow of the American Association for the Advancement of Science, American Geophysical Union, and American Meteorological Society (president, 2010). She is the recipient of an Editor's Award of the AMS *Journal of Atmospheric Sciences* (1989), the NCAR Education Award (1995), and the AMS Charles Anderson Award (2004).

Further Resources

National Corporation for Atmospheric Research. "Margaret (Peggy) LeMone." http://box.mmm.ucar.edu/individual/lemone/.

Leopold, Estella Bergere

b. 1927
Paleoecologist

Education: B.A., University of Wisconsin, 1948; M.S., University of California, Berkeley, 1950; Ph.D., botany, Yale University, 1955

Professional Experience: assistant research hydrologist, Laboratory of Tree Ring Research, University of Arizona, 1951; mycologist, Forest Products Laboratory, Madison, Wisconsin, 1952; research assistant, Genetics Experiment Station Research, Smith College, 1952; teaching assistant, plant science and zoology, Yale University, 1952–1954; research botanist, Paleontology and Stratigraphy Branch, U.S. Geological Survey, Denver, Colorado, 1955–1976; director, Quaternary Research Center, University of Washington, Seattle, 1976–1982; professor, Department of Botany and College of Forest Resources, University of Washington, Seattle, 1982–1999, emeritus

Concurrent Positions: adjunct professor, biology, University of Colorado, 1967–1976; visiting professor, Department of Botany and Institute for Environmental Studies, University of Wisconsin, Madison, 1971–1972; member and chair, Aldo Leopold Foundation, 1996–2004 (president 1996–1998 and 2004)

Paleoecologist, Estella Leopold.
(Courtesy of University of Washington/
UnivPhoto)

Estella Leopold is one of the leading authorities on paleoecology, which is the study of prehistoric organisms and their environments. She describes her work as comparing the pollen and spores that exist today with those found in rocks for a particular earlier time period. In this way, researchers try to determine the landscape and climate represented by fossils, which are probably the most important evidence of environments of the past. In her research in the Rocky Mountains, she found that extinction and evolution are highest in the middle of the continent because of the variable seasonal changes, while the coastal areas, which have more moderate climates, are able to sustain older species, such as the giant redwood. She was one of the leaders in the successful campaign to save Colorado's Florissant fossil beds, and in 1962, the National Park Service decided to designate the fossil beds as a national monument, but did not enact legislation. Meanwhile, developers started building recreational subdivisions in the park. In 1969, the Defenders of Florissant, Inc. persuaded the U.S. Congress to enact legislation to designate 6,000 acres for the national monument. She was a past director of the Quaternary Research Center at the University of Washington, Seattle (the Quaternary period, the present period of the Earth's history, originated about 2 million years ago).

Leopold developed her interest in ecology in her childhood under the tutelage of her father, the conservationist and writer Aldo Leopold. Estella received an undergraduate degree in botany from the University of Wisconsin, Madison, where her father taught wildlife management. Growing up, the family regularly spent weekends on a farm, where they planted tree seedlings and restored an old cornfield back to a tall-grass prairie. All five Leopold children followed careers in science, and Estella and her two brothers, Starker and Luna, are all members of the National Academy of Sciences.

Leopold was elected to membership in the National Academy of Sciences in 1974. She has served on many distinguished scientific committees on conservation and ecology, and is a fellow of the American Association for the Advancement of Science (president, 1995) and of the Geological Society of America. She is a member of the American Quaternary Association (president, 1982–1984), Botanical Society of America, Ecological Society of America, and American Academy of Arts and Sciences. She is also on the board and past president of the Aldo Leopold Foundation, which works for ecological and environmental awareness and protection in her father's name.

Further Resources

University of Washington, Seattle. "Pollen and Seed Laboratory." http://protist.biology.washington.edu/eleopold/.

The Aldo Leopold Foundation. http://www.aldoleopold.org.

Lesh-Laurie, Georgia Elizabeth

b. 1938
Developmental Biology

Education: B.S., Marietta College, 1960; M.S., University of Wisconsin, 1961; Ph.D., biology, Case Western Reserve University, 1966

Professional Experience: instructor, biology, Case Western Reserve University, 1965–1966; assistant professor, biological science, State University of New York, Albany, 1966–1968; assistant professor, Case Western Reserve University, 1969–1973, associate professor, 1974–1977, assistant dean, 1973–1976; professor, biology, Cleveland State University, 1977–1990, dean, College of Graduate Studies, 1981–1986, dean, College of Arts and Sciences, 1986–1990, interim provost, 1989–1990; vice chancellor of academic affairs, University of Colorado, Denver, 1990–1995, interim chancellor, 1995–1997, chancellor, 1997–2003

Georgia Lesh-Laurie is renowned for her research on a drug that can be used in place of digitalis for the treatment of congestive heart failure. Digitalis, made from the purple foxglove plant, increases the heart's pumping power without increasing oxygen demand, but patients with kidney problems are unable to use it. Lesh-Laurie's stimulant is a protein found in the toxin of the hydra, a small freshwater cousin of the jellyfish, and the protein was discovered after people stung by jellyfish noticed a sudden neurological and cardiovascular response. Sponsored by the American Heart Association, she continued work in the 1980s on developing a drug incorporating the protein.

Early in her career, Lesh-Laurie assumed administrative responsibilities in addition to her teaching and research. She served as assistant dean for three years at Case Western Reserve, and at Cleveland State University, she was department chair, dean of the College of Graduate Studies, dean of the College of Arts and Sciences, and interim provost for a year. She moved to the University of Colorado as vice chancellor of academic affairs in 1990 and then served as chancellor of that institution for seven years before retiring in 2003. Lesh-Laurie has been a member of the American Association for the Advancement of Science, American Society of Zoologists, Society for Developmental Biology, New York Academy of Sciences, and American Society for Cell Biology.

L'Esperance, Elise Depew Strang

1878–1959
Pathologist

Education: M.D., Woman's Medical College of New York, 1900

Professional Experience: intern, New York Babies Hospital, 1900; physician, private practice, 1901–1908; physician and instructor, Cornell University Medical Center, 1910–1920, assistant professor, 1920–1932; director, Kate Depew Strang

Tumor Clinic, New York Infirmary, 1933–1941; associate professor to professor, Cornell University Medical Center, 1942–1950

Concurrent Positions: editor, *Medical Woman's Journal*, 1936–1941; editor, *Journal* of the American Medical Women's Association, 1946–1948

Elise L'Esperance was a physician who established family clinics and promoted the early detection and treatment of cancer. Her research focused on the pathology and treatment of malignant tumors. Because their mother died from cancer, L'Esperance and her sister, May Strang, used an inheritance to open the first of three clinics in New York City devoted to the detection of cancer in 1933. The clinic offered complete physical examinations to apparently healthy women and provided referral service for any sign of cancer. Several new techniques were developed at the Strang clinics, such as the Pap smear for the diagnosis of cervical cancer. She staffed the clinic entirely with women physicians, and she conducted an extensive campaign of public education. Later, she opened other clinics where the services were expanded to men and children. Other groups in other cities built upon this model; the value of early detection became more widely accepted both by the public and the medical profession. She also worked in the fields of tuberculosis and Hodgkin's disease.

L'Esperance was a member of the last class to graduate from the Women's Medical College of New York in 1899, but, due to an attack of diphtheria, did not receive her degree until the next year. After serving her internship, she engaged in private practice in New York and Detroit. She became increasingly interested in pathology, and she accepted a position at Cornell's medical college. She left Cornell to direct the Kate Depew Strang Tumor Clinic in New York for eight years, but returned to teaching, eventually advancing to full professor right before she retired in 1950.

In addition to focusing her efforts on women's health, and serving as editor of two women's health journals, L'Esperance actively promoted careers in medicine for women. She received numerous awards, the most prestigious of which was the Albert Lasker Award of the American Public Health Association (1951). She also received the Elizabeth Blackwell Citation in 1950 for her achievements in pathology and cancer detection. She was elected a fellow of the New York Academy of Medicine, and she was named an honorary member of the American Radiologists Society. She was president of the American Medical Women's Association in 1948. Her other memberships included the American Medical Association, the American Association of Pathologists and Bacteriologists, the American Association of Immunologists, the American Radium Society, the Harvey Society, and the American Cancer Society.

Levelt-Sengers, Johanna Maria Henrica

b. 1929
Physicist

Education: B.Sc., physics and chemistry, University of Amsterdam, 1950, M.S., 1954, Ph.D., physics, 1958

Professional Experience: research associate, Van der Waals Laboratory, University of Amsterdam, 1958–1959; research physicist, 1959–1963; physicist, Heat Division, U.S. National Bureau of Standards, 1963–1984, physicist and senior fellow, National Bureau of Standards/National Institute of Standards and Technology, 1984–1995, emeritus

Concurrent Positions: research associate and instructor, theoretical chemistry, University of Wisconsin, 1958–1959; lecturer, University of Nijmegen, Netherlands, 1962–1963; visiting professor, University of Louvain, Belgium, 1971; visiting research scientist, Instituut voor Theoretische Fysica, Amsterdam, 1974–1975

Johanna Levelt-Sengers is renowned for her research on critical phenomena and fluid mixtures. Her research included thermodynamic properties of fluids and fluid mixtures; critical phenomena in fluids; equation of state, theoretical and experimental; and supercritical aqueous systems. She has been involved in establishing indexes or standards on water and steam properties and power for the International Association for the Properties of Water and Steam (IAPWS) and the American Society of Mechanical Engineers (ASME) Steam Tables. Her husband, Jan V. Sengers, is also a physicist, and the two came to the United States in 1963 to work for the National Bureau of Standards. The Sengers have collaborated and published numerous papers together, and in 1992, the couple were awarded honorary doctorates from the Technical University of Delft in their home country of The Netherlands. In 1995, she retired from a more than 30-year career at the National Institute of Standards and Technology (formerly National Bureau of Standards), but has remained active as a conference organizer, committee member on the ASME International Steam Tables, and author of a book on thermodynamics, *How Fluids Unmix*, published in 2002. She has also co-chaired the InterAcademy Council's advisory panel on promoting women in science and technology careers.

Levelt-Sengers is one of the few women scientists who has been elected a member of both the National Academy of Engineering (1992) and the National Academy of Sciences (1996). She has received numerous other awards, such as the Edward Uhler Condon Award (1975), Special Achievement Award (1977), and Certificate of Recognition (1978), all from the National Bureau of Standards;

the Department of Commerce Silver Medal (1972) and Gold Medal (1978) awards; the Interagency Committee for Women in Science and Engineering's WISE Award (1985); the Alexander von Humboldt Award (1991); the L'Oréal-UNESCO Women in Science Award (2003); and ASME's Yeram S. Touloukian Award (2006). She is a fellow of the American Physical Society and a member of the American Society of Mechanical Engineers, American Institute of Chemical Engineers, American Chemical Society, and International Association for the Properties of Water and Steam (president, 1991–1991; U.S. national representative, 1990–2004). She is also a member of the Royal Netherlands Academy of Arts and Sciences and the Royal Holland Society of Sciences and Humanities.

Further Resources

National Institute of Standards and Technology. "Johanna M. H. Anneke Levelt Sengers (Scientist Emeritus)." http://www.boulder.nist.gov/div838/ProfilesSengers.html.

Leverton, Ruth Mandeville

1908–1982
Nutritionist

Education: B.S., home economics, University of Nebraska, 1928; M.S., nutrition, University of Arizona, 1932; Ph.D., nutrition, University of Chicago, 1937

Professional Experience: high school teacher, 1928–1930; teaching fellow, home economics, University of Arizona, 1930–1932, assistant, experiment station, 1932–1934; assistant professor, home economics, University of Nebraska, 1937–1940; associate specialist, Bureau of Home Economics, U.S. Department of Agriculture (USDA), 1940–1941; associate professor, home economics and director, human nutrition research, University of Nebraska, 1941–1949, professor, 1949–1953; professor, home economics and assistant director, agricultural experiment station, Oklahoma Agricultural and Mechanical College, 1954–1957; assistant director, human nutrition research division, USDA, 1957–1958, associate director, institute of home economics, 1958–1961, assistant director of administration, 1961–1971, science advisor, 1971–1974

Concurrent Positions: Fulbright professor, University of the Philippines, 1949–1950

Ruth Leverton was a nutritionist whose research included human metabolism and requirements of minerals, nutritive value of food products, and blood regeneration and prevention of anemia. Her research had an important impact on American food practices at mid-century, including decisions about wartime food rationing and nutrition, the development of a system of Recommended Dietary Allowances, the fortification of grains, and food assistance programs. Following a pattern of many educated women of her generation, she taught school for several years after she received her undergraduate degree. But Leverton decided to continue her own education, earning a master's degree in nutrition at the University of Arizona before moving to the University of Nebraska for a doctorate. Her research focused on women's iron needs and made connections between insufficient protein intake and iron-deficiency anemia. Leverton's research was among the first to highlight the differences between men's and women's dietary and nutritional needs.

Leverton worked briefly for the U.S. Department of Agriculture at the beginning of World War II, but returned to academia at Nebraska where she rose from associate professor to professor of home economics between 1941 and 1953. She accepted a position at Oklahoma Agricultural and Mechanical College (now Oklahoma State University) as professor of home economics and assistant director of the agricultural experiment station. She returned to the USDA in 1957 as assistant director of Human Nutrition Research, the highest-ranking woman at the USDA at that time. She remained at the USDA in various positions until her retirement in 1974. Throughout this time, she also traveled extensively throughout Asia, Africa, Latin America, and Europe. She represented the United States on the International Rice Commission and the International Congress of Nutrition, and lectured on nutritional health at conferences worldwide.

Leverton published more than 200 academic papers, was the author of the classic book *Food Becomes You* (1952), and was the co-author of *Your Diabetes and How To Live with It* (1953). Leverton received the Borden Award for Dairy Foods Research (1942 and 1953), the Distinguished Service Award from the USDA (1972), the Conrad A. Elvehjem Award of the American Institute of Nutrition (1973), the Federal Woman's Award (1977), and a Medallion Award of the American Dietetic Association (1977). She was the first woman to receive an honorary doctorate of science from her alma mater, University of Nebraska, in 1961. She was a member of the American Dietetic Association, the American Home Economics Association, the American Public Health Association, the American Institute of Nutrition, and the American Association for the Advancement of Science.

Further Resources

Hampl, Jeffrey S. and Marylynn I. Schnepf. "Ruth M. Leverton (1908–1982)." http://jn.nutrition.org/cgi/content/full/129/10/1769.

Leveson, Nancy G.

Aerospace Engineer, Computer Scientist

Education: B.A., mathematics, University of California, Los Angeles, 1965, M.S., Graduate School of Management, 1967, Ph.D., computer science, 1980

Professional Experience: systems engineer, IBM, 1967–1970; assistant professor, information and computer science, University of California, Irvine, 1980–1985, associate professor, 1985–1990, professor, 1990–1993; Boeing Professor of Computer Science and Engineering, University of Washington, Seattle, 1993–1998; professor, aeronautics and astronautics, and professor, engineering systems, Massachusetts Institute of Technology (MIT), 1998–

Concurrent Positions: visiting professor, Laboratory for Computer Science, MIT, 1988–1989; adjunct professor, computer science, University of British Columbia, 1993–; Hunsaker Visiting Professor, aeronautics and astronautics, MIT, 1997–1998

Nancy Leveson is an aerospace engineer who pioneered a new research field in software safety systems, which involves using computer programs to prevent and analyze safety situations where property or life are at risk. Her research has focused particularly in the area of air and space flight, and involves creating accident models that take into account the role of computers as well as human decision-making in risk management. She has published over 200 scientific papers and articles, and her system for aircraft collision avoidance has been adopted by the Federal Aviation Administration (FAA) for use in commercial airlines. She has been a distinguished invited guest lecturer at national and international universities and a consultant or advisory council member for numerous industry and government organizations related to software-related safety issues in nuclear power plants, transportation, air traffic control, and aerospace systems and accidents, including authoring an analysis of the *Columbia* space shuttle explosion of 2003. Leveson completed her undergraduate and graduate education at the University of California, Los Angeles, including a doctorate in computer science in 1980. She taught at the University of California, Irvine and at the University of Washington, Seattle before joining the faculty at MIT, where she holds joint appointments in the departments of Aeronautics and Astronautics, and in Engineering Systems.

Leveson was elected to the National Academy of Engineering in 2000. She is a fellow of the Association for Computing Machinery (ACM) and the International Association for the Advancement of Space Safety (IAASS), and has been on the Board of Directors of the Computing Research Association (CRA), International Council on System Engineering, and Geisinger Institute on Electronic Health

Records Safety. She was the recipient of the AIAA Information Systems Award (1995), ACM Allen Newell Award (1999), CRA Habermann Award (2004), ACM SIGSOFT Outstanding Software Engineering Research Award (2004), and System Safety Society Professional Achievement Award.

Further Resources

Massachusetts Institute of Technology. Faculty website. http://sunnyday.mit.edu/.

Levi-Montalcini, Rita

b. 1909
Neuroembryologist

Education: M.D., University of Turin, 1936, 1940

Professional Experience: research associate, zoology, Washington University, St. Louis, 1947–1951, associate professor, 1951–1958, professor, 1958–1981

Concurrent Positions: director, Research Center of Neurobiology of the Consiglio Nazionale delle Ricerche (CNR), Rome, 1961–1969, Laboratory of Cellular Biology, 1969–1978

Rita Levi-Montalcini is a neurologist who shared the Nobel Prize in Physiology or Medicine in 1986 with colleague Stanley Cohen for their discovery of the nerve growth factor (NGF), responsible for the rapid growth of immature cells implicated in diseases such as cancer and Alzheimer's. Born and educated in Italy, she conducted the early stages of her prize-winning research beginning in 1952 while on the faculty at Washington University in St. Louis, Missouri. Her research was focused on the effect of a nerve growth factor isolated from the mouse salivary gland on the sympathetic nervous system and of an antiserum to the nerve growth factor. In order to advance the work more quickly, she smuggled two tumor-infected mice on the plane to Rio de Janeiro to consult with a colleague about the process of growing tissues *in vitro*. She spent the next six years on the project until she achieved success. With a National Science Foundation grant in 1961, she set up a small research unit in Rome so she could be close to her family. After a few years, when she received grants from the Italian government to establish an independent research institute, she alternated six months in Rome and six months in the United States.

As teenagers in Italy, she and her twin sister Paola were sent to a finishing school until, at age 20, she finally convinced her father that she would never marry, so he hired tutors in mathematics, science, Latin, and Greek to prepare her for university entrance examinations. After completing her medical degree, she continued research at the University of Turin. There she learned a new technique of staining embryonic chick neurons with chrome silver to make nerve cells stand out in the smallest detail. She continued using this technique in her private research when she was dismissed from her position at the University of Turin because her family was Jewish. She was unable to practice medicine, use the university library, or even visit friends at the university. During World War II, she set up a laboratory in her home and hid her experiments from the authorities. Since she was unable to publish her papers in Italian journals, she received international attention when they were published in Swiss and Belgian journals that

Rita Levi-Montalcini shared the 1986 Nobel Prize for Physiology or Medicine with Stanley Cohen for her discovery of nerve growth factor, the protein that promotes cell growth in the peripheral nervous system. (Nobel Foundation)

could be read in the United States. After the war, she returned to the laboratory at the University of Turin until she was invited to join a research group in 1947 at Washington University after the director read the papers she had published. She spent 30 years in St. Louis, returning to Italy permanently upon her retirement in 1981. In 2009, Levi-Montalcini celebrated her hundredth birthday.

Levi-Montalcini was elected to the National Academy of Sciences in 1968, and in 1987, she was awarded the National Medal of Science, the highest scientific honor in the United States. With her collaborator, Stanley Cohen, she also jointly received the Louisa Gross Horwitz Prize from Columbia University (1983) and the Albert Lasker Award for Basic Medical Research (1986). In 2001, she was named an honorary Senator for Life in the Italian Senate. She has received honorary degrees from Polytechnic University of Turin (2006) and Complutense University of Madrid, Spain (2008). She has been a member of the American Association for the Advancement of Science, Society for Developmental Biology, American Association of Anatomists, and Pontifical Academy of Sciences. In 1988, she published an autobiography, *In Praise of Imperfection: My Life and Work.*

Further Resources

McGrayne, Sharon Bertsch. 1993. *Nobel Prize Women in Science: Their Lives, Struggles, and Momentous Discoveries*. Secaucus, NJ: Birch Lane Press.

Wasserman, Elga. 2002. *The Door in the Dream: Conversations with Eminent Women in Science*. Washington, D.C.: Joseph Henry Press.

Lewis, Margaret Adaline Reed

1881–1970
Embryologist

Education: Woods Hole Marine Biological Laboratory, 1900; A.B., Goucher College, 1901; Bryn Mawr College, 1902–1903, 1908–1909; Columbia University, 1903–1906; University of Zurich, 1906; University of Paris and University of Berlin, 1908

Professional Experience: assistant, zoology, Bryn Mawr College, 1901–1902; lecturer, physiology, New York Medical College for Women, 1904–1907; lecturer, Barnard College, 1907–1909; instructor, anatomy and physiology, training school for nurses, Johns Hopkins Hospital, 1911–1912; collaborator, department of embryology, Carnegie Institution, 1915–1927, research associate, 1927–1940; member, Wistar Institute, 1940–1958, emeritus member, 1958–1964

Concurrent Positions: preparator in zoology, Columbia University, 1903–1906; lecturer, New York Medical College, 1904–1905

Margaret Lewis was a world-renowned authority on tumors, with expertise in the chemotherapy of cancer, the cytology of living cells in tissue cultures, the origin of epithelioid cells, and the relation of white blood cells to tumors. While working in Berlin, she may have conducted the first known successful *in vitro* mammalian tissue culture experiment. She and her husband, Warren H. Lewis, perfected the technique to develop clear solutions on special slides. This technique is known as the Lewis culture, and the medium is called the Locke-Lewis solution. In later years, they studied the chemotherapy of dyes in cancer. As early as 1915, they were able to provide a reasonably complete description of a number of living cells microscopically. By 1917, they had begun to determine some physiological activities. Later, at the Carnegie Institution, she added important studies of the effects of acidity on these processes. She published nearly 150 scientific papers, often co-authored with her husband.

Lewis received her undergraduate degree from Goucher College in 1901 and studied at a number of universities in the United States and Europe without completing a graduate degree. She held brief appointments at several U.S. colleges before joining the Carnegie Institution department of embryology. In 1940, she was elected a member of the Wistar Institute, where she held emeritus status for several years after her retirement. She and her husband jointly received the Gerhard Gold Medal of the Pathological Society of Philadelphia and an honorary degree from Goucher College in 1938. She was an honorary life member of the Tissue Culture Society and a member of the American Association of Anatomists. She is identified in some sources as "Margaret Reed" or as "Mrs. Warren H. Lewis."

Libby, Leona Woods Marshall

1919–1986
Physicist

Education: B.S., University of Chicago, 1938, Ph.D., chemistry, 1943

Professional Experience: research associate, metallurgical laboratory, Manhattan Project, 1942–1944; physicist, Hanford Engineering Works, Washington, 1944–1946; fellow, Institute for Nuclear Studies, University of Chicago, 1946–1947, research associate, 1947–1953, assistant professor, physics, 1953–1960; associate professor, physics, New York University, 1960–1962, professor, 1962–1963; associate professor, University of Colorado, Boulder, 1963; staff member, Rand Corporation, California, 1963–1970; staff member, R&D Associates, California, 1970–1976

Concurrent Positions: consulting physicist, E. I. du Pont de Nemours & Company, 1944–1946; fellow, Institute for Advanced Studies, Princeton, New Jersey, 1957–1958; visiting scientist, Brookhaven National Laboratory, 1958–1960; visiting scientist, Brookhaven National Laboratory, 1958-1986; visiting adjunct professor, University of California, Los Angeles, 1973–1986

Leona Marshall Libby was a physicist whose research focused on high-energy nuclear physics, nuclear reactions, fundamental particles, astrophysics, and stable isotopes in tree thermometers. She discovered that historical climate could be measured from the isotope ratios in tree rings. She also conducted early research on neutron and proton scattering. She was a member of the Manhattan Project, the group that built the first and second Argonne reactors, the Oak Ridge reactor, and the three Hanford reactors. She worked with the most important scientists in this field in the mid-twentieth century, including several Nobel Laureates. For her doctorate, she studied with Robert Mulliken, who won the Nobel Prize in Chemistry in 1966; even before completing her Ph.D., she became the first female researcher at the Chicago Metallurgical Laboratory working with Enrico Fermi, who had won the Nobel Prize in Physics in 1938. Their top-secret work on the first

Nuclear physicist Leona Woods Marshall Libby worked on the Manhattan Project during World War II. (AP/Wide World Photos)

nuclear fission reactor and development of the atomic bomb became later known as the Manhattan Project. Many women scientists were able to secure government contracts and positions during World War II, and numerous women physicists and chemists were ultimately involved in the Manhattan Project. Libby (in this early period known professionally by her first married name, Marshall) spent several years as a researcher and then assistant professor with the Institute for Nuclear Studies at the University of Chicago, where she studied nuclear explosions and neutron diffusion.

In 1958, Leona Marshall moved to the Brookhaven National Laboratory in New York and later taught atomic and nuclear physics at New York University. She left New York in 1964 to teach physics at the University of Colorado, Boulder, and in 1966 married her second husband, Willard Frank Libby, another prominent chemist who had recently received the Nobel Prize in 1960 for his work on radio-carbon dating. In 1972, the couple relocated to California, where Leona worked first as a visiting professor and then adjunct instructor at University of California, Los Angeles. There she continued her research in particle physics and began her work on environmental engineering, tree rings and ancient climates, publishing two books in the 1970s on environmental issues. After Willard Libby's death in 1980, Leona Marshall Libby collected and edited his papers and published *The Life Work of Nobel Laureate Willard Frank Libby* in 1982.

In the 1970s and 1980s, Leona Marshall Libby remained an outspoken advocate for nuclear power in the face of increasing public criticism. In 1979, she published an autobiography of her early work in nuclear physics entitled *The Uranium People*. She was elected a fellow of the American Physical Society and the Royal Geographical Society, and was a member of the National Science Foundation Postdoctoral Fellowship Evaluation Board.

Further Resources

Howes, Ruth and Carolyn L. Herzenberg. 1999. *Their Day in the Sun: Women of the Manhattan Project*. Philadelphia, PA: Temple University Press.

Linares, Olga Frances

b. 1936
Anthropologist

Education: B.A., anthropology, Vassar College, 1958; Ph.D., anthropology, Harvard University, 1964

Professional Experience: instructor, anthropology, Harvard University, 1965; lecturer, anthropology, University of Pennsylvania, 1966–1971; research scientist to senior scientist, Smithsonian Tropical Research Institute, 1973–2008, emerita

Concurrent Positions: research curator, Center for American Archaeology, Peabody Museum, Harvard University, 1974–

Olga Linares is an anthropologist recognized for her research on the rural populations of western Africa and Central America. Her research centers on the agrarian practices and political economy of western African and Central American rural populations, and on human adaptations to the tropical forest, past and present. She is working in the area of economic anthropology among primarily agrarian populations and looks not only at the types of crops that are grown and marketed but also at the sexual division of labor. She examines the social, spatial, and temporal relations in archaeological perspective.

The strength of her research can be seen in the book *Power, Prayer, and Production: The Jola of Casamance, Senegal* (1992). The central thesis is that ideology and production are part of the same system and any consideration of the division of labor—whether by gender, age, status, or ethnic identity—must take into account the influence of ideology. She compares three communities that are engaged in intensive wet-rice cultivation but structure their agriculture very differently. One is a non-Muslim community in which both men and women commune with spirit shrines, and relations between the generations and the sexes tend to be reciprocal and cooperative. Another community has adopted Islam and has divided production along territorial, generational, gender, and kinship lines. The third community also is Islamic and there is a strong Islamic community nearby; this group has more extreme inequality and social separation between the sexes and the generations. In each case, Linares examined the same set of factors: marriage and residence patterns, cropping and land tenure arrangements, the role of ritual and religious powers and duties, the organization of labor, the effects of introduced technologies, and the dynamics of social power and conflict. After an appointment as a lecturer in anthropology at the University of Pennsylvania, Linares secured joint appointments with the Smithsonian Tropical Research Institute, first as a research scientist and then as a senior scientist, and with the Center for American Archaeology at the Peabody Museum of Harvard University as a research curator. She has written numerous journal articles and book chapters.

Linares was elected to membership in the National Academy of Sciences in 1992. She is a fellow of the American Association for the Advancement of Science and a member of the American Anthropological Association, African Studies Association, Royal Anthropological Association, and Latin American Studies Association.

Further Resources

Smithsonian Tropical Research Institute. "Olga F. Linares." http://www.stri.org/english/scientific_staff/staff_scientist/scientist.php?id=24.

Lippincott, Sarah Lee

b. 1920
Astronomer

Education: student, Swarthmore College, 1938–1939; B.A., University of Pennsylvania, 1942; M.A., astronomy, Swarthmore College, 1950

Professional Experience: research assistant, astronomy, Swarthmore College, 1942–1951, research associate, 1952–1972, lecturer, 1961–1976, director, Sproul Observatory, 1972–1981, professor, 1977–1981, emerita

Concurrent Positions: visiting astronomer, Lick Observatory, University of California, Santa Cruz, 1949; visiting astronomer, California Institute of Technology, 1978

Sarah Lippincott is known for her research in astrometry, which is the branch of astronomy that deals with the measurement of the positions and motions of celestial bodies. One of her projects was to look for extrasolar planets or planetlike companions to nearby stars. The Sproul Observatory at Swarthmore College has had a long-term program of tracing the motions of stars within five parsecs of the Earth to look for such perturbations. The data, going back an average of 50 years, are on photographic plates containing images of those stars; these are in the archives of the observatory. Lippincott found three stars that were candidates for having unseen companions. In addition to her work at Swarthmore, she was a visiting astronomer at major West Coast observatories. She held a Fulbright fellowship in France and was a member of the French solar eclipse expedition to Oland, Sweden, in 1954. Lippincott spent her entire professional career at Swarthmore College, beginning as a research assistant and eventually becoming professor and director of the Sproul Observatory. She trained many female astronomers at Swarthmore, including well-known cosmologist **Sandra Faber**.

Lippincott held a master's degree, but received an honorary doctorate from Villanova University in 1973. She published numerous papers in scientific journals and is co-author of the book *Point to the Stars*, of which three editions were published between 1963 and 1976. She is a member of the American Astronomical Society and International Astronomical Union (president, 1973–1976). In some sources she is identified by her married name, Zimmerman.

Liskov, Barbara Huberman

b. 1939
Computer Scientist

Education: B.A., mathematics, University of California, Berkeley, 1961; M.S., computer science, Stanford University, 1965, Ph.D., 1968

Professional Experience: applications programmer, Mitre Corporation, 1961–1962; programmer, language translation project, Harvard University, 1962–1963; graduate research assistant, artificial intelligence, Stanford University, 1963–1968; member of technical staff, computer science research and development, Mitre Corporation, 1968–1972; assistant to associate professor, computer science and electrical engineering, Massachusetts Institute of Technology (MIT), 1972–1980, professor, 1980–, NEC Professor of Software Science and Engineering, 1986–1997, Ford Professor of Engineering, 1997–, associate head, computer science, 2001–

Barbara Liskov is recognized as an expert on computer software, and her research on programming methodology, distributed computing, programming languages, and operating systems has been at the forefront of the field of software and computer operating systems. She has been instrumental in designing software that has formed the basis of widely used programming languages such as C++ and Java. She worked for Mitre Corporation for several years before joining MIT as a faculty member in 1972. As a member of the Programming Methodology Group of the Computer Science and Artificial Intelligence Laboratory at MIT, her work has focused on distributed systems, object-oriented databases, programming languages, software design and upgrades, and, most recently, systems operations plans due to computer failure or hacking, an important area of research in the Internet age. In addition to her academic duties, she has consulted for major computer companies such as Digital Equipment, Hewlett-Packard, NCR, Prime Computers, Cadence, Intermetrics, BBN Corporation, and Cisco Systems.

Liskov received her Ph.D. from Stanford University in 1968, the first woman to earn a doctorate in a computer science program. She was elected to membership in the National Academy of Engineering (1988) and received an honorary doctorate from the Swiss Federal Institute of Technology (ETH) in Zurich (2005). She is the recipient of an Achievement Award from the Society of Women Engineers (1996), the John von Neumann Medal of the Institute of Electrical and Electronics Engineers (IEEE) (2004), the Association for Computing Machinery (ACM) SIGPLAN Lifetime Achievement Award (2008), and the A. M. Turing Award of the Association for Computing Machinery (2009), one of the highest awards in computer science, for her contribution to "virtually every modern computing-related convenience in people's daily lives." In 2002, she was profiled as a top scientist in both *Popular Science* and *Discover* magazines. She is a member of the IEEE and the ACM, and a fellow of the American Academy of Arts and Sciences.

Further Resources

Massachusetts Institute of Technology. "MIT's Magnificent Seven: Women Faculty Members Cited as Top Scientists." http://web.mit.edu/newsoffice/2002/women.html.

Massachusetts Institute of Technology. Faculty website. http://www.pmg.csail.mit.edu/~liskov/.

Lochman-Balk, Christina

1907–2006
Geologist, Paleontologist

Education: A.B., Smith College, 1929, A.M., geology, 1931; Ph.D., paleontology, Johns Hopkins University, 1933

Professional Experience: assistant geologist, Smith College, 1929–1931; instructor, Mount Holyoke College, 1935–1940, assistant to associate professor, 1940–1947;

lecturer, physical science, University of Chicago, 1947; lecturer, life sciences, New Mexico Institute of Mining and Technology, 1954, professor, geology, 1955–1972

Concurrent Positions: strategic geologist, New Mexico State Bureau of Mines and Mineral Resources, 1955–1957

Christina Lochman-Balk was a prominent geologist whose research area was the Cambrian paleontology and stratigraphy of the western United States, Mexico, and Newfoundland. In particular, she studied Cambrian trilobites and published several important papers and updates on invertebrate paleontology in North America. She held positions at several universities and eventually rose through the ranks to full professor in a predominantly male profession. After receiving her doctorate, she accepted a position at Mount Holyoke in 1935, advancing to assistant professor and associate professor. After her marriage in 1947, she followed her husband Robert Balk to the University of Chicago, where he was appointed a professor of geology; there, she could only get a position as a lecturer. The couple relocated to New Mexico where, again, he was a professor and she a lecturer until she was promoted after his death in 1955. She remained at the New Mexico Institute of Mining and Technology (New Mexico Tech) until her retirement in 1972, also serving a two-year appointment as a strategic geologist for the New Mexico State Bureau of Mines.

During her tenure at the New Mexico Institute of Mining and Technology, she was renowned as a teacher and as a researcher, and helped expand the program's offerings for doctoral studies in the earth sciences. She supervised numerous doctoral students who went on to make important geological discoveries of their own. As Dean of Women, she was particularly interested in promoting the careers of female scientists. She also established two fellowship opportunities for student research in geology and earth sciences. Lochman-Balk was elected a fellow of both the American Association for the Advancement of Science and the Geological Society of America, and was a member of the Paleontological Society, which awarded her its President's Citation in 1996.

Further Resources

New Mexico Institute of Mining and Technology. Faculty website. http://www.ees.nmt.edu/balk/.

Loeblich, Helen Nina Tappan

1917–2004
Paleontologist

Education: B.S., University of Oklahoma, 1937, M.S., 1939; Ph.D., geology, University of Chicago, 1942

Professional Experience: assistant geologist, University of Oklahoma, 1937–1939; instructor, Tulane University, 1942–1943; geologist, U.S. Geological Survey (USGS), 1943–1945, 1947–1959; research associate, paleontology, Smithsonian Institution, 1954–1957; lecturer, geology, University of California, Los Angeles (UCLA), 1958–1965, associate research geologist, 1961–1963, senior lecturer, geology, 1965–1966, professor, 1966–1984

Helen Loeblich was a renowned researcher in micropaleontology whose research focused on living and fossil foraminiferans, tintinnids, the camoebians, and organic-walled siliceous and calcareous phytoplankton. She was an assistant geologist at the University of Oklahoma before taking over her the teaching responsibilities of her husband, fellow paleontologist Al Loeblich, at Tulane University while he was on active military duty during World War II. After the war, she held positions with the USGS working at the Naval Petroleum Reserve in Alaska, and, with Al and their children, traveled to Europe on a Guggenheim Fellowship to conduct research on historical collections of forminifera for the Smithsonian Institution. The family relocated to California in 1957, where Al worked as a researcher for Chevron Oil and Helen returned to academia as a lecturer in geology at UCLA, where she eventually advanced to full professor and remained until her retirement in 1984.

With Al Loeblich she co-published more than 200 papers, articles, and books, and helped update the 1964 edition of *Treatise on Invertebrate Paleontology*. She received high praise for her 1980 book *The Paleobiology of Plant Protists*, and their joint 1987 two-volume work *Foraminfiera Genera and Their Classification* was designated the best geography and earth science book of 1988 by the Association of American Publishers. Helen Loeblich was also an accomplished scientific artist who, in 1976, designed a stamp for the fiftieth-anniversary celebration of the Society of Economic Paleontologists and Mineralogists (now the Society for Sedimentary Geology).

Among the awards Helen Loeblich received were the Woman of Science Award from UCLA (1982), Paleontological Society Medal (1983), Raymond C. Moore Medal for Excellence in Paleontology (1984), and Woman of the Year Award in Natural History from the Palm Springs Desert Museum (1987). The Loeblich's were named honorary directors of the Cushman Foundation of Foraminiferal Research in 1982 and Helen Loeblich was elected president of the Paleontological Society in 1985. She was also a fellow of the Geological Society of America, an honorary member of the Society for Sedimentary Geology, and a member of the American Microscopical Society.

Further Resources

"In Memoriam: Helen Nina Tappan Loeblich." University of California, Los Angeles. http://www.universityofcalifornia.edu/senate/inmemoriam/helenninaloeblich-tappan.html.

Long, Irene (Duhart)

b. 1951
Aerospace Physician

Education: B.A., biology, Northwestern University, 1973; M.D., St. Louis University School of Medicine, 1977; M.S., aerospace medicine, Wright State University School of Medicine, Ohio, 1981

Professional Experience: medical resident, Ames Research Center, National Aeronautics and Space Administration (NASA), 1981-1982, John F. Kennedy Space Center, 1982; chief, Occupational Medicine and Environmental Health Office, NASA John F. Kennedy Space Center, 1982–1994, director, Biomedical Office, 1994–2000, Chief Medical Officer and Associate Director of Spaceport services, 2000–

Irene Long is one of the highest-ranking professional women at NASA. She was the first black female chief of the Occupational Medicine and Environmental Health Office, and is responsible for overseeing not only the health of the astronauts but also the health of some 18,000 workers, civil servants, and contractors at the Kennedy Space Center. She works with a team of physicians to provide medical services to the astronauts in emergency cases, such as an aborted mission, and she oversees inspecting workspaces at the Kennedy Space Center to protect employees from exposure to various possible hazards—toxic chemicals, fire, or decompression, for example—when a spacecraft is launched. She coordinates the efforts of the Department of Defense, environmental health agencies, and the astronaut office when they work together to stage successful launches, as well as to prepare for emergency situations. In her own research, Long has found that lower oxygen levels do not impede the flow of blood in people with the sickle-cell trait, and so they should not be banned from flying. She has also used the Johnson Space Center's collection of medical data to research the physical condition of astronauts, including the effects of space on the individuals' physiology and the consequences of weightlessness.

Long has also worked to encourage women and minorities to have careers in science and engineering through the Space Life Sciences Training Program. Participants in the program spend six weeks at the Kennedy Space Center studying space physiology in plants, animals, and humans, learning how to develop experiments, and becoming acquainted with the basic concepts of teamwork. Long received the Presidential Award of the Society of NASA Flight Surgeons (1995) and later served as president of the society (1998). She also received an Outstanding Achievement Award from Women in Aerospace (1998) and the Lifetime Achievement Award from the National Women of Color Technology Awards Conference (2005). Long is a member of the Aerospace Medical Association.

Further Resources

National Aeronautics and Space Administration. "Irene Duhart Long, M.D.: Chief Medical Officer and Associate Director, Center Operations." http://www.nasa.gov/centers/kennedy/about/biographies/long.html.

Long, Sharon (Rugel)

b. 1951
Developmental Biologist, Molecular Biologist

Education: B.S., biochemistry, California Institute of Technology, 1973; Ph.D., cell and developmental biology, Yale University, 1979

Professional Experience: research fellow, biology, Harvard University, 1978–1981; assistant to associate professor, biological science, Stanford University, 1982–1992, professor, 1992–

Concurrent Positions: investigator, Howard Hughes Medical Institute, 1994–2001; dean, School of Humanities and Sciences, Stanford Univeristy, 2001–2007

Developmental and molecular biologist, Sharon Long. (Courtesy of Stanford University News Service Library)

Sharon Long is renowned for her studies in plant genetics. Her research includes genetics and developmental biology of symbiotic nitrogen fixation in legumes, the role of plasmids in symbiosis, plant cell biology, and plant molecular biology. She identified and cloned the genes that allow bacteria to locate and enter certain plants; she has worked with the rhizobium bacterium that invades the roots of such legumes as alfalfa, soybeans, and peas, and lives symbiotically with the plant, receiving moisture and protection from it and producing nitrogen for the plant's growth. Her specific contribution is to genetically alter the bacterium to make better invaders. Her research involves allowing the bacterium to invade other major food crops, which will enable farmers to reduce the amounts of nitrogen fertilizer that are spread on food crops and eventually are washed off by rain into streams and rivers.

Long was elected to membership in the National Academy of Sciences in 1993 and has had a distinguished career as a teacher, researcher, and administrator. After a postdoctoral fellowship at Harvard, she joined the faculty at Stanford University in 1982, serving as full professor since 1992, and was dean of the School of Humanities and Sciences for six years. At Stanford, she has twice received the Dean's Award for Distinguished Teaching (1988 and 1992). Among her prestigious national honors are a Presidential Young Investigators Award of the National Science Foundation (1984–1989) and a MacArthur fellowship (1992–1997). She has also been the recipient of a Shell Foundation Research Award, a Charles A. Schull Award from the American Society of Plant Physiology, a National Science Foundation Faculty Award for Women, and the Wilbur Cross Medal for alumni from Yale University (2002). She is a fellow of the American Association for the Advancement of Science, American Academy of Arts and Sciences, and American Philosophical Society. She is a member of the American Academy of Microbiology and a member of the Genetics Society of America, American Society of Plant Physiologists, American Society for Microbiology, and Society for Developmental Biology.

Further Resources

Stanford University. "Sharon R. Long Lab." http://cmgm.stanford.edu/biology/long/.

Love, Susan M.

b. 1948
Surgeon

Education: B.S., Fordham University, 1970; M.D., State University of New York Medical Center, 1974; M.B.A., Anderson School of Business, University of California, Los Angeles, 1998

Professional Experience: surgical intern, Beth Israel Hospital, Boston, Massachusetts, 1974–1975, surgical resident, 1975–1979, surgical coordinator, 1979, clinical fellow, pathology, 1980, assistant in surgery, Beth Israel Hospital, 1980–1987, director, Breast Clinic, 1980–1988, associate surgeon, 1987–1992, director of research, Faulkner Breast Center, 1992; clinical fellow, surgery, Harvard Medical School, Boston, 1977–1978, clinical instructor, 1980–1987, assistant clinical professor, surgery, 1987–1992; associate professor, clinical surgery, University of California, Los Angeles, 1992–1996, Revlon Chair in Women's Health, 1995–1996, director, Revlon/UCLA Breast Center, 1992–1996, adjunct associate

Surgeon Susan Love, 1996. She is the founder and director of the Dr. Susan Love Research Foundation for breast cancer research. (AP/Wide World Photos)

professor, 1996–1997, adjunct professor, general surgery, 1997–; president and medical director, Dr. Susan Love Research Foundation (formerly Santa Barbara Breast Cancer Institute), 1996–

Concurrent Positions: visiting registrar, Guy's Hospital, London, England, 1977–1978; clinical associate, Dana Farber Cancer Institute, 1981–1992

Susan M. Love is a surgeon who retired from medical practice to advocate for breast cancer research. She had a distinguished career as a physician, rising to become the first female general surgeon at Beth Israel Hospital in Boston and a professor at Harvard Medical School. She founded the National Breast Center Coalition in 1990 to bring together the latest research and political advocacy for greater awareness and more funding dedicated to breast cancer. She moved to California in 1992 as a professor at the University of California, Los Angeles and served as director of the Revlon/UCLA Breast Center before founding the Dr. Susan Love Research Foundation in 1996. Love's work focuses on the lack of research funding, misinformation among doctors and patients, and concern about drastic and unnecessary treatment methods offered to women in the past, such as radical mastectomies. Her message to women diagnosed with breast cancer is that they should do their own research, be informed, get second opinions, and be their own advocates in the battle against the disease.

Love has published two books: *Dr. Susan Love's Breast Book* (1990; 4th ed., 2005) and *Dr. Susan Love's Menopause and Hormone Book* (1998; 2nd ed., 2003). She has contributed to numerous medical textbooks and has been a member of editorial or review boards of medical journals. Love's high-profile research center and popular books have made her a prominent public figure, and she has made several television appearances on Discovery Health channel, Lifetime, the *Oprah Winfrey Show*, *Good Morning America*, the *Today Show*, and other news programs. She lectures often for women's health groups and has served as a

Breast Cancer Research

Breast cancer is said to strike one in eight American women and, although more women die of heart disease and lung cancer each year, breast cancer is seen as a particularly insidious and dreaded disease among women. The fear of the disease stems, in part, from the historically brutal nature of the treatment. The first radical mastectomies were performed in the nineteenth century, with removal of the breasts, lymph nodes, and chest muscles. By the late twentieth century, as greater understanding of the course and spread of the disease was achieved, a greater range of treatments were offered, including surgery, radiation, and chemotherapy, but also new experimental drug options. Emotional support for the disease has also increased as patient advocacy groups have created support networks and called for earlier diagnosis, more humane treatments, increased research funding, and public campaigns such as the pink ribbon crusade and October as National Breast Cancer Awareness Month. Women scientists who have died of breast cancer include early-nineteenth-century British paleontologist Mary Anning, environmental scientist **Rachel Carson**, mathematician **Carol Karp**, and American physician **Jerri Nielsen**, who attracted worldwide media attention in 1998 after performing a lumpectomy on her own breast while stranded at the South Pole research center.

medical advisor or board member for the National Alliance of Breast Cancer Organizations, Lesbian Health Foundation, Wellness Community, International Breast Cancer Research Foundation, President's National Action Plan on Breast Cancer, and numerous other organizations. Her own research on breast ducts led to the co-founding in 1998 of a medical device company, now known as Pro-Duct Health, for which she remains a consultant. In 2008, her own Dr. Susan Love Research Foundation joined with the Avon Foundation's "Army of Women" to support breast cancer awareness and research.

Love's numerous awards and honors include, but are not limited to, the following: Women Who Have Made a Difference by the International Women's Forum (1991), Achievement Award of the American Association of Physicians for Human Rights (1992), Women of Distinction by the National Council on Aging (1994), Spirit of Achievement Award from Albert Einstein College of Medicine, Yeshiva University (1995), Alumni Achievement Award from the State University of New York (SUNY) College of Medicine (1999), Radcliffe Medal of the Radcliffe College Alumnae Association (2000), Humanitarian of the Year Award of Western University of Health Sciences (2001), Excellence in Cancer Awareness Award from Congressional Families for Cancer Awareness (2002), Women Inspiring Hope and Possibility from the National Women's History Project (2004), Director's Award of the National Cancer Advisory Board and National Cancer

Institute (2004), and induction into the International Women's Forum Hall of Fame (2006). She has also received honorary degrees from several universities.

Love has been a member of the North American Menopause Society, American Medical Women's Association (branch president, 1987), American College of Surgeons, American Society of Clinical Oncology, American Association of Physicians for Human Rights, American Society of Preventive Oncology, Society for the Study of Breast Disease, American Association for Cancer Research, Association of Women Surgeons, American College of Women's Health Physicians (founding member), American College of Physicians Executives, Doctors against Abuse from Steroid Sex Hormones (DASH), Longmire Surgical Society, Massachusetts Medical Society, Los Angeles Medical Society, and Los Angeles Academy of Medicine. She is married to California surgeon Dr. Helen Sperry Cooksey.

Further Resources

Dr. Susan Love Research Foundation. http://www.dslrf.org/.

Stabiner, Karen. 1998. *To Dance with the Devil: The New War on Breast Cancer.* NY: Delta Books.

Lubchenco, Jane

b. 1947
Marine Ecologist, Conservation Biologist

Education: B.A., biology, Colorado College, 1969; M.S., zoology, University of Washington, 1971; Ph.D., marine ecology, Harvard University, 1975

Professional Experience: assistant professor, ecology, Harvard University, 1975–1977; assistant to associate professor, zoology, Oregon State University, 1977–1988, professor and department chair, 1988–1992, distinguished professor, 1993–2009; Under Secretary of Commerce and Administrator, National Oceanic and Atmospheric Administration (NOAA), 2009–

Concurrent Positions: principal investigator, National Science Foundation, 1976–; visiting professor, University of the West Indies, Kingston, Jamaica, 1976; visiting professor, Discovery Bay Marine Lab, 1977; research associate, Smithsonian Institution, 1978–1984; visiting professor, Universidad Catolica, Santiago, Chile, 1986; visiting professor, Institute of Oceanography, Qingdao, China, 1987; visiting professor, University of Canterbury, Christchurch, New Zealand, 1995–1996, 1999–2000, and 2002–2003.

Jane Lubchenco is a marine ecologist interested in biodiversity and sustainable ecological systems, and is active in national and international studies in ecology and global climate change. In 2009, she was chosen by President Obama as head of NOAA, the first woman to hold that post. She had previously served on the National Science Board under President Clinton from 1996 to 2006 and advised the president, vice president, and U.S. Congress on issues related to climate change. Her work has focused on marine plant–herbivore interactions, chemical ecology, predator–prey interactions, algal ecology, and life histories. She is also interested in biodiversity and sustainable ecological systems. She conducted her early field research in Panama from 1977 to 1983. She helped draft the Sustainable Biosphere Initiative of the Ecological Society of America in 1991 and co-authored a 1997 article entitled "Human Domination of Earth's Ecosystems," in which the authors warned that human alteration of the Earth was substantial and growing.

She began her career at Harvard University, but after she and her husband, marine biologist Bruce A. Menge, had been married for several years, they sought joint employment at a research university where they could combine family and career. Oregon State University allowed them to split one tenure-track position into two separate, half-time tenure-track positions so that each of them could engage in research and work toward tenure while their children were young. As their children grew older, the couple were able to gradually move into full-time status, a model they actively endorse as an alternative for faculty with families. Some of her early papers were published under her married name, Jane Menge.

Lubchenco was elected to membership in the National Academy of Sciences in 1996 and is an elected member of the American Academy of Arts and Science (president, 1997), American Philosophical Society, and Royal Society. She is also a member of the Phycological Society of America, American Society of Naturalists, American Institute of Biological Sciences, International Council for Science, and Ecological Society of America (president, 1992–1993). She has served as advisor for numerous marine organizations including the Ocean Trust Fund, Environmental Defense Fund, and Monterey Bay Aquarium. She has received numerous honorary degrees as well as the Mercer Award of the Ecological Society of America (1979), a MacArthur Foundation "genius" fellowship (1993–1998), a Pew fellowship (1993), the Heinz Award for the Environment (2002), the American Association for the Advancement of Science (AAAS) Award for Public Understanding of Science and Technology (2005), and the Zayed International Prize for the Environment (2008).

Further Resources

Wasserman, Elga. 2002. *The Door in the Dream: Conversations with Eminent Women in Science*. Washington, D.C.: Joseph Henry Press.

National Oceanic and Atmospheric Administration. "NOAA Leadership: Dr. Jane Lubchenco." http://www.noaa.gov/lubchenco.html.

Lubic, Ruth (Watson)

b. 1927
Nurse-Midwife

Education: diploma, School of Nursing Hospital, University of Pennsylvania, 1955; B.S., Teachers College, Columbia University, 1959, M.A., applied anthropology, 1961; certificate, nurse-midwifery, State University of New York at Brooklyn, 1962; Ed.D., applied anthropology, Columbia, 1979

Professional Experience: faculty member, School of Nursing, New York Medical College, and Maternity Center Association, State University of New York School of Nurse-Midwifery, Downstate Medical Center, 1955–1958; nurse, Memorial Hospital for Cancer and Allied Diseases, New York, 1955–1958; clinical associate, Graduate School of Nursing, New York Medical College, 1962–1963; general director, Maternity Center Association, New York, 1970–1995; founder, president, and co-CEO, District of Columbia Developing Families Center, 2000–

Concurrent Positions: consultant, midwifery, nursing, and maternal and child health, Office of Public Health and Science, U.S. Department of Health and Human Services, 1995–

Ruth Lubic is known for her contributions to the public health field, particularly those related to childbearing women, and has been a driving force behind the expansion of the midwifery profession in the United States. During her nursing training, Lubic observed that doctors often treated maternity patients with condescension and insensitivity, and that women often did not receive the prenatal and postnatal information they needed. These observations contrasted with her own experience in 1959, when her obstetrician allowed her husband to be present in the delivery room and to remain there with her and their newborn child for an hour after birth. For more than 25 years, she was director of the Maternity Century Association of New York, which was founded in 1918 as a nonprofit health agency dedicated to the advancement of education about childbearing and improving the care given to women during pregnancy and birth, and after delivery. Under Lubic's direction, the Maternity Center Association open the nation's first freestanding birth centers.

Lubic's father was a pharmacist; after his death, her mother ran the pharmacy, and Ruth worked there to save money to enter nursing school at the age of 25.

Martha Ballard, Early American Midwife

Martha Moore Ballard (1735–1812) was a midwife and healer who, for more than 25 years, kept a detailed diary of her medical practice and domestic work in the frontier town of Hallowell, Maine. In addition to raising her own nine children and assisting her husband in the family business, Ballard contributed to her family income and community life as a midwife who delivered hundreds of babies and treated a variety of illnesses in her small town. Ballard's diary (which was recovered and published by historian Laurel Thatcher Ulrich in the 1991 Pulitzer Prize–winning book *The Midwife's Tale* and also inspired a documentary film and a student research website, http://dohistory.org/) reveals important information about childbirth and medicine in the late eighteenth and early nineteenth century, in particular highlighting the differences in obstetrical practice between midwives such as Ballard and a new class of professionally trained male doctors.

Interestingly, medicine seemed to be part of the female family legacy, as Martha Ballard's diary was eventually handed down to and preserved by a great-great-granddaughter, Mary Hobart, who in 1884 was one of the first American women to earn a medical degree and was the first woman admitted to the Massachusetts Medical Society.

She graduated from college in 1959 and went on to receive a certificate in nurse-midwifery from State University of New York at Brooklyn in 1962, but while working for the center, she realized that her limited knowledge of different cultures was barring her from responding adequately to the needs of some of her clients. Therefore, she entered the graduate program in applied anthropology at Columbia University's Teachers College and earned an educational doctorate (Ed.D.) in 1979. She had already become director of the Maternity Center Association in 1970. In 1993, she was the first nurse ever to be honored with a MacArthur Foundation grant (1993), which she used to open the District of Columbia Developing Families Center in 2000. The Center's stated goal is "to meet the primary health care, social service, and child development needs of underserved individuals and childbearing and childrearing families through a collaborative that builds on their strengths and promotes their empowerment."

Lubic received the Rockefeller Public Service Award from Princeton University (1981), the Lillian D. Wald Spirit of Nursing Award from the Visiting Nurse Service of New York (1994), and the Gustav O. Lienhard Award of the Institute of Medicine (2001). She is co-author of *Childbearing: A Book of Choices* (1987). She is a member of the Institute of Medicine of the National Academy of Sciences, the American Public Health Association, and the American College of Nurse-Midwives. She was founder of the National Association of Childbearing Centers and served as that organization's president from 1983 to 1992.

Further Resources

Institute of Medicine of the National Academies. "Past Recipients of the Gustav O. Lienhard Award." http://www.iom.edu/Activities/Quality/Lienhard/Past-Recipients.aspx.

DC Developing Families Center. http://www.developingfamilies.org/.

Andrews, Wyatt. "The Midwife on a Mission." http://www.cbsnews.com/stories/2008/09/08/eveningnews/main4428250.shtml.

Lubkin, Gloria (Becker)

b. 1933
Physicist

Education: B.A., physics, Temple University, 1953; M.A., physics, Boston University, 1957

Professional Experience: mathematician, Aircraft Division, Fairchild Stratos Corporation, 1954, and Letterkenny Ordnance Depot, U.S. Department of Defense, 1955–1956; physicist, technical research group, Control Data Corporation, 1956–1958; acting chair, physics, Sarah Lawrence College, 1961–1962; vice president, Lubkin Associates, 1962–1963; associate to senior editor, *Physics Today*, 1963–1984, editor, 1985–1994, editorial director, 1994–2000, editor at large, 2001–2003

Gloria Lubkin has contributed to the physics profession in her 40-year career as editor of *Physics Today*, the publication of the American Institute of Physics. Her research includes nuclear physics and the history of physics, and in the 1960s, she began conducting oral histories of famous physicists. She is also an expert on science policy and has conducted several roundtables on issues in science that have been published in the journal, including issues of funding and scientists' relationship to government and to industry. Lubkin came to the journal with a solid background of experience. While working on her master's degree, she worked as a mathematician for Fairchild Stratos Corporation and the U.S. Department of Defense, and she was a physicist with Control Data Corporation before serving as acting chair of the physics department at Sarah Lawrence College. She joined the staff of *Physics Today* as an associate editor and rose through the ranks to editor and then editorial director before retiring emeritus in 2003.

Lubkin has served on numerous commissions and has received appointments to significant committees. She was a member of the Nieman Advisory Committee of

Harvard University (1978–1982) after being a recipient of a Nieman fellowship (1974–1975). In the American Physical Society, she has been a member of the executive commission of the Forum of Physics and Society (1977–1978) and a member of the executive committee of the History of Physics Division. She was also a consultant for the Center for the History and Philosophy of Physics of the American Institute of Physics (1966–1967). She was co-chair of the advisory commission for and co-founder of the Theoretical Physics Institute of the University of Minnesota (1987–1988), which now has a Gloria Becker Lubkin professorship of Theoretical Physics named in her honor.

Lubkin is a fellow of the American Physical Society and the American Association for the Advancement of Science, and a member of the New York Academy of Science and the National Association of Science Writers.

Luchins, Edith Hirsch

1921–2002
Mathematician

Education: B.A., Brooklyn College, 1942; M.S., New York University, 1944; Ph.D., mathematics, University of Oregon, 1957

Professional Experience: inspector, Sperry Gyroscope Company, New York, 1942–1943; instructor, mathematics, Brooklyn College, 1944–1946, 1948–1949; assistant, applied mathematics laboratory, New York University (NYU), 1946; research fellow and research associate, mathematics, University of Oregon, 1957–1958; research associate to associate professor, mathematics, University of Miami, Florida, 1959–1962; associate professor, Rensselaer Polytechnic Institute, New York, 1962–1970, professor, 1970–1992

Concurrent Positions: visiting professor, mathematics, U.S. Military Academy, West Point, 1991–1992, adjunct professor, cognitive sciences, 1994

Edith Luchins was recognized for her research on Banach algebras, functional analysis, and mathematical psychology. She was particularly interested in cognitive processes in mathematical problem solving, as well as the role of gender in learning and teaching mathematics. She especially wanted to encourage more women to pursue mathematics as a field of study. Luchins (then Hirsch) had emigrated to the United States from Poland when she was just six years old and, although (or because) neither parent had been formally educated, they stressed the importance of an education for Edith. In her New York City high school, she excelled in math, even tutoring other students and assisting teachers with grading.

Not only her family but her future spouse supported her education; Abraham Luchins insisted that she complete her undergraduate degree before they were married in 1942. She completed her bachelor's degree and then master's at NYU in quick succession, also teaching at Brooklyn College and working during World War II as a government chemist inspector of anti-aircraft equipment at Sperry Gyroscope. Female mathematicians and scientists were in great demand in government and industry to fill in for men during the war.

Luchins had begun doctoral work at NYU but eventually took several years off from her studies to raise children and follow her husband's career to Montreal, Canada, and then to Oregon, where she finally received her doctorate from the University of Oregon in 1957 before giving birth to her fifth child. Her years in Canada were also important to her career, however, as she worked closely with her husband, an educational psychologist, and developed an interest in mathematics education that would influence her commitment to teaching and learning as well as research. Her collaborations with her husband also led the two to co-author several books. She taught at the University of Miami for four years before being appointed associate professor at Rensselaer Polytechnic in 1962, and in 1970 became the first woman promoted to full professor there. She formally retired in 1992 but remained active in her research until her death in 2002. She was a member of the Mathematical Association of America, the American Mathematical Society, the Society for Industrial and Applied Mathematics, the American Education Research Association, and the American Association for the Advancement of Science.

Further Resources

Murray, Margaret Anne Marie. 2000. *Women Becoming Mathematicians: Creating an Identity in Post–World War II America.* Cambridge, MA: MIT Press.

Rensselaer Polytechnic Institute. "Obituary: Edith Luchins." http://www.rpi.edu/web/Campus.News/dec_02/dec_2/luchins.html.

Lucid, Shannon (Wells)

b. 1943
Biochemist, Astronaut

Education: B.S., chemistry, University of Oklahoma, 1963, M.S., 1970, Ph.D., biochemistry, 1973

Professional Experience: teaching assistant, chemistry, University of Oklahoma, 1963–1964; senior laboratory technician, Oklahoma Medical Research Foundation,

Astronaut Shannon Lucid exercises on a treadmill which has been assembled in the Russian Mir space station Base Block module, 1996. (NASA)

1964–1966; chemist, Kerr-McGee, 1966–1968; graduate assistant, biochemistry and molecular biology, University of Oklahoma Health Science Center, 1969–1973; research associate, Oklahoma Medical Research Foundation, 1974–1978; astronaut, National Aeronautics and Space Administration (NASA), 1978–1996; chief scientist, Solar System Exploration Division, Jet Propulsion Laboratory, 2002–2003, CAPCOM, mission control, Johnson Space Center, 2005–, management, Astronaut Office, 2008–

Shannon Lucid is a biochemist and astronaut who set the record for the most hours in space of any U.S. astronaut after she stayed aboard the Russian space station *Mir* for 179 days in 1996. Her task on the *Mir* was to conduct biomedical experiments on the effects of long-term space flight on humans. In addition her work on the space station, she flew as a mission specialist on the space shuttles *Discovery* (1985), *Atlantis* (1989), *Atlantis* (1991), and *Columbia* (1993). In 2002, Lucid became chief scientist of NASA's Solar System Exploration program, directing future space research and explorations and communicating NASA's missions to the public. Since 2005 she has served as CAPCOM (capsule communicator) for several space shuttle missions.

Lucid was born in Shanghai, China to missionary parents and was raised in Oklahoma. She earned both her undergraduate and graduate degrees in chemistry and biochemistry from the University of Oklahoma, receiving her doctorate in 1973. She then worked as a laboratory research associate at the Oklahoma Medical Research Foundation before joining the astronaut training program as part of the first group of women to be selected for the space program in 1978, along with **Judith Resnik**, **Sally Ride**, and others; Lucid was the only mother among the original group of female astronauts. During the 1980s, there was much publicity about the women astronauts, and their photos and interviews appeared in numerous magazines.

In 1996, Lucid was the first female astronaut to be awarded the Congressional Space Medal of Honor. She was recognized by Russian President Yeltsin in 1997 with the highest honor given to noncitizens, the Order of Friendship Medal.

Further Resources

Kevles, Bettyann H. 2003. *Almost Heaven: The Story of Women in Space*. New York: Basic Books.

National Aeronautics and Space Administration. "Shannon W. Lucid (Ph.D.)." http://www.jsc.nasa.gov/Bios/htmlbios/lucid.html.

Lurie, Nancy (Oestreich)

b. 1924
Anthropologist

Education: B.A., University of Wisconsin, Madison, 1945; M.A., University of Chicago, 1947; Ph.D., anthropology, Northwestern University, 1952

Professional Experience: instructor, anthropology and sociology, University of Wisconsin, Milwaukee, 1947–1949, 1951–1952; instructor, anthropology, University of Colorado, 1950; research associate, Peabody Museum, Harvard University, 1954–1956; lecturer, anthropology, Rackham School, University of Michigan, 1957–1959, lecturer, School of Public Health, 1959–1961, assistant professor, 1961–1963; associate professor, anthropology, University of Wisconsin, Milwaukee, 1963–1967, professor, 1967–1972; Anthropology Curator, Milwaukee Public Museum, 1972–1994, emerita

Concurrent Positions: American Association for the Advancement of Science grant, National Archives, 1953–1954; adjunct faculty member, University of Wisconsin, Milwaukee, 1972–1994

Nancy Lurie is a cultural anthropologist known for her studies of North American Indians and her work in applied and action anthropology aimed at identifying and solving community problems. Her work centered on the Winnebago or Ho-Chunk tribe of Wisconsin, and she was adopted by a member of that tribe, Mitchell Redcloud, Sr., whom she interviewed during the course of her graduate research. Her adoption gave her an entrée to Redcloud's family when she later conducted extensive research into the role of Native American women, who she felt were ignored in most histories of Native Americans. Her book, *Mountain Wolf Woman, Sister of Crashing Thunder* (1961), is the autobiography of one of Redcloud's family members, as Mountain Wolf Woman dictated it to Lurie, her adopted niece. As part of her activist anthropology, Lurie has also consulted with and served as an expert witness and researcher for Indian clients before the U.S. Indian Claims Commission on issues related to tribal identities, boundaries, land use, and occupancy.

As a child, Lurie's father took her to learn about American Indians at the Milwaukee Public Museum. As soon as she was old enough to ride the public transportation alone, she spent many hours at the museum and worked in the anthropology department as a volunteer. She conducted her first fieldwork among the Winnebago while still an undergraduate. There was very little information about this group available at the time, and she continued her research in graduate school. Her doctoral thesis compared cultural change in the Nebraska and Wisconsin enclaves of the Winnebago. During the late 1950s and early 1960s, she collaborated on research with **June Helm** on the northern Athabaskan Indians, studying the Dogrib settlements in the Canadian Northwest. Among her action anthropology projects, several involved the Winnebago and Menominee tribes. In 1972, she left the university and spent the next 20 years of her career as curator and head of the anthropology section of the Milwaukee Public Museum, the first woman to head one of the museum's scientific sections.

Lurie has received numerous honors and awards and is a fellow of the American Anthropological Association (president, 1983–1985) and a member of the American Association for the Advancement of Science, American Ethnological Society, and Society for Applied Anthropology. In 2004, the Ho-Chunk Nation formally recognized Lurie for her work on behalf of their people by presenting her with a custom-made blanket.

Further Resources

Gacs, Ute et al. 1988. *Women Anthropologists: Selected Biographies*. Westport, CT: Greenwood Press.

M

Maccoby, Eleanor (Emmons)

b. 1917
Psychologist

Education: student, Reed College, 1934, 1936; B.S., University of Washington, Seattle, 1939; M.A., University of Michigan, 1949, Ph.D., psychology, 1950

Professional Experience: study director, Division of Program Surveys, U.S. Department of Agriculture, 1943–1946; study director, Survey Research Center, University of Michigan, 1946–1948; lecturer and researcher, Laboratory of Human Development, Department of Social Relations, Harvard University, 1950–1958; associate professor, psychology, Stanford University, 1958–1966, professor, 1966–1987, emeritus

Eleanor Maccoby is a developmental and social psychologist whose studies of the social behavior of young children continue to influence research and theories of gender differences. She edited and wrote a chapter for *The Development of Sex Differences* (1966) on the differences in the development of male and female children, and, in particular, the reasons boys and girls perform differently on intellectual tests. In *The Psychology of Sex Differences* (1974), she and her co-author, Carol Nagy Jacklin, examined the research on gender differences and theorized that gender-typed behavior is a joint product of biological predispositions, social shaping, and cognitive self-socialization processes, and that there was no evidence for many widely held beliefs about the differences between boys and girls. The book was immediately controversial, but it was a first step toward more objective scientific investigations of sex differences. In *Social Development* (1980), she examined family socialization and argued that children's development is influenced by the nature and effect of parent–child interactions.

In the late 1980s, Maccoby and her team of researchers began a long-term study of the effect of divorce on young children. They followed 500 divorcing families for the book *Dividing the Child: Social and Legal Dilemmas of Custody* (1992; co-authored with legal scholar Robert Mnookin). Maccoby and her researchers followed up four years later when the children were adolescents and found that they did well as long as there was minimal parental conflict involved in joint-custody arrangements, results published in *Adolescents After Divorce* (1996; co-authored with Christy M. Buchanan and Sanford M. Dornbusch). Maccoby again turned to her interest in sexual identity and gender differences in *The Two Sexes: Growing Up Apart, Coming Together* (1998), which explores how individuals express their sexual identity at successive periods of their lives and in different social contexts.

Maccoby's interest in psychology began in 1940 when she obtained a position doing public-opinion surveys for the Department of Agriculture. There she gained experience in applied psychology by conducting surveys of wartime programs such as fuel oil rationing and the sale of war bonds. She conducted research for her doctoral thesis in B. F. Skinner's laboratory at Harvard. Through the Laboratory of Human Development, she conducted interviews of mothers for a socialization study on childrearing practices. When the major investigator left the department, she was assigned to teach his courses in child psychology. After moving to Stanford University, she began her work on gender studies.

Maccoby was elected to membership in the National Academy of Sciences in 1993. She has received numerous prizes, such as the Distinguished Scientific Contributions Award of the American Psychological Association (1988), the Kurt Lewin Memorial Award (1991), and the Gold Medal Award for Life Achievement in Psychological Science of the American Psychological Association (1996). She is a member of the Society for Research in Child Development (president, 1981–1983), the American Psychological Association, the Social Science Research Council, and the American Academy of Arts and Sciences.

Macklin, Madge Thurlow

1893–1962
Geneticist

Education: A.B., Goucher College, 1914; M.D., Johns Hopkins University, 1919

Professional Experience: instructor, embryology, University of Western Ontario, 1921–1930, assistant professor, 1930–1945; research associate, Ohio State University, 1945–1959

Madge Macklin performed pioneering research in medical genetics, and she campaigned to include genetics in the standard medical school curriculum. Eventually, she was able to convince her contemporaries of the clinical importance of the family history in diagnosis, therapy, prognosis, and prevention of disease. She demonstrated that both environment and hereditary factors are significant in specific cancers, such as those of the stomach and breast. After her marriage in 1918 and receiving her M.D. from Johns Hopkins in 1919, she moved to the University of

Western Ontario as a lecturer in embryology classes for first-year medical students. Despite her significant work, she received only successive one-year appointments at Western Ontario, perhaps due to her controversial views on eugenics. She viewed eugenics as a branch of preventive medicine in that physicians should determine which people are physically and genetically qualified to be parents of the next generation. She advocated sterilization of people with certain mental diseases. Another factor for her short appointments was that her husband taught at the university, and many institutions were reluctant to hire both husband and wife as faculty, although they did not specifically forbid it.

Macklin was meticulous in her research in preparing carefully controlled experiments and data analysis. The contributions she made in applying sound statistical techniques to genetics were of great significance. In 1945, when she was notified that her contract at Western Ontario would not be renewed, she accepted a position at Ohio State as a National Research Council associate and as a lecturer in medical genetics. Her husband remained at Western Ontario. Macklin received an honorary degree from Goucher College in 1938, and the Elizabeth Blackwell Medal from the American Medical Women's Association in 1957. She was elected president of the American Society of Human Genetics in 1959.

MacLeod, Grace

1878–1962
Nutritionist

Education: B.S., chemistry, Massachusetts Institute of Technology, 1901; A.M., Columbia University, 1914, Ph.D., 1924

Professional Experience: teacher, Massachusetts public schools, 1901–1910; teacher, chemistry and physics, Pratt Institute, 1910–1917; assistant editor, *Industrial and Engineering Chemistry*, 1917–1919; instructor, nutrition, Teachers College, Columbia University, 1919–1924, assistant professor to professor, 1924–1944

Concurrent Positions: cooperating investigator, nutrition laboratory, Carnegie Institution, 1922–1928

Grace MacLeod was recognized by her contemporaries for her work in nutrition. Her research involved utilization of calcium and other supplements, efficiency of proteins, energy metabolism of children, and availability of iron. She spent nearly 25 years at Teachers College, Columbia University, building one of the outstanding nutrition programs in the United States. After the retirement of her colleague and former professor, **Mary Swartz Rose**, in 1940, MacLeod was the head of the nutrition program. In 1944 and 1956, she helped revise and then co-author two new editions of Rose's book, *Foundations of Nutrition*, and she co-authored the fifth edition of Rose's *Laboratory Handbook for Dietetics*. During World War II, she worked on food and nutrition guidelines, becoming chair of the Food and Nutrition Council of Greater New York. After her formal retirement in 1944, she continued to consult for government agencies (such as the U.S. Department of Agriculture [USDA]) on children's nutritional and energy needs.

Born in Scotland, MacLeod came to the United States when she was only four years old. She majored in chemistry at the Massachusetts Institute of Technology (MIT) after being encouraged in science and math by her high school teachers. She went on to Columbia, but her career followed the general pattern for a woman of her generation. She taught public school for more than 10 years after receiving her undergraduate degree and then taught college chemistry and physics while working on her master's degree at Columbia. She spent two years as an assistant editor of a major journal in chemistry, *Industrial and Engineering Chemistry*, before joining the staff of Columbia University while she completed her doctorate. She rose through the academic ranks to full professor and did significant work in the field of nutrition, which was just being recognized as a profession, thanks in large part to the work of her team at Columbia. Her sister, Florence MacLeod, was also a nutritionist, working at the University of Tennessee.

MacLeod published numerous papers and articles, and was on the editorial board of the *Journal of Nutrition* and the *Journal of the American Dietetic Association*. She was a member of the American Association for the Advancement of Science, American Society of Biological Chemistry, American Chemical Society, Society of Biological Chemists, Society for Experimental Biology and Medicine, American Institution of Nutrition, American Dietetic Association, and American Home Economics Association.

Macy-Hoobler, Icie Gertrude

1892–1984
Chemist

Education: A.B., English and music, Central College for Women, Missouri, 1914; B.S., chemistry, University of Chicago, 1916; A.M., chemistry, University of Colorado, Boulder, 1918; Ph.D., physiological chemistry, Yale University, 1920

Professional Experience: assistant chemist, University of Colorado, Boulder, 1916–1917, physiological chemist, school of medicine, 1917–1918; assistant biochemist, Western Pennsylvania Hospital, 1920–1921; instructor, University of California, Berkeley, 1921–1923; director, Nutrition Research Laboratory (later Research Laboratory of the Children's Fund of Michigan), Merrill-Palmer School, Detroit, Michigan, 1923–1954; staff, Merrill-Palmer Institute of Human Development and Family Life, 1954–1959, consultant, 1959–1974

Icie Macy-Hoobler was one of the most influential physiological chemists of the early twentieth century for her research on nutrition, mineral metabolism in human pregnancy, lactation, and growth, and the chemistry of red blood cells in health and disease. Her most important work was on the effect of nutrition on both mother and child. She studied the nutritional requirements of women and children and proved that malnutrition in women had a significant effect upon birth defects, and upon infant health and growth. As a graduate student at Yale, she began research on cottonseeds, which, during World War I, were being substituted for wheat flour. She found that animals that had been fed cottonseeds became ill due to gossypol, a poison present in the plant. She held a series of short-term positions at various schools while completing her advanced degrees and, after receiving her Ph.D., was offered the directorship of nutrition research at the Merrill-Palmer School in Detroit, which in 1931 became the Research Laboratory of the Children's Fund of Michigan. In this position, she mentored biochemistry and nutrition graduate students from the University of Chicago and other schools, and oversaw numerous important projects. Her research group was instrumental in showing the need for vitamin D and in encouraging the irradiation of milk. She studied amino acids in foods and the standardization and minimum daily requirements of vitamins B and C. Her work led to public health campaigns to disseminate new scientific information on nutrition to mothers and children. She contributed to hundreds of scientific papers and several books, including *Nutrition and Chemical Growth in Childhood* (3 vols., 1942–1951), *Hidden Hunger* (1945; co-authored with H. H. Williams), and *Chemical Anthropology: A New Approach to Growth in Children* (1957; co-authored with H. J. Kelly).

Icie Macy attended Central College for Women in Lexington, Missouri, where she studied English and received certification as a music teacher in order to please her parents. She went on to earn another bachelor's degree, this time in chemistry, from the University of Chicago in 1916, where she studied with Nobel Prize–winning physicist Robert A. Millikan. She earned a master's degree in chemistry from the University of Colorado, Boulder, where she taught inorganic and physiological chemistry. A professor encouraged her to continue on for a doctorate, and she enrolled at Yale University, where in 1920 she was one of the earliest women to earn a Ph.D. in physiological chemistry. She married late in life, to pediatrician Raymond Hoobler, in 1938.

Macy-Hoobler was the first woman to chair a division of the American Chemical Society—the biochemistry division (1930–1931). She was active in establishing the Women's Award of the American Chemical Society, later known as the Garvan Medal, which she received in 1946. She also was awarded the Borden Award (1939), the Osborn and Mendel Award (1952), and the Modern Medicine Award (1955). She was a member of the American Association for the Advancement of Science, American Chemical Society, American Society of Biological Chemists, Michigan Academy of Arts, Sciences and Letters, American Institute of Chemists, and American Institute of Nutrition (president, 1944). She received honorary degrees from Wayne State University (1945) and Grand Valley State College (1971), and was inducted into the Michigan Women's Hall of Fame. Macy-Hoobler published an autobiography, *Boundless Horizons: Portrait of a Woman Scientist* (1982), in which she related some of the difficulties she encountered as a woman in her long career.

Further Resources

Williams, Harold H. 1984. "Icie Gertrude Macy Hoobler (1892–1984): A Biographical Sketch." *The Journal of Nutrition*. American Institute of Nutrition. 1351–1362. (30 April 1984). http://jn.nutrition.org/cgi/reprint/114/8/1351.pdf.

Makemson, Maud Worcester

1891–1977
Astronomer

Education: Radcliffe College, 1908–1909; A.B., astronomy, University of California, Berkeley, 1925, A.M., 1927, Ph.D., astronomy, 1930

Professional Experience: newspaper reporter, *Review* (Bisbee, Arizona) and *Gazette* (Phoenix, Arizona), 1917–1923; public school teacher, 1925–1926; research assistant, astronomy, University of California, Berkeley, 1926–1929, instructor, 1930–1931; assistant professor, astronomy and math, Rollins College, 1931–1932; assistant professor, Vassar College, 1932–1957, director of observatory, 1936–1957; research astronomer, University of California, Los Angeles, 1959–1964, lecturer, astronomy, 1960–1964

Concurrent Positions: Fulbright fellow, astronomy, Ochanomizu Women's University, Tokyo, Japan, 1953–1954; consultant, Consolidated Lockheed–California, 1961–1963, General Dynamics, Fort Worth, Texas, 1965

Maud Makemson was recognized for her research on astrodynamics. Her research centered on celestial mechanics and astrodynamics, and on cultural topics such as Polynesian astronomy, navigation, and the Mayan calendar, subjects on which she published in anthropological journals. She began her college career at Radcliffe, but after more than 10 years as a housewife and then newspaper reporter, she entered the University of California in 1923 as a divorced mother of three young children to complete her undergraduate degree. With the help of relatives to watch the children, she continued on to receive a Ph.D. in astronomy at Berkeley in 1930, with a focus on celestial mechanics. She taught briefly at Rollins College in Florida before accepting a faculty position at Vassar College in 1932. Makemson was the first professor of astronomy at Vassar who had not been a student of Maria Mitchell. Makemson remained at Vassar for 25 years, during which time she also became director of the observatory. She continued to work well into her seventies even after formal retirement. She held a position as a research astronomer and lecturer in astronomy at the University of California, Los Angeles after retirement from Vassar, and also consulted for Consolidated Lockheed in California and General Dynamics in Texas, where she moved in 1964 to be near one of her children. At the time of her death, she was still busy at work on a translation of a 1645 Latin astronomy text.

Makemson was co-author of *Introduction to Astrodynamics* (1961, 1967). In addition to her scientific papers, she wrote two other books: *The Morning Star Rises* (1941), on Polynesian astronomy, and *Book of the Jaguar Priest* (1951), for which she had received a Guggenheim fellowship to work on the translation and study of an ancient Mayan calendar. Both of these projects brought her prestige and recognition within the field. In 1953, she received a Fulbright fellowship to teach astronomy at Ochanomizu Women's University in Japan. She was elected a fellow of the American Association for the Advancement of Science and was a member of the American Astronomical Society, the American Institute of Aeronautics and Astronautics, and the American Association of Variable Star Observers.

Further Resources

Lankford, John and Ricky L. Slavings. 1997. *American Astronomy: Community, Careers, and Power, 1859–1940*. University of Chicago Press.

Maling, Harriet Mylander

1919–1987
Pharmacologist

Education: A.B., Goucher College, 1940; A.M., Radcliffe College, 1941, Ph.D., medical science and physiology, 1944

Professional Experience: assistant pharmacologist, Harvard Medical School, 1944–1945, instructor, 1945–1946; assistant professor, medical school, George Washington University, 1951–1952, assistant research professor, 1952–1954; pharmacologist, National Heart and Lung Institute, National Institutes of Health, 1954–1962, head, physiology section, 1962–retirement

Harriet Maling was a pharmacologist whose research focused on autonomic and cardiovascular drugs. She began her career plan as an undergraduate at Goucher College in Maryland, where she planned to combine her interests in medicine and biological research. She worked as an assistant pharmacologist and instructor at Harvard Medical School after receiving her doctorate in medical science and physiology from Radcliffe. During her final year of doctoral study, Maling received a fellowship from the American Association of University Women. She then took a five-year break from seeking employment, during which time she married and gave birth to four children. She returned to teaching at the medical school of George Washington University as assistant professor and later assistant research professor. She joined the National Heart and Lung Institute as a chemical pharmacologist in 1954, and began her research on the effects of different drugs on heart function. She was named the head of the division of physiology in 1962. She published dozens of scientific papers and was a member of the editorial board of the *Journal of Pharmacology and Experimental Therapeutics* from 1962 to 1965.

Maling was a member of the American Association for the Advancement of Science, the Society for Experimental Biology and Medicine, the American Society for Pharmacology and Experimental Therapeutics, and the New York Academy of Science.

Maltby, Margaret Eliza

1860–1944
Physicist

Education: A.B., Oberlin College, 1882, A.M., 1891; B.S., physics, Massachusetts Institute of Technology, 1891; Ph.D., University of Göttingen, 1895

Professional Experience: high school teacher, 1884–1887; instructor, physics, Wellesley College, 1889–1893, department head, 1896; instructor, physics and mathematics, Lake Erie College for Women, Ohio, 1897–1898; research assistant, Physikalisch-Technische Reichsantalt, 1898–1899; instructor, chemistry, Barnard

College, 1900–1903, adjunct professor, physics, 1903–1910, associate professor and department head, physics, 1910–1931

Margaret Maltby was one of the earliest researchers in physical chemistry and the first American woman to receive a degree in physics from a German university. Her areas of research were measurement of high electrolytic resistances, measurement of periods of rapid electrical oscillations, conductivity of very dilute solutions of certain salts, and radioactivity. She was also interested in music and acoustics and is believed to have offered the first course in the physics of sound. During her lifetime, physics was almost exclusively a male profession, and Maltby worked at various institutions before securing a tenure-track faculty position at Barnard. After she graduated from Oberlin College, she attended the Art Students' League for a year before returning to Ohio to teach high school for four years. She entered the Massachusetts Institute of Technology (MIT) in 1887 to study physics and received her undergraduate degree in 1891, the same year that Oberlin granted her a master's degree. Besides earning these degrees, she also studied theoretical physics for a year at Clark University. She taught physics at Wellesley College for four years while continuing her graduate studies at MIT. She then was awarded a traveling fellowship to the University of Göttingen, where she received her doctorate in 1895. She returned to Wellesley as department head before accepting a position at Lake Erie College as instructor in physics and mathematics.

In 1900, Maltby came to Barnard to teach chemistry and physics, and spent the last 20 years of her career as head of the physics department. Although she had early success as a researcher, she never advanced to full professor at Barnard, perhaps because her teaching and administrative duties left little time for further research. She did, however, have a tremendous influence on a generation of female students of science, and committed a significant amount of time and energy to securing funding for equipment and for her students. She had an even greater role in developing career opportunities for women as chair of the fellowship committee of the American Association of University Women (AAUW) from 1913 to 1924, which gave funds to women for advanced study both in the United States and abroad. In 1926, the AAUW established a fellowship in her name. She contributed a chapter on "The Physicist" for a 1920 guide to *Careers for Women* (edited by Catherine Filene).

Maltby was elected a fellow of the American Physical Society, and she also was a member of the American Association for the Advancement of Science.

Further Resources

Byers, Nina and Gary A. Williams. 2006. *Out of the Shadows: Contributions of Twentieth-Century Women to Physics*. New York: Cambridge University Press.

Marcus, Joyce

Archaeologist

Education: B.A., University of California, Berkeley, 1969; M.A., Harvard University, 1971, Ph.D., anthropology, 1974

Professional Experience: visiting lecturer, anthropology, University of Michigan, Ann Arbor, 1973–1975, visiting assistant professor, 1975–1976, assistant professor, anthropology, 1976–1981, assistant curator, Latin American Archaeology, University of Michigan Museum of Anthropology, 1978–1981, associate professor and associate curator, 1981–1984, professor, anthropology, and curator, Latin American Archaeology, 1985–, Elman R. Service Professor of Cultural Evolution, 1998–2005, Robert L. Carneiro Distinguished University Professor of Social Evolution, 2005–

Joyce Marcus is an archaeologist whose research interests include the social, political, and economic development of ancient societies in Latin America. Her primary research has focused on the Zapotec, Maya, and pre-Inca societies of ancient Mexico, Guatemala, and Peru. She has combined archaeological fieldwork with hieroglyphic and ethnohistoric sources to analyze how these civilizations evolved over time, and has documented the emergence of the earliest villages, hereditary inequality, social stratification, warfare, and state religion. She and her colleagues have excavated one of Mexico's earliest agricultural villages and the earliest appearance of hieroglyphic texts circa 700 to 650 BC.

Born in California, Marcus completed her undergraduate education at the University of California, Berkeley and went on to receive her doctorate in anthropology from Harvard University in 1974. She has spent her entire teaching career at the University of Michigan in Ann Arbor, where she is also the Curator of Latin American Archaeology for the Museum of Anthropology. She has published numerous articles and books, including *The Cloud People: Divergent Evolution of the Zapotec and Mixtec Civilizations* (1983, with Kent Flannery; new ed., 2003), *Mesoamerican Writing Systems: Propaganda, Myth, and History in Four Ancient Civilizations* (1992, an Honorable Mention for Outstanding Book in the Social Sciences and Humanities by the Latin American Studies Association), *Zapotec Civilization: How Urban Society Evolved in Mexico's Oaxaca Valley* (1996, with K. Flannery), *Women's Ritual in Formative Oaxaca: Figurine-Making, Divination, Death and the Ancestors* (1998), *La Civilización Zapoteca: Como Evolucionó La Sociedad Urbana en el Valle de Oaxaca* (2001, with K. Flannery, recipient of the Premio Caniem 2001 en el Arte Editorial award in Mexico), *Agricultural Strategies* (2006, with Charles Stanish), *Monte Albán* (2008), *Excavations at*

Cerro Azul, Peru: The Architecture and Pottery (2008, awarded the Cotsen Book Prize in archaeology), *The Ancient City* (2008, with Jeremy A. Sabloff), and *Andean Civilization* (2009, with Ryan Patrick Williams).

Marcus has been an invited lecturer at institutions both in the United States and abroad, and has been a consultant to the American Museum of Natural History in New York; University Museum, University of Pennsylvania; Cotsen Institute of Archaeology, University of California, Los Angeles; and Harvard University's Peabody Museum. She has received numerous grants for her research, including from the Ford Foundation, the National Endowment for the Humanities, the American Association of University Women, and the National Science Foundation.

Marcus was elected to the National Academy of Sciences in 1997 and was the first archaeologist to be elected to the Council of the National Academy of Sciences (2005–2008). She is an elected fellow of the American Academy of Arts and Sciences, American Philosophical Society, and Institute of Andean Studies, and a member of the American Anthropological Association, Society for American Archaeology, American Society for Ethnohistory, Midwest Andeanist Society, and Midwest Mesoamericanist Society. The University of Michigan has acknowledged her research and teaching with the Henry Russel Award for Scholarly Research (1979), a Literature, Science, and Arts Excellence in Research Award (1995), and a Distinguished Faculty Achievement Award (2007). The Universidad Autónoma de Campeche awarded Marcus a "special recognition" Reconocimiento (2003), and she received a Mentor Recognition Award from the University of California, San Diego (2007).

Further Resources

University of Michigan. Faculty website. http://www.lsa.umich.edu/anthro/faculty_staff/marcus.html.

Margulis, Lynn (Alexander)

b. 1938
Cell Biologist, Microbiologist

Education: B.A., University of Chicago, 1957; M.S., University of Wisconsin, 1960; Ph.D., genetics, University of California, Berkeley, 1965

Professional Experience: postdoctoral researcher, Brandeis University, 1963–1965; assistant professor, biology, Boston University, 1966–1971, associate professor, 1971–1977, professor, 1977–1988, Distinguished University Professor of

Cell biologist and microbiologist, Lynn Margulis. (Courtesy of the University of Massachusetts)

Biology, 1986–1988; Distinguished University Professor of Geosciences, University of Massachusetts, Amherst, 1988–

Concurrent Positions: chair, Space Science Board Committee on Planetary Biology and Chemical Evolution, National Academy of Science, 1977–1980

Lynn Margulis has been called the most gifted theoretical biologist of her generation, and she has questioned accepted truths about evolution, heredity, and cell biology. Her research on evolutionary links between cells containing nuclei and cells without nuclei led her to formulate a symbiotic theory of evolution in the 1970s that finally is becoming more widely accepted in the scientific community. Prior to her work, scientists held that evolution was based on natural selection. Her theory of symbiosis proposes that eukaryotes (cells with nuclei) evolved when different kinds of prokaryotes (cells without nuclei) formed symbiotic systems to enhance their chances for survival. The first such symbiotic fusion would have taken place between fermenting bacteria and oxygen-using bacteria. All cells with nuclei, she theorizes, are derived from bacteria that formed symbiotic relationships with other primordial bacteria some 2 billion years ago. She argues that the primary mechanism driving biological change is symbiosis and that competition plays a secondary role. The manuscript in which she first presented her symbiotic theory was rejected or lost by 15 journals before it was published in the *Journal of Theoretical Biology* in 1966. Her comprehensive exposition of the theory is presented in the book *The Origin of Eukaryotic Cells* (1970), and the revised version was published as *Symbiosis in Cell Evolution* (1981). By the time the second book was published, the scientific establishment had finally accepted the idea that mitochondria and chloroplasts evolved symbiotically.

Another of Margulis's current theories that has not yet been accepted is that the Earth as a whole is alive. This idea is popularly known as "the Gaia hypothesis," named for the Greek goddess of the Earth and first proposed by the chemist James Lovelock. Margulis provided evidence for this theory in her research on protozoa, algae, seaweeds, molds, and microbes that prompted *Omni* magazine to dub her "the wizard of ooze" in 1985. When Margulis was elected to membership in the National Academy of Sciences in 1983, she viewed the honor as an indication that her theories were being accepted by the scientific community.

Margulis has published numerous books, including *Symbiotic Planet: A New Look at Evolution* (1998), and several books with her son, Dorion Sagan, such as *What Is Life?* (1995), *What Is Sex?* (1997), and *Acquiring Genomes: A Theory of the Origins of Species* (2002). In 2006, she and Dorion founded their own publishing company, Sciencewriters Books. Her 2007 book, *Mind, Life, and Universe: Conversations with Great Scientists of Our Time* (co-edited with Eduardo Punset), includes profiles of several other women scientists.

In 1999, Margulis was awarded both the National Medal of Science and the Proctor Prize for scientific achievement. In 2009, she was awarded the Darwin-Wallace Medal of the Linnean Society of London for "major advances in evolutionary biology." She is a fellow of the American Association for the Advancement of Science and a member of the International Society for the Study of the Origin of Life and International Society for Evolutionary Protistology ("protistology" refers to the taxonomic kingdom of protists such as protozoans, eukaryotic algae, and slime molds). She was married to physicist and popular author Carl Sagan, and her early work was published under the name Lynn Sagan. She was married a second time, to crystallographer Thomas Margulis, but the couple later divorced.

Further Resources

University of Massachusetts. Faculty website. http://www.geo.umass.edu/faculty/margulis/.

Marlatt, Abby Lillian

1869–1943
Home Economist and Educator

Education: B.S., Kansas State Agricultural College, 1888, M.S., chemistry, 1890

Professional Experience: head, domestic economy, Utah State Agricultural College, 1890–1894; high school teacher, home economics, 1894–1909; director, home economics department, University of Wisconsin, 1909–1935

Abby Marlatt helped develop the profession of home economics by insisting upon broad training and high standards. Her department at the University of Wisconsin became the model for other home economics programs around the country. After she received her master's degree from Kansas State Agricultural College, Marlatt was invited to establish a program in domestic economy at Utah State. In 1894, she accepted a position to establish a program at the Manual Training High School in Providence, Rhode Island. She took advantage of the location by enrolling in advanced studies at Clark University and Brown University. In 1909, the dean of agriculture at the University of Wisconsin invited her to revitalize the school's home economics program. Under her management, the department rapidly expanded in number of courses, students, and faculty. She also established high academic standards for her students, requiring courses in English, foreign languages, and science, and offering technical courses including bacteriology, physiology, and journalism. Her program greatly broadened the training available to home economics majors beyond the domestic skills courses that many colleges offered.

During World War I, Marlatt served in the food-conservation division of the U.S. Department of Agriculture and on several other federal committees. In 1903, she was chair of the Lake Placid Conference on Home Economics and was instrumental in continuing the conference series. After the group was established as the American Home Economics Association, she was vice president from 1912 to 1918. She also directed two fundraising campaigns for the association. She was a member of the American Chemical Society and the American Association for the Advancement of Science. She received honorary degrees from Kansas State in 1925 and from Utah State in 1938.

Marrack, Philippa Charlotte

b. 1945
Immunologist

Education: B.A., Cambridge University, 1967, Ph.D., biological sciences, 1970

Professional Experience: postdoctoral fellow, molecular biology, Cambridge University, 1970–1971; postdoctoral fellow, biology, University of California, San Diego, 1971–1973; postdoctoral fellow, microbiology, University of Rochester, New York, 1973–1974, associate, 1974–1975, assistant professor, microbiology, 1975–1976, assistant professor, oncology, microbiology and cancer center, 1976–1979, associate professor, 1979–1982; associate professor, biophysics, biochemistry, and genetics, University of Colorado Health Science Center, 1980–1985, professor, biochemistry and molecular genetics, 1985–, professor, microbiology and immunology, 1988–1994, professor, immunology, 1994–

Concurrent Positions: investigator, American Heart Association, 1976–1981; member, Department of Medicine, National Jewish Health Center, 1979–; head, Division of Basic Immunology, National Jewish Center of Immunology and Respiratory Medicine, 1988–1990 and 1998–1999; investigator, Howard Hughes Medical Institute, 1986–; advisory head, research in allergy/asthma, National Jewish Health Center, 2004–2006

Philippa Marrack is renowned for her research on the body's immune system and the intricate web of defenses it raises against viruses, bacteria, and other trespassers. Her particular interest is how the body accepts or rejects its own tissues, which is the study of the "T cells" formed in the thymus gland that control the immune system, and her work has implications for the development of vaccines. Very little was known about the T cells in the immune system until the late 1960s, and Marrack and her husband, John Kappler, who have worked together for more than 30 years, became the leading scientists conducting this research.

Born in England, Marrack began her research while still a graduate student at Cambridge University. After receiving her doctorate, she moved to the University of California, San Diego as a fellow in immunology. She joined the cancer research laboratory of R. W. Dutton, who had recently learned to grow cultures of T lymphocytes, and there she met John Kappler, who also was working in the laboratory. They married in 1974 and moved to the University of Rochester, where she was a postdoctoral fellow in immunology. After she won an American Heart Association investigatorship to do basic research, she was recognized as an equal partner with Kappler and the two began to pursue joint projects and publish together. They established a system whereby the person who performed the principal experiments is always the first listed author, and the one who primarily wrote the paper is named second. They moved to Denver, Colorado, in 1979, where they both hold joint appointments with the University of Colorado Health Sciences Center and National Jewish Health. They are also investigators for the Howard Hughes Medical Institute.

Marrack and Kappler have received numerous awards and honors from international organizations and universities for their research; most recently, she is the recipient of the American Association of Immunologists Lifetime Achievement Award (2002), the L'Oréal-UNESCO Women in Science Award (2004), and the National Jewish Health Presidential Award (2004). She is a member of the Royal Society, the American Association of Immunologists (vice president, 1999–2000; president, 2000–2001), and the Science Council of the American Heart Association. She was elected to the National Academy of Science in 1989 and to the Institute of Medicine in 2008.

Further Resources

Howard Hughes Medical Institute. "Philippa Marrack, Ph.D." http://www.hhmi.org/research/investigators/marrack_bio.html.

Martin, Emily

b. 1944
Anthropologist

Education: B.A., anthropology, University of Michigan, B.A., 1966; Ph.D., Cornell University, 1971

Professional Experience: assistant professor, anthropology, Program in Comparative Culture, University of California, Irvine, 1971–1972; assistant professor, anthropology, Yale University, 1972–1974; associate professor, Johns Hopkins University, 1974–1976, professor, 1976–, Mary Elizabeth Garrett Professor of Arts and Sciences, 1981–1994; professor, anthropology, Princeton University, 1994–2001; professor, anthropology, New York University, 2001–

Emily Martin is a cultural anthropologist whose research interests include religion, ideology, politics, models and explanations in social anthropology, political economy of health, gender, anthropology of science, rationality, psychiatry, the unconscious, anthropology of science and medicine, gender, cultures of the mind, emotion and rationality, history of psychiatry and psychology, and both Chinese and U.S. culture and society. Martin began her anthropological research in China and published several books on Chinese religion, ritual, and rural society. She became interested in issues related to science and gender, and her 1987 book, *The Woman in the Body: A Cultural Analysis of Reproduction*, won the Eileen Basker Memorial Prize. She continued this work in a pathbreaking 1991 article, "The Egg and the Sperm: How Science Has Constructed a Romance Based on Stereotypical Male-Female Roles," which highlighted how the language of gender influences scientific research and findings; the article has become a classic in feminist science criticism. Her interest in the cultural anthropology of science and medicine led to the publication in 1994 of *Flexible Bodies: Tracking Immunity in American Culture from the Days of Polio to the Age of AIDS*, and her 2007 book, *Bipolar Expeditions: Mania and Depression in American Culture*.

Martin taught at Yale, Johns Hopkins, and Princeton universities; since 2001, she has been a professor of anthropology at New York University (NYU) and is affiliated with the NYU Institute for the History of the Production of Knowledge. Always interested in the connections between science and culture, Martin has been involved in a variety of projects linking academic anthropology to broader issues of interest to the general public and to practitioners in other sciences. She is founding editor of the magazine *Anthropology Now* and is one of the organizers of an interdisciplinary program called the Psycences Project, which brings together scholars in the humanities and social sciences with clinicians in psychology and psychiatry to discuss research on the human mind across these fields.

Martin has been a distinguished invited lecturer and visiting scholar at many institutions and is a member of the American Anthropological Association, American Ethnological Society, Royal Anthropological Institute, Society for Medical Anthropology, and Association for Feminist Anthropology. Some of her early works were published under the name Emily Martin Ahern.

Further Resources

New York University. Faculty website. http://anthropology.as.nyu.edu/object/emilymartin .html.

Marvin, Ursula Bailey

b. 1921
Planetary Geologist

Education: B.A., history, Tufts University, 1943; M.S., geology, Harvard University–Radcliffe, 1946, Ph.D., 1969

Professional Experience: assistant silicate chemist, University of Chicago, 1947–1950; mineralogist, Union Carbide Ore Company, 1953–1958; instructor, mineralogy, Tufts University, 1958–1961; geologist, Smithsonian Astrophysical Observatory (now the Harvard-Smithsonian Center for Astrophysics), Harvard University, 1961–1998, senior geologist emeritus

Concurrent Positions: lecturer, Tufts University, 1968–1969; lecturer, geology, Harvard University, 1974–1977

Ursula Marvin has been a prominent planetary geologist whose research interests include mineralogy and petrology of meteorites and lunar samples, history of geology, and geological mapping of Galilean satellites. She received a master's degree in 1946, and for many years, she and her husband, Thomas Crockett Marvin, were independent geologists and mineralogists who worked throughout the world. She did not return to Harvard to complete her Ph.D. until 23 years after completing the required course work at Harvard. One of her early projects involved traveling to Brazil to locate manganese oxide to be used in batteries manufactured by the Union Carbide Company. She also worked as a chemist at the University of Chicago and as an instructor in mineralogy at Tufts University. She has had concurrent positions with the Smithsonian Astrophysical Observatory (or SAO; now the Harvard-Smithsonian Center for Astrophysics, or CfA), the Harvard College Observatory, and as a lecturer with Tufts and Harvard, where she was the first female faculty member in geology. She began studying meteorites at the Harvard-Smithsonian CfA in 1961, and remained there until her retirement in 1998.

At the CfA, Marvin worked on a NASA study of the mineral makeup of lunar rocks brought back by the Apollo astronauts. Her space work also included the geological mapping of Jupiter's largest satellite or moon, Ganymede. During two polar expeditions as part of the National Science Foundation's (NSF) Antarctic Search Meteorites Team in 1978–79 and 1981–82, she collected and studied meteorites, sending samples to other scientists around the world. She returned to Antarctica in 1985 as part of an NSF team to research the boundary and impact of the meteor that may have led to the extinction of the dinosaurs 65 million years ago. Her work on these various projects was acknowledged with both an asteroid (1991) and an ice mountain, or "nunatak," (1992) named in her honor.

Even after her formal retirement in 1998, Marvin has remained affiliated with the CfA as a consultant, senior geologist emeritus, and consulting expert. She has been particularly involved in advancing the careers of women in science. Between 1974 and 1977, she was the first coordinator of the Federal Women's Program (now the Women's Program Committee) at the SAO. She served on the American Geological Institute's Committee of Women in the Geosciences, for which she compiled and edited the annual *Roster of Women in the Geosciences Professions*. She has dedicated her time to a variety of academic and professional committees dedicated to science education, including as a trustee of Tufts University (1975–1985), a trustee of the Universities Space Research Association (USRA) (1979–1984), and secretary-general of the International Commission on the History of Geological Sciences (1989–1996; vice president for North America, 1996).

Marvin was the honored recipient of the History of Geology Award in 1986 from the Geological Society of America (GSA). In 1997, Marvin received Lifetime Achievement Awards from Women in Science and Engineering (WISE) and the Harvard-Smithsonian Center for Astrophysics. In 2005, she received the Sue Tyler Friedman Award of the GSA for work in recording the history of geology. She has been a member of the Mineralogical Society of America, the Meteoritical Society (president, 1975 and 1976), History of Earth Sciences Society (president, 1991), American Association for the Advancement of Science, and American Geophysical Union.

Further Resources

"Ursula Marvin Honored by 'Wise' Award for Lifetime Achievement in Science." *CfA Almanac*. (July 1997). http://www.cfa.harvard.edu/lib/online/almanac/797.htm.

Mathias, Mildred Esther

1906–1995
Botanist

Education: A.B., Washington University, St. Louis, 1926, M.S., 1927, Ph.D., botany, 1929

Professional Experience: assistant, Missouri Botanical Garden, 1929–1930; research associate, New York Botanical Garden, 1932–1936; research associate, University of California, Berkeley, 1937–1942; herbarium botanist, University of California, Los Angeles (UCLA), 1947–1951, lecturer, botany, 1951–1955, assistant professor to professor, 1955–1974; director, Botanical Garden, UCLA, 1956–1974

Concurrent Positions: assistant specialist, experiment station, UCLA, 1951–1955, assistant plant systematist, 1955–1957, associate plant systematist, 1957–1962

Mildred Mathias was a prominent, award-winning botanist whose research included classification of plants of the western United States, subtropical ornamental plants, and tropical medicinal plants. She was an expert on Umbelliferae, or the carrot family, of which she discovered 100 new species or combinations; the genus *Mathiasella* is named in her honor. Mathias had originally planned to study mathematics, but few courses were available to women when she began her studies in the early 1920s. She went on to complete her bachelor's, master's, and doctorate in botany at Washington University in St. Louis. She worked at the Missouri Botanical Garden and the New York Botanical Garden before moving to California as a research associate at Berkeley in 1937. She began her work at UCLA in 1947 as a herbarium botanist, then serving on the faculty until her retirement in 1974. Mathias was married and the mother of four children, which was unusual for a career woman of her generation.

Beyond her research and teaching, Mathias was committed to educating the public about horticulture and to conservation. She was director of the UCLA botanical garden for almost 20 years, and in 1979, it was renamed the Mildred E. Mathias Botanical Garden in her honor. In addition to her work at the botanical garden, she co-hosted a weekly gardening show on television and wrote articles on horticulture and gardening for popular magazines. She worked to protect lands from development at the local, national, and international level, helping to establish the U.S. Natural Reserve System and founding the Organization for Tropical Studies to preserve lands in Costa Rica. She was the author of *Color for the Landscape: Flowering Plants for Subtropical Climates* (1973).

Mathias was named Woman of the Year by the *Los Angeles Times* in 1964, and her numerous other honors include the Nature Conservancy National Award, the California Conservation Council Merit Award, the UCLA Medical Auxiliary Woman of Science Award, a Merit Award from the Botanical Society of America (1973), the Liberty Hyde Bailey Medal from the American Horticultural Society (1980), a Medal of Honor from the Garden Club of America (1982), and the UCLA Emeritus of the Year Award (1990). She served as executive director of the American Association of Botanical Gardens and Arboretums, and was a member and president of both the American Society of Plant Taxonomists (president, 1964) and the Botanical Society of America (president, 1984). She was also a member of the Society for the Study of Evolution, American Association for the Advancement of Science, and the American Society of Naturalists. In 1996, the Botanical Garden at UCLA produced a video about her life and work entitled *Mildred Mathias: A Lifetime of Memories.*

Further Resources

University of California, Los Angeles. "Mildred E. Mathias Botanical Garden." http://www.botgard.ucla.edu/bg-home.htm.

Matson, Pamela Anne

b. 1953
Soil Scientist, Environmental Scientist

Education: B.S., biology and English, University of Wisconsin, Eau Claire, 1975; M.S., environmental science, Indiana University, 1980; Ph.D., forest ecology, Oregon State University, 1983

Professional Experience: postdoctoral fellow, entomology, North Carolina State University, 1983; research scientist, Ames Research Center, National Aeronautics and Space Administration (NASA), 1983–1993; professor, ecosystem ecology, University of California, Berkeley, 1993–1997; professor, geological and environmental studies, Stanford University, 1997–

Pamela Matson is renowned for her pioneering research into the role of land-use changes on global warming. She has analyzed greenhouse gas emissions resulting from tropical deforestation and investigated the negative effects of intensive agriculture on the atmosphere, especially the effects of tropical agriculture and cattle ranching. After receiving her doctorate in biology in 1975, her early research focused on forest ecology and then broadened to include many other areas in the global environment. She worked as a research scientist for NASA for 10 years before entering academia as a full professor at the University of California, Berkeley. In 1997, she moved to Stanford as professor of environmental studies, where she has served as Dean of Earth Sciences since 2002.

Matson has served on numerous boards and committees dedicated to conservation, ecology, and the study of the environment and global climate change. She is the founding editor-in-

Environmental scientist Pamela Anne Matson. (Courtesy of Stanford University News Service Library)

chief of the *Annual Review of Environment and Resources* (2002–present) and has served on the editorial board of numerous other journals, including *Ecosystems, Biogeochemistry*, and *Global Change Biology*. She has published nearly 100 papers and book chapters, and is co-editor of *Biogenic Trace Gases: Measuring Emission from Soil and Water* (1995) and *Principles of Terrestrial Ecosystem Ecology* (2002).

In 1994, she was elected to the National Academy of Sciences, and in 1995 she received the prestigious five-year MacArthur fellowship. Among her numerous other awards and honors are the NASA Exceptional Service Award (1993), the University of Wisconsin, Eau Claire Distinguished Alumni Award (1996), the Oregon State University Distinguished Alumni Award (1998), and a McMurtry Fellowship in Undergraduate Education at Stanford (2002). She has served on numerous professional boards and government committees on sustainability and environmental issues, and has been named a National Associate of the National Academy of Sciences (2002). She is a member of the American Academy of Arts and Sciences, Ecological Society of America (president, 2001), American Association for the Advancement of Science, American Geophysical Union, American

Institute of Biological Sciences, and American Association of University Women. In 2000, she was named a Fellow of the Aldo Leopold Leadership Program at Stanford University.

Further Resources

Wasserman, Elga. 2002. *The Door in the Dream: Conversations with Eminent Women in Science*. Washington, D.C.: Joseph Henry Press.

Stanford University. Faculty website. http://pangea.stanford.edu/research/matsonlab/members/Matson.htm.

Matthews, Alva T.

b. 1933
Engineer

Education: B.S., Columbia University, 1955, M.S., 1957, Ph.D., engineering science, 1965

Professional Experience: design and research engineer, Weidlinger Associates, 1957–1983; senior research engineer, Rochester Applied Science Associates, 1957–1983; independent consultant, 1983–

Concurrent Positions: instructor, civil engineering, Columbia University; adjunct associate professor, mechanical and aerospace sciences, University of Rochester; instructor, engineering, Swarthmore

Alva Matthews is a research engineer recognized for her work in the field of structural analysis and wave propagation in solids. She designed helicopter blades and satellite-tracking antennae, and analyzed auto accidents in order to learn how to build safer cars. As a wave-propagation specialist, she has studied earthquakes to see how shock waves are transmitted through soil and rocks, and how buildings can be designed to withstand earth tremors. Her work had implications for studies of the effect of nuclear weapons on structures such as buildings. After receiving her master's degree, she was employed at Weidlinger Associates, a construction engineering firm in New York. She was concurrently a senior research engineer with Rochester Applied Science Associates in Rochester, New York. At the same time, she was an instructor of civil engineering at Columbia and lectured in the evenings at the University of Rochester.

Matthews decided to become an engineer while still a teenager when she accompanied her father, an industrial builder, to construction sites. She began her college education at Middlebury College in Vermont, then transferred to Barnard before moving on to Columbia University. As a student worker in a contractor's field office, she was prohibited from entering the tunnels with the men, and an early advisor at Middlebury told her engineering was too difficult for a girl and she would never find a job. She ignored this advice, and went on to earn three engineering degrees from Columbia, including becoming the first woman to receive a doctorate in civil engineering from that institution. She dedicated herself to promoting engineering education and careers for women; in 1973, she spoke before the Society of Women Engineers on "Engineering as an Ideal Woman's Career." After starting her own family in the late 1970s, Matthews retired from her full-time engineering position, but remained a private consultant and adjunct instructor of engineering sciences at various colleges.

Matthews is a member of the American Society of Mechanical Engineers and has been honored with the Society of Women Engineers Achievement Award (1971) and the Engineering Award of the Federation of Engineering and Scientific Societies of Drexel University (1976). In 2005, she was appointed to the Dean's Engineering Council at Columbia. She is identified in some sources by her married name, Alva Matthews Solomon.

Further Resources

Alva Matthews. The Society of Women Engineers. http://societyofwomenengineers .swe.org/index.php?option=com_content&task=view&id=49&Itemid=55.

Hatch, Sybil E. 2006. *Changing Our World: True Stories of Women Engineers*. Reston, VA: American Society of Civil Engineers.

Maury, Antonia Caetana de Paiva Pereira

1866–1952
Astronomer

Education: B.A., Vassar College, 1887

Professional Experience: staff member, Harvard College Observatory, 1888–1896; teacher and lecturer, physical science and astronomy, various institutions, 1896–1918; staff member, Harvard College Observatory, 1918–1935; curator, Draper Park Observatory Museum, 1935–1938

Antonia Maury was an astronomer whose research interests were spectra of bright Northern stars and spectroscopic binaries. She was one of the first women to receive a professional appointment at the Harvard College Observatory. She started working at Harvard in 1888 after she graduated with honors from Vassar. In her research, she developed a new, two-dimensional system of stellar classification that included the width and sharpness of lines. It turned out that the differences in width and sharpness resulted from differences in the size and luminosity of stars. During that time, she also confirmed the observatory director Edward C. Pickering's discovery of a double star and then discovered a second star system. She left the observatory in 1896 due to conflicts with Pickering, who wanted his staff to gather data quickly under another system, while Maury wanted to develop a classification that yielded a wider range of data. Maury lectured and taught at several schools in the interim, but returned to the observatory in 1918 after Pickering retired. Her other significant work was on spectroscopic binaries, including some very complex systems, but she did not work steadily on this area of research. After retiring from Harvard in 1935, she spent three years as curator of the Draper Park Observatory Museum in New York.

Although Maury's contributions were not fully appreciated at Harvard, they had a significant influence on scientists elsewhere. Her early studies are now widely recognized as an essential step in the development of theoretical astrophysics, and she received the Annie J. Cannon Prize of the American Astronomical Society in 1943. Although she worked chiefly as an astronomer, she also was an active ornithologist, a naturalist, and a conservationist who participated in the campaign to save the redwood forests. She was a member of the American Astronomical Society, the Royal Astronomical Society, and the National Audubon Society. Her younger sister was paleontologist **Carlotta Maury**.

Maury, Carlotta Joaquina

1874–1938
Paleontologist

Education: Radcliffe College, 1891–1894; Ph.B., Cornell University, 1896, Ph.D., Cornell University, 1902

Professional Experience: high school teacher, 1900–1901; assistant, paleontology, Columbia University, 1904–1906; paleontologist, Louisiana Geological Survey, 1907–1909; lecturer, geology, Barnard College, 1909–1912; professor, geology and zoology, University of the Cape of Good Hope, 1912–1915; paleontologist, Brazil Survey, 1918–1938

Carlotta Maury was a paleontologist whose research interests were in the recent and Pleistocene eras of New York and the Gulf of Mexico, the Tertiary period of Florida and the West Indies, and stratigraphy of Venezuela. She received degrees from Radcliffe and Cornell, and studied at the Jardin des Plantes in Paris for one year before receiving her doctorate from Cornell University in 1902.

Maury participated in several research expeditions and published numerous reports of her work on Antillean, Venezuelan, and Brazilian stratigraphy and fossil faunas. Many of these reports were sent to the American Museum of Natural History. She was the author of *A Comparison of the Oligocene of Western Europe and the Southern United States* (1902). She was the paleontologist for a geological expedition to Venezuela in 1910 and 1911, organized and conducted the Maury expedition to the Dominican Republic in 1916, was consulting paleontologist and stratigrapher for the Venezuelan division of the Royal Dutch Shell Petroleum Company from 1910 to 1938, and was official paleontologist to Brazil from 1918 to 1938.

Maury was elected a fellow of both the Geological Society of America and the American Geographical Society. She also was a member of the American Association for the Advancement of Science and a corresponding member of the Brazilian Academy of Sciences. Her older sister was astronomer **Antonia Maury**.

McCammon, Helen Mary (Choman)

b. 1933
Geologist, Marine Biologist

Education: B.Sc., University of Manitoba, 1955; M.S., University of Michigan, 1956; Ph.D., geology, Indiana University, 1959

Professional Experience: research technician, stratigraphy, Manitoba Department of Mines and Natural Resources, 1952–1959; lecturer, geography, University of North Dakota, 1961; assistant professor, geology, Department of Earth Science, University of Pittsburgh, 1963–1968, associate professor, 1968; visiting associate professor, geology, Department of Geology, University of Illinois, Chicago, 1968–1970; research associate, geology, Field Museum of Natural History, Chicago, 1969–1972; director of environmental science, Environmental Protection Agency, 1973–1976; senior oceanography and marine scientist, Environmental Research Division, U.S. Department of Energy, 1977–1979, director, 1979–1991, deputy director, Environmental Science Division, 1991–retired

Helen McCammon is a geologist who is also known for her work in marine physiology and ecology. Like many environmental scientists, her research has required broad interdisciplinary knowledge and includes the impact of energy activities in coastal and terrestrial environments ranging from arctic tundra to temperate forest and desert regions, as well as marine physiology and ecology. As deputy director of the Environmental Science Division of the U.S. Department of Energy, her responsibilities included overseeing ecological research and overseeing the division's budget. Her background also includes considerable experience in research and teaching geology in several universities, plus a stint as a geologist at the Field Museum of Natural History in Chicago. However, her research included studies of living animals as well as terrestrial and marine physical environments, the usual subjects of geologists.

After she completed her dissertation on paleontology for her doctorate in geology, McCammon decided to study how marine organisms live today and began to research living invertebrates, especially brachiopods. These lampshells were common 300 million years ago and can still be found in New Zealand, the Antarctic, and other cold-water regions. Her fieldwork took her to many of these places, but because geology is a male-dominated field, she had problems in obtaining funding and having her papers published. She also faced discrimination at the University of Illinois, Chicago when the department head decided to discontinue her salary in order to hire a man, and asked her to continue teaching as an unsalaried faculty member. Her husband was a faculty member at the university at the time, but her position had been only that of a visiting associate professor. She refused to accept the arrangement and accepted a position at the Field Museum instead. Her own husband was supportive of her research, sharing household duties, involving the children in their sample-collecting and research, and supporting their long-distance relationship when he still worked in Chicago and Helen took a position with the U.S. government in Boston and then Washington, D.C. In 1973, she joined the Environmental Protection Agency, beginning her long career with the federal government as chief scientist and administrator in various departments.

McCammon is a fellow of the American Association for the Advancement of Science and a member of the American Geological Institute, American Society of Zoologists, Oceanic Society, and American Society of Limnology and Oceanography.

McClintock, Barbara

1902–1992
Geneticist

Education: B.S., botany, Cornell University, 1923, M.A., botany, 1925, Ph.D., botany, 1927

Professional Experience: instructor, botany, Cornell University, 1927–1931, research associate, 1934–1936; fellow, National Research Council, 1931–1933; fellow, Guggenheim Foundation, 1933–1934; assistant professor, botany, University of Missouri, 1936–1941; staff member, Carnegie Institution of Washington, Cold Spring Harbor Laboratory, 1942–1967, distinguished member, 1967–1992; Andrew White Professor at Large, Cornell University, 1965–1992

Barbara McClintock received the Nobel Prize in Physiology or Medicine in 1983 for her pioneering work on the mechanism of genetic inheritance. She discovered early on that genes can move from one area on the chromosomes to another, a finding known as "jumping genes" that now helps molecular biologists identify, locate, and study genes. She observed the changes in color patterns in kernels of Indian corn and correlated these changes with changes in the chromosome structure. She received the Nobel Prize for this discovery more than 32 years after publishing her findings. In spite of early recognition for her work, she was relatively unknown in the scientific community for decades until she was awarded the Nobel Prize. Her work was largely outside the mainstream of science at that time, and few were able to comprehend the significance of her research until other scientists' work on DNA in the 1960s was used to verify her experiment and support her theories.

After she received her doctorate in botany in 1927, McClintock stayed on at Cornell as an instructor for five years, and then worked in research for another six years under fellowships from the National Research Council and the Guggenheim Foundation. Since Cornell was not appointing women to faculty positions, she had to find other sources for income. Research positions were very scarce for women during the Depression, but she accepted an appointment as assistant professor of botany at the University of Missouri for five years. That was the last teaching position she held, as she preferred to focus exclusively on research. Starting in 1942, she worked at the Cold Spring Harbor Laboratory on Long Island, where she maintained a small apartment on the grounds of the laboratory. In 1981, she was awarded a lifetime tax-free annual fellowship of $60,000 from the MacArthur Foundation, and she continued to work her accustomed schedule of long hours seven days a week in the lab until just shortly before her death.

McClintock was elected a member of the National Academy of Sciences in 1944 and was the first woman president of the Genetics Society of America in 1945. She received the Kimber Genetics Award (1967), the National Medal of Science (1970), the Rosenstiel Award (1978), and the Lasker Award (1981). She was a member of the American Association for the Advancement of Science, the American Academy of Arts and Sciences, the American Philosophical Society, the American Society of Naturalists, and the Royal Society of England. In 2005, the U.S. Postal Service issued a postage stamp featuring McClintock as part of their *American Scientists* series.

Further Resources

Keller, Evelyn Fox. 1983. *A Feeling for the Organism: The Life and Work of Barbara McClintock*. San Francisco, CA: W.H. Freeman.

Fedoroff, Nina V. and David Botstein. 1992. *The Dynamic Genome: Barbara McClintock's Ideas in the Century of Genetics.* Cold Spring Harbor, NY: Cold Spring Harbor Laboratory Press.

The Barbara McClintock Papers, Profiles in Science, National Library of Medicine. http://profiles.nlm.nih.gov/LL/.

McCoy, Elizabeth Florence

1903–1978
Soil Microbiologist

Education: B.S., University of Wisconsin, 1925, M.S., 1926, Ph.D., bacteriology, 1929

Professional Experience: postdoctoral fellow, National Research Council, Rothamsted Experimental Station, England, and Botanical Institute, Karlova University, Prague, Czechoslovakia, 1929–1930; assistant to associate professor, agricultural bacteriology, University of Wisconsin, 1930–1943, professor, 1943–1973

Elizabeth McCoy was a microbiologist whose research included anaerobes, serology, freshwater bacteria, water quality and waste disposal, and industrial fermentations. She was notable for her work in soil microbiology and for detecting a high-yielding strain of *penicillium* as part of a World War II–era government project with the Office of Scientific Research and Development. She discovered another antibiotic, oligomycin, which is still used to research and treat fungal diseases in plants and was also under development at that time by the pharmaceutical company Pfizer. McCoy was granted patents for her methods of isolating oligomycin and for processing butyl alcohol. She gained national attention as a female scientist when the *New York Times* reported in 1946, "Wisconsin University Girl Wins Patent on an Industrial Solvent"; at the time, McCoy was hardly a "girl," as she was a 43-year-old professor of bacteriology.

McCoy was born and raised on a Wisconsin farm, and she developed an early interest in agricultural science and diseases. She received her bachelor's, master's, and doctoral degrees in agricultural bacteriology at the University of Wisconsin, where she was then employed as a faculty member for more than 40 years, one of the few women of that time to advance to full professor. Over the years, she conducted research experiments in California, Puerto Rico, England, and Czechoslovakia, and was involved in notable studies related to botulinum food poisoning, vaccine development, and microbial ecology (pollution) of rivers and lakes. She oversaw bacteriological and water-quality research projects at Trout Lake Station in northern Wisconsin, at nearby Lake Mendota, and at Lake Michigan. She published numerous scientific papers and articles, and authored or co-authored books on *Root-nodule Bacteria and Leguminous Plants* (1932) and *Anaerobic Bacteria and Their Activities in Nature and Disease* (1939).

McCoy was elected a fellow of the American Public Health Association and was a member of the American Association for the Advancement of Science, the American Academy of Microbiologists, the Society for Experimental Biology and Medicine, and the Wisconsin Academy of Sciences, Arts and Letters, for which she served as president. She served as editor of *Biological Abstracts* and of the *Journal of Bacteriology*. She received a posthumous honorary doctorate from the University of Wisconsin, Milwaukee. After her death, her patents, as well as her family farm, were donated to the Wisconsin Alumni Research Foundation.

Further Resources

Fisher, Madeline. 2003. "Discovery Provides Reminder of Bacteriology Prof and WARF Inventor." Wisconsin Alumni Research Foundation. (November 5, 2003). http://www.warf .org/news/news.jsp?news_id=138.

McCracken, (Mary) Isabel

1866–1955
Entomologist

Education: A.B., Stanford University, 1904, A.M., 1905, Ph.D., 1908; University of Paris, 1913–1914

Professional Experience: teacher, public schools, 1890–1900; assistant in physiology and entomology, Stanford University, 1903–1904, instructor, entomology and bionomics, 1904–1909, assistant professor, entomology and zoology, 1909–1918, associate professor, 1918–1930, professor, 1930–1931; research associate, California Academy of Sciences, 1931–1945

Isabel McCracken was an entomologist who conducted research on a variety of topics, including bees, beetles, birds, mosquitoes, and silkworms, and taught in the area of economic entomology. Her career followed the pattern of many women of her generation in that she first taught in the public schools of her native Oakland, California, for more than a decade before entering college. She attended Stanford University and was employed as a staff assistant while working toward her advanced degrees, with studies in physiology, natural history, and entomology.

Entomologist Isabel McCracken. (National Library of Medicine)

She conducted field research and published scientific papers on the genetics of beetles and on birds of the Sierra Nevada mountains. She received her Ph.D. in 1908, at the advanced age of 42. She became an assistant professor of entomology and zoology after receiving her doctorate, and spent the remainder of her career at Stanford, finally advancing to full professor in 1930, just one year before she retired. After her retirement from teaching, McCracken worked as a research associate at the California Academy of Sciences. Her research there concentrated on her long-term interest in birds and their relationship to insects.

In addition to her scientific papers, McCracken co-authored a textbook called *The Animals and Man* (1911). She was elected a fellow of the California Academy of Sciences, and she was a member of the Entomological Society of America.

McFadden, Lucy-Ann Adams

b. 1952
Astronomer, Geophysicist

Education: B.A., natural sciences, Hampshire College, Massachusetts, 1974; M.S., earth and planetary science, Massachusetts Institute of Technology, 1977; Ph.D., geology and geophysics, University of Hawaii, 1983

Professional Experience: research associate, geography, University of Maryland and Goddard Space Flight Center, National Aeronautics and Space Administration (NASA), 1983–1984, research associate, astronomy, University of Maryland, 1984–1986, assistant research scientist, astronomy, 1986–1987; assistant research physicist, California Space Institute, University of California, San Diego, 1987–1991, associate research physicist, 1991–1995; associate research scientist, graduate faculty, astronomy, University of Maryland, 1996–2007, research professor, 2007–

Concurrent Positions: National Science Foundation visiting professor, University of Maryland, 1992–1995; Near-Earth Asteroid Rendezvous (NEAR) Mission Science Team Multispectral Imager/Near-Infrared Spectrometer, 1994–2000; visiting scientist, Space Telescope Science Institute, 1995; faculty director, College Park Scholars, Science, Discovery & the Universe program, 1997–2001; Deep Impact Discovery Mission Co-Investigator, Education/Outreach Manager, NASA, 1999–2006; Dawn Discovery Mission, 2002–2015; Deep Impact Extended Mission, 2007–2011

Lucy-Ann McFadden is a planetary scientist who has specialized in searching for Earth-approaching asteroids and dead comets. She has estimated that between 1,500 and 2,000 asteroids and dead comets roam space near the Earth. Most asteroids and comets pass Earth at high speed millions of miles away, but those that come closer provide astronomers with insight into the solar system's past. The small bodies in the inner solar system contain primarily rock and metal, while those in the outer solar system contain ices and dark, carbon-based compounds. The difference in composition among various types indicates how material was spread through the solar system while it was forming, and by bouncing radar beams off their surfaces, astronomers can determine the sizes, shapes, and compositions of the objects. McFadden's research involves determining the surface composition of asteroids to understand their nature, source, and evolution. She uses the Hubble Space Telescope to study the relationship between asteroids and comets based on the composition of solid components and the reflectance properties of meteorites. She has published numerous papers on the characteristics of the objects that she has studied and also participated in the observations of the Shoemaker-Levy comet (named for Gene and **Carolyn Shoemaker**, who first identified the comet that impacted Jupiter).

In addition to her position as a faculty research scientist at the University of Maryland, McFadden has been a principal investigator and educational and outreach director of several NASA missions that have included observations of Mars, the moon, and other planets. Through her work at NASA, in her local community, and through the Internet, she has been heavily involved in promoting educational programs that inspire students to pursue careers in the sciences. She is also the co-editor of the *Encyclopedia of the Solar System* (2006).

McFadden has received numerous awards and honors from NASA and other institutions. She is a member of the American Association for the Advancement of Science, American Astronomical Society, American Geophysical Union, and Meteoritical Society.

Further Resources

University of Maryland. Faculty website. http://www.astro.umd.edu/~mcfadden/.

McNutt, Marcia Kemper

b. 1952
Marine Geophysicist

Education: B.A., physics, Colorado College, 1973; Ph.D., earth science, Scripps Institution of Oceanography, University of California, San Diego, 1978

Professional Experience: postdoctoral research associate, Scripps Institution of Oceanography, 1978; visiting assistant professor, University of Minnesota, 1978–1979; geophysicist, tectonphysics, Office of Earthquake Studies, U.S. Geological Survey (USGS), 1979–1982; assistant professor, geophysics, Massachusetts Institute of Technology (MIT), 1982–1986, associate professor, 1986–1988, professor, 1989–1998; president and chief executive officer, Monterey Bay Aquarium Research Institute, 1997–2009; director, USGS, 2009–

Concurrent Positions: secretary, John Muir Geophysical Society, 1979–1983; associate editor, *Journal of Geophysical Research*, 1980–1983; associate director, MIT SeaGrant College, 1993–1995; director, Joint Program in Oceanography and Applied Ocean Science and Engineering, MIT and Woods Hole Oceanographic Institution, 1995–1997; affiliated professor, geophysics, Stanford University, 1998–; affiliated professor, earth sciences, University of California, Santa Cruz, 1998–

Marcia McNutt is renowned for her research on plate tectonics using a variety of techniques, including the Geosat global-positioning satellite. She is particularly known for her work on mapping the ocean floor and measuring the depth of the ocean. Her research includes studies of long-term rheology of the Earth's crust and upper mantle using gravity and topography data, isotasy, paleomagnetism of seamounts, and thermal modeling of the lithosphere. Plate tectonics, a science that developed around 1965 to 1970, is the theory of global tectonics in which the lithosphere is divided into a number of crustal plates, each of which moves on the plastic asthenosphere more or less independently to collide with, slide under, or move past adjacent plates. The study of plate tectonics seeks to explain how the continents were formed and to predict how the plates will move into new patterns. Although geologists have mapped much of the land portion of the Earth, data on the oceans are still being revealed and much of it has remained classified by the U.S. government for strategic reasons. The data that have been released can indicate new locations for fishing or for oil drilling as well as predicting the future activity of underwater volcanoes.

McNutt has worked particularly on mapping areas of the southern oceans, which had remained uncharted because they are far from shipping lanes and not of strategic importance from a military standpoint. The standard method has been to use echo sounders by deploying entire arrays of acoustic transceivers on the hulls of ships to measure the ocean depth in a swath several kilometers wide. McNutt has improved this research by using highly sensitive radar altimeters in Earth orbit to sense minute changes in sea level caused by the gravitational attraction of topography on the seafloor; the radar altimeters measure the water density by hitting the water/rock interface on the ocean floor. The measurements obtained by the echo sounders and by the radar altimeters provide similar readings, but the orbiting altimeters provide a much higher resolution than the echo sounders. In addition to her faculty appointments, in 1997, McNutt became Director of the Monterey Bay Aquarium Research Institute (MBARI) in Moss Landing, California, a position she held for more than 10 years. In 2009, she was chosen by President Obama as the new head of USGS and science advisor to the U.S. Secretary of the Interior.

McNutt was elected to the National Academy of Sciences in 2005. She has been a member of the National Aeronautical and Space Administration (NASA) Science Steering Group Geopotential Research Mission, the Committee on Geodesy of the National Research Council, and the Geodynamics Committee, and was chair of the President's Panel on Ocean Exploration under President Clinton. She is a fellow of the American Geophysical Union (president, 2000–2002), Geological Society of America, American Association for the Advancement of Science, American Academy of Arts and Sciences, and International Association of Geodesy. McNutt has received numerous awards and honors for her research, including the James B. MacElwane Medal of the American Geophysical Union (1988), Scientist of the Year award from the ARCS Foundation (2003), and Maurice Ewing Medal from the Society of Exploration Geophysicists (2007).

Further Resources

U.S. Geological Survey. "Marcia McNutt, Director, U.S. Geological Survey." http://www.usgs.gov/aboutusgs/organized/bios/mcnutt.asp.

McSherry, Diana Hartridge

b. 1945
Medical Physicist, Computer Scientist

Education: B.A., physics, Harvard University, 1965; M.A., Rice University, 1967, Ph.D., nuclear physics, 1969

Professional Experience: fellow, nuclear physics, Rice University, 1969; research physicist in ultrasonics, Digicon, Inc., 1969–1974, executive vice president of medical ultrasound, 1974–1977; president, cardiology analysis systems, Digisonics, Inc., 1977–1982; vice president, Digicon, Inc., 1980–1987; chief operating officer, Cogniseis Development, Inc., 1987–1995, president, 1995–; president, Digisonics

Concurrent Positions: chair, Information Products Systems, Houston, 1982–1986

Diana McSherry is a research biophysicist known for her development of computer-based cardiology analysis systems and has worked in the specific areas of echocardiology, ventriculography, and hemodynamics. Echocardiology uses reflected ultrasonic waves to examine the structure and functioning of the heart; ventriculography involves examining the ventricles of the heart, which are the lower chambers on each side of the heart that receive blood from the atria and in turn force it into the arteries; and hemodynamics is the branch of physiology dealing with the forces involved in the circulation of the blood. The system she developed uses ultrasonic waves and computer processing to produce images of the heart and circulation system, and it was a major breakthrough in the 1970s, when scientists were just beginning to develop the software for medical applications. Her product permits physicians to view the inside of a patient's body without making an incision; the ultrasound is reflected from the heart, producing an image that is refined after being fed into a computer.

After receiving her doctorate from Rice University, McSherry was a fellow in nuclear physics there for one year and then spent the rest of her career working in corporations. She started as a research physicist in ultrasonics at Digicon, Inc., and became executive vice president of medical ultrasound and then president of cardiology analytical systems when the company was acquired by Digisonics, Inc. She is currently president and manager of Digisonics, which creates ultrasound equipment and systems for use in cardiology, radiation, and OB/GYN applications. She also served on the board of directors as chair for Information Products Systems, Houston.

McSherry is a member of the Institute of Electrical and Electronics Engineers, American Institute of Ultrasound in Medicine, American Physical Society, and American Heart Association.

McWhinnie, Mary Alice

1922–1980
Biologist

Education: B.S., DePaul University, 1944, M.S., biology, 1946; Ph.D., biology, Northwestern University, 1952

Professional Experience: assistant, biological sciences, DePaul University, 1944–1950, instructor, 1950–1952, assistant to associate professor, biology, 1952–1960, professor, 1960–1980

Mary McWhinnie was one of the first two women scientists to winter in Antarctica to study krill. Her research involved crustacean metabolism, with special reference to carbohydrates during the molt cycle, and her findings highlighted the importance of krill to the ocean food chain. As a child growing up in Illinois, she developed an interest in nature and, especially, fishing, and went on to study biology at DePaul University. She worked as an assistant in biological sciences at the university while also completing her master's degree, and went on to receive her Ph.D. from Northwestern University.

She spent her entire teaching career at DePaul University, advancing through the ranks from instructor to professor. It was while studying crayfish in Chicago that she became interested in comparing them to their cold-water cousins, krill. She prepared a proposal for the National Science Foundation (NSF) and, in 1962, embarked on a two-month research cruise as the first American woman scientist assigned to the U.S. Antarctic Research program. She returned again in 1972, this time as the ship's chief scientist, and in 1974, she and her female research assistant were the first women to spend the winter at the McMurdo research station on Antarctica.

Over the course of her career, McWhinnie made 11 trips to Antarctica to study krill and became the worldwide expert on krill as an ocean food source. In the late 1970s, she became an ecological spokesperson advocating for the protection of krill against overfishing. McWhinnie fell ill while preparing for another trip to Antarctica in the winter of 1979–1980 and died at the age of 57 of undiagnosed cancer of the lungs and brain.

McWhinnie received numerous NSF grants to carry on her polar biology research, as well as other funding, such as an assistantship at Woods Hole Marine Biological Laboratory in 1952, summers as a faculty fellow of the American Physiological Society in 1957, and fellow of the Lalor Foundation in 1958. She was a member of the Panel on Biological and Medical Sciences of the National Academy of Science, and the National Science Foundation Committee on Polar Research. She was elected a fellow of the American Physiological Society and was a member of the American Association for the Advancement of Science and the Biophysical Society.

Further Resources

Land, Barbara. 1981. *The New Explorers: Women in Antarctica*. New York: Dodd, Mead.

Chipman, Elizabeth. 1986. *Women on the Ice: A History of Women in the Far South*. Carlton, Vic.: Melbourne University Press.

Mead, Margaret

1901–1978
Anthropologist

Education: B.A., Barnard College, 1923; M.A., psychology, Columbia University, 1924, Ph.D., anthropology, 1929

Professional Experience: assistant to associate curator, ethnology, American Museum of Natural History, 1926–1964, curator, 1964–1969; instructor and adjunct professor, anthropology, Columbia University, 1947–1978

Concurrent Positions: instructor, Vassar College, New York University, Fordham University, and others; director, Columbia University Research in Contemporary Cultures, 1948–1950

Anthropologist Margaret Mead in Samoa, in a photo sent to colleague **Ruth Benedict**, 1926. (Library of Congress)

Margaret Mead was perhaps the foremost anthropologist of the twentieth century. Through such bestselling books as *Coming of Age in Samoa* (1928), *Sex and Temperament in Three Primitive Societies* (1935), and *Male and Female* (1949), she changed anthropology from an esoteric discipline to a subject that was fascinating to the public at large. Her expeditions to Samoa, New Guinea, and Bali, and her work with Native American tribes, provided material for more than 1,500 books, articles, films, and occasional pieces. She was the first anthropologist to compare childrearing practices and roles of women in various cultures, topics that had not been of interest to male anthropologists. She was a founder of a new school of anthropology that examines the ways a culture shapes an individual's personality. Along with her third husband, Gregory Bateson, she pioneered the use of photography and eventually film and video to document vanishing cultures, and thus her work was spread to the general public in new ways.

Mead's father was a faculty member at the University of Pennsylvania, her mother a sociologist, and her paternal grandmother a pioneer child psychologist. Mead received her master's degree in psychology from Columbia in 1924 and then spent six months studying adolescents in Samoa. She came to the controversial conclusion, explained in *Coming of Age in Samoa*, that people were a product of their environment more than heredity. In later years, she conceded that she was too inexperienced as a field investigator at the time she made the study, but she never revised the book or returned to Samoa. She went on to conduct fieldwork on the Manus tribe of the Admiralty Islands and, with her second husband, Reo Fortune, visited three native American tribes—the Arapesh, the Mundugumor, and the Tcambuli—to study the social conditioning of the two sexes. Later, with Bateson, she engaged in fieldwork in Bali and New Guinea. Her book *And Keep Your Powder Dry* (1942) studied American character against the background of seven other cultures.

In addition to her faculty appointments, Mead was affiliated with American Museum of Natural History for much of her career and established a Hall of Peoples of the Pacific there. She was the first female president of the Society for Applied Anthropology in 1949, and also served as president of the American Anthropological Association in 1960 and the American Association for the Advancement of Science in 1975. She was elected a member of the National Academy of Sciences in 1975 and the American Philosophical Society in 1977. She also was a member of the American Academy of Arts and Sciences. Her autobiography is *Blackberry Winter: My Earlier Years* (1972), and there are numerous biographies, including a family history by her daughter, cultural anthropologist **Mary Catherine Bateson**.

Further Resources

Banner, Lois W. 2003. *Intertwined Lives: Margaret Mead, Ruth Benedict, and Their Circle*. New York: Random House.

Bateson, Mary Catherine. 1984. *With a Daughter's Eye: A Memoir of Margaret Mead and Gregory Bateson*. New York: W. Morrow.

Medicine, Beatrice A.

1924–2005
Anthropologist

Education: B.S., education, art, and history, South Dakota State University, Brookings, 1945; M.A., sociology and anthropology, Michigan State University, 1953; Ph.D., cultural anthropology, University of Wisconsin, Madison, 1983

Professional Experience: lecturer, sociology and anthropology, University of Montana, Missoula, 1967–1968; director, American Indian Research, Oral History Project, and assistant professor, anthropology, University of South Dakota, Vermillion, 1968–1969; assistant professor, anthropology, San Francisco State University, 1969–1970, associate professor, 1970–1971; predoctoral lecturer, anthropology, University of Washington, Seattle, 1971–1973; visiting professor, anthropology, Native American Studies, Dartmouth College, 1973–1974; visiting professor, anthropology, Colorado College, 1974–1975; visiting associate professor, anthropology, Stanford University, 1975–1976; associate professor, anthropology, and coordinator, Interdisciplinary Program in Native American Studies, California State University, Northridge, 1982–1985; professor, anthropology, and director, Native Centre, University of Calgary, Canada, 1985–1988

Concurrent Positions: assistant professor, Teacher Corps, University of Nebraska, Omaha, summer 1969; fellow, Center for the History of American Indians, Newberry Library, Chicago, 1972–1973; visiting professor, educational anthropology, University of New Brunswick, Canada, summer 1976; visiting professor, Education Policy Sciences, University of Wisconsin, Madison, summer 1979; visiting professor, Graduate School of Public Affairs, University of Washington, Seattle, spring 1981; lecturer, Standing Rock College, North Dakota, 1989; visiting professor, anthropology, Humboldt State University, California, 1991; visiting professor, Colorado College, 1991; visiting professor, Saskatchewan Indian Fed. College, Canada, 1991; visiting distinguished professor, women's studies, University of Toronto, 1992; research coordinator, Women's Perspectives, Royal Commission on Aboriginal Peoples, Ottawa, Ontario, Canada, 1993–1994; visiting professor, rural sociology, South Dakota State University, Brookings, 1993; visiting scholar, Museum of Anthropology, University of British Columbia, Vancouver, 1995; adjunct professor, Department of Educational Foundations, University of Alberta, Edmonton, 1995–2005; Buckman Professor, Department of Human Ecology, University of Minnesota, Twin Cities, 1996; Stanley Knowles Distinguished Professor, Brandon University, Manitoba, Canada, 1998, visiting professor, 1999–2005

Beatrice Medicine was a recognized expert on the study of tribal traditions among the Dakota Indians. She was one of the few Native American women to earn an advanced degree in anthropology, and she worked to dispel anthropological myths that have tended to oversimplify and homogenize Native American cultures. In her writing and teaching, she established a more realistic picture of the plurality and diversity of Native American life from the real and complex Native American perspectives. Her research centered on the changing Native American family and on women's roles, real and perceived, past and present. Much erroneous information exists because the first narratives and histories were written by white men who were the product of a patriarchal society, and they largely ignored or incorrectly reported the role of women in Native American society. She had an already long career as a visiting professor and fellow at numerous colleges and universities before attending the University of Wisconsin at Madison as an Advanced Opportunity Fellow to complete her doctorate in anthropology in 1983. She went on to direct Native American studies programs at several universities, including at California State University, Northridge, and the University of Calgary, Alberta, Canada. Even after her formal retirement in 1988, she continued to be a visiting or adjunct professor at several institutions in both the United States and Canada.

Although she spent a career in academia, Medicine maintained strong ties to her reservation home. She was born and raised on the Standing Rock Sioux Reservation in northern South Dakota, and her family stressed maintaining tribal traditional cultural identity. In addition to her research on her own people, Medicine was involved in work with the aboriginal peoples of New Zealand, Australia, and Canada. She was extensively involved in the field of mental health, focusing on issues such as alcohol and drug abuse among Native Americans. The title of her doctoral thesis was "An Ethnography of Drinking and Sobriety among the Lakota Sioux." She was an advocate for Indian leadership and helped establish a network of Indian social service centers in urban areas. In her role as head of the Women's Branch of Canada's Royal Commission on Aboriginal Peoples, she helped draft legislation protecting the legal rights of native families.

Medicine published more than 60 articles and chapters in books, including *Native American Women: A Perspective* (1978) and *The Hidden Half: Studies of Plains Indians Women* (1983). A collection of her writings, entitled *Learning to Be an Anthropologist and Remaining "Native,"* was published in 2001. She was a member of the American Anthropological Association and Society for Applied Anthropology.

Further Resources

Medicine, Beatrice and Sue-Ellen Jacobs, eds. 2001. *Learning to Be an Anthropologist and Remaining "Native": Selected Writings*. Urbana: University of Illinois Press.

Meinel, Marjorie Pettit

1922–2008
Astronomer

Education: B.A., Pomona College, 1943; M.A., astronomy, Claremont College, 1944

Professional Experience: researcher and associate editor, rocket programs, California Institute of Technology, 1944–1945; research associate, solar energy, University of Arizona, 1974–1984; visiting scientist, optics, Jet Propulsion Laboratory, California Institute of Technology, 1984–2000

Concurrent Positions: consultant, Office of Technological Assessment, U.S. Congress, 1974–1980; consultant, Arizona Solar Energy Research Commission, 1975–1981

Marjorie Meinel was recognized for her work on solar energy applications, upper atmospheric phenomena, volcanic eruptions, solar and variable stars, and astronomical optics. As a graduate teaching assistant during World War II, she taught navigation to Army airmen. After receiving her master's degree from Claremont College, she obtained employment on secret military rocket programs at the California Institute of Technology. Throughout her career, she conducted collaborative research with her husband, astronomer Aden Meinel, on solar optics, solar energy, volcanic eruptions, and cosmic radiation. They co-authored several papers and books and, in their later post-retirement years, the couple continued to expand their diverse interests into topics such as paleoanthropology and global warming, pursuing their research and giving public lectures.

Both of Meinel's parents were pioneering astronomers; her father, Edison Pettit, was one of the founding astronomers at Mt. Wilson Observatory in Los Angeles, and her mother, Hanna Steele Pettit, was one of the first women to receive a doctorate in astronomy from the University of Chicago. Marjorie met Aden Meinel when they both enrolled in a program for gifted high school students at Pasadena Junior College. Aden Meinel became the first Director of the Kitt Peak National Observatory (1958–1960) and was the founder and first Director of the Optical Sciences Center at the University of Arizona. Marjorie took time off from her career to raise seven children, but she remained professionally active in collaboration with and support of Aden's research. Once her children were grown, she spent 10 years as a research associate in solar energy at the University of Arizona and was active in the 1970s as a member of state and national solar energy committees. In 1984, the couple returned to California as Distinguished Visiting Scientists at the Jet Propulsion Laboratory, where they researched solar optics and helped launch the Hubble Space Telescope.

Among Meinel's numerous awards and acknowledgements, she was named one of five outstanding "Women in Physics" by the American Physical Society (1980) and received the Goddard Award (1984), the George van Biesbroeck Award for Services in Astronomy (1990), the National Aeronautics and Space Administration (NASA) Exceptional Scientific Achievement Medal (1993), and the Kingslake Medal of the Optical Society of America (1994 and 2001). The Meinels also received many awards for their joint work, including a Gold Medal Award of the Society of Photographic Instrumentation Engineers (SPIE) (1997). She was a member of SPIE and of the New York Academy of Sciences and the Society of Photo-Optical Instrumentation Engineers.

Further Resources

LaFee, Scott. "Astronomers Link Human Evolution, Cosmic Radiation." http://www .signonsandiego.com/news/science/20060607-9999-lz1c07meinel.html.

Mendenhall, Dorothy Reed

1874–1964
Research Physician

Education: B.L., Smith College, 1895; student, chemistry and physics, Massachusetts Institute of Technology; M.D., Johns Hopkins University, 1900

Professional Experience: fellow, Johns Hopkins University, 1901–1902; resident physician, Babies Hospital, New York, 1903–1906; lecturer, home economics, University of Wisconsin, 1914–1936

Concurrent Positions: medical officer, United States Children's Bureau, 1917–1936

Dorothy Reed Mendenhall was recognized first for her early work on Hodgkin's disease and later for her pioneering efforts in obstetrics. In 1900, she was one of the first women to receive a medical degree at Johns Hopkins after the school lifted its ban on admitting women students. More than 50 years later, she would establish a scholarship fund at Johns Hopkins for female medical students. While working as an intern and fellow in pathology and bacteriology at Johns Hopkins, she earned an international reputation for her recognition of the Reed (or Reed-Sternberg) cell, named in her honor, as the distinctive characteristic of Hodgkin's disease. Prior to her work, Hodgkin's disease was believed to be a form of tuberculosis. Since there were few opportunities for women to advance at Johns Hopkins, she moved to New York, where in 1903 she was appointed the first resident physician at Babies Hospital. After losing her own first child at birth, she changed her research interests to improving obstetrics and infant mortality in the United States.

In 1906, she moved to Madison, Wisconsin, where her husband was a faculty member in physics. After an interval of several years to care for her growing family, she became a lecturer in home economics at the University of Wisconsin and began researching infant mortality, nutrition, and public health. After conducting a campaign of lectures and pamphlets, she organized Wisconsin's first infant welfare clinic in 1915. In 1917, she served as a medical officer for the U.S. Children's Bureau while her husband was on war duty in Washington, D.C. She continued her affiliation with the Children's Bureau until 1936 while maintaining her position at the University of Wisconsin. She studied European countries with low infant mortality rates and used the information to produce numerous bulletins on nutrition and childcare for the university, the Wisconsin State Board of Health, and the U.S. Department of Agriculture. She advocated the use of midwives for healthy pregnancies and specialized training in obstetrics for physicians. Her work also led to the creation of height and weight standards to improve infant and child nutrition and health. She focused not only on the role of doctors, however, also reaching out to women and prospective mothers with correspondence courses on nutrition and hygiene.

Further Resources

National Institutes of Health. "Dr. Dorothy Reed Mendenhall." Changing the Face of Medicine: Celebrating America's Women Physicians. National Library of Medicine, National Institutes of Health. http://www.nlm.nih.gov/changingthefaceofmedicine/ physicians/biography_221.html.

Menken, Jane Ava (Golubitsky)

b. 1939
Demographer, Sociologist

Education: A.B., mathematics, University of Pennsylvania, 1960; M.S., biostatistics, School of Public Health, Harvard University, 1962; Ph.D., sociology and demography, Princeton University, 1975

Professional Experience: assistant, biostatistics, School of Public Health, Harvard University, 1962–1964; mathematical statistician, National Institute of Mental Health, 1964–1966; research associate, biostatistics, School of Public Health and Administrative Medicine, Columbia University, 1966–1969; research staff, Office of Population Research, Princeton University, 1969–1971, research demographer, 1975–1980, assistant to associate director, 1978–1987; associate professor, sociology, Princeton University, 1977–1980, professor, sociology and public affairs, 1980–1987, visiting professor, public and international affairs, 1987–1988; professor, social sciences, and research associate, Population Studies

Center, University of Pennsylvania, 1987–2001, director, 1989–1995; professor, sociology, University of Colorado, Boulder, 1997–, and director, Institute of Behavioral Science, 2001–

Concurrent Positions: fellow, Center for Advanced Study in the Behavioral Sciences, Stanford University, 1995–1996; honorary professor, School of Public Health, University of the Witwatersrand, Johannesburg, South Africa, 2006–

Jane Menken is recognized as one of the top demographers in the United States. Demography is the science of vital and social statistics, such as birth, death, diseases, and marriage. It differs from statistics, which is the science that deals with numerical facts or data, in that demography is centered on the populations—the people—and the interpretation and forecasting of trends. Menken's work has often involved controversial social issues, such as the effects of government policy on fertility, breastfeeding rates, and birth control and abortion access. In particular, her work has focused on women's and children's health and on the study of aging. In addition to her numerous scholarly articles and papers, she is the co-author of the book *Mathematical Models of Conception and Birth* (1973), and co-editor of *Natural Fertility* (1979), *Teenage Sexuality, Pregnancy, and Childbearing* (1981), *World Population and U.S. Policy: The Choices Ahead* (1986), and, most recently, *Aging in Sub-Saharan Africa* (2006). She is a founding member of the editorial board of two journals, *Demographic Research* and *Southern African Journal of Demography.*

Menken worked for the federal government for a short time as a statistician for the National Institute of Mental Health, but has primarily been employed by major universities with distinguished reputations in demography. She has been a visiting scholar, invited lecturer, and conference participant at institutions around the world, and has served on numerous panels and commissions including, but not limited to, as a member of the population advisory committee for the Rockefeller Foundation (1981–1993), member of the committee on AIDS research of the National Academy of Sciences (1987–1994), chair of Family Health International's project on *The Impact of Family Planning Programs on Women's Lives* (1994–1998), and member of the World Health Organization Study on Global Aging and World Health (2004–2006). She has also served in numerous advisory roles for National Research Council and National Institutes of Health (NIH) studies on population and fertility.

Menken is an elected member of the National Academy of Sciences (1989) and the Institute of Medicine (1995). She is a former Guggenheim fellow (1992–1993), a fellow of the American Association for the Advancement of Science, and a member of the American Academy of Arts and Sciences, Population Association of America (president, 1985), American Public Health Association,

Sociological Research Association (president, 1995–1996), American Sociological Association, American Statistical Association, Society for the Study of Social Biology, and International Union for the Scientific Study of Population.

Further Resources

University of Colorado. Faculty website. http://www.colorado.edu/ibs/PP/menken/.

Michel, Helen (Vaughn)

b. 1932
Nuclear Chemist

Education: B.S., chemistry, University of California, Berkeley, 1955; student, Indiana University, 1955–1956

Professional Experience: chemist, University of California, Berkeley, Radiation Laboratory (later Lawrence Radiation Laboratory), 1956–1990

Helen Michel is a nuclear chemist who achieved great success and worldwide recognition for her expertise in operating the complex electronic instruments in the Lawrence Radiation Laboratory in Berkeley, California. Her analyses led to many important scientific discoveries in the fields of nuclear science, geochemistry, plant biology, and archaeometry, the dating of archaeological specimens through specific techniques such as radiocarbon dating. One study of the mid-1970s that attracted a vast amount of publicity involved her group's role in determining the authenticity of an artifact called the Plate of Brass. Historical evidence indicated it had been left by English explorer Sir Walter Drake in the sixteenth century when he landed on the coast of what is now California. The plate had been discovered in 1936 in San Francisco, and was kept at Berkeley. After examining samples of the metal as well as the plate itself by x-ray fluorescence, atomic absorption, and emission spectroscopy, Michel determined that the Plate of Brass was not authentic and that it had probably been made in the last half of the nineteenth century or the early part of the twentieth. Similar studies conducted at Oxford University verified her conclusions. She was also involved in another news story of the 1980s when her expertise in chemical soil analysis was crucial in a long-term project that substantiated the theory that an asteroid impact resulted in the extinction of the dinosaurs some 65 billion years ago. Michel's contributions to the work that proved the asteroid theory are described in Luis Alvarez's book, *Adventures of a Physicist* (1987), and Walter Alvarez's *T-Rex and the Crater of Doom* (1997), among other reports.

Michel decided on a career in chemistry while still in elementary school, but by the time she completed her college education in the 1950s, it was difficult for a woman to obtain any job, much less in the sciences. She secured a part-time job at the Radiation Laboratory in Berkeley in the Division of Nuclear Chemistry while she was still an undergraduate. She spent a year in graduate work at Indiana University before returning to Berkeley for a full-time position as a chemist. Even without an advanced degree, she earned the respect of her colleagues and was always included as a co-author on all of the papers describing research in which she participated.

In the 1960s, she and her husband expanded their scientific interests to the hobby of breeding orchids. They established a business as part of the Orchid Ranch in Livermore, California, and after retiring from Lawrence Laboratory in 1990, Helen took over much of the daily supervision of the business.

Further Resources

Rogers, Phila W. 1979. "Investigating a Mass Extinction Occurring 65 Million Years Ago." http://www.lbl.gov/Science-Articles/Archive/fingerprinting-past.html.

Micheli-Tzanakou, Evangelia

b. 1942
Neurophysicist, Biomedical Engineer, Biophysicist

Education: B.S., University of Athens, 1968; M.S., Syracuse University, 1974, Ph.D., physics, 1977

Professional Experience: fellow, biophysics, Syracuse University, 1977–1980, consultant, 1980–1981; assistant to associate professor, biomedical engineering, Rutgers University, 1981–1990, professor and department chair, 1990–2000, professor and director, Computational Intelligence Laboratories, Department of Biomedical Engineering, 2000–

Concurrent Positions: consultant, Eye Defect and Engineering Research Foundation, 1978–1980; adjunct instructor, University of Medicine and Dentistry of New Jersey

Evangelia Micheli-Tzanakou is a physicist who does extensive research on brain function, including pattern recognition; digital signal processing of biological signals; neural networks, data compression, and image reconstruction; hearing aids; and neural network modeling of the brain. She is renowned for her research in using optimization techniques to understand to problems of brain functions and dysfunctions, and she has pursued a multiphase quest in order to gain this understanding. Some of the methods she developed are used in cardiology to predict the prognosis of heart-attack patients, and she has compared people who age normally with patients who have Alzheimer's and Parkinson's diseases. She has developed a set of algorithms for modeling the visual system and applied this technique to other functions of the nervous system and to research other brain functions, such as pattern recognition. In 1994, her research indicated that people with advanced educational and occupational attainment are able to cope longer before the onset of Alzheimer's, a finding that has proved controversial.

Biomedical engineering, or bioengineering, is a relatively new discipline that developed in the 1960s and involves the application of engineering principles and techniques to problems of medicine and biology. In many institutions, it is an interdisciplinary effort on the part of physicians, biophysicists, electrical engineers, and computer scientists. Some researchers have expertise in several or all of these disciplines. In Micheli-Tzanakou's work, information processing by the visual system is examined by computer-controlled techniques, and recordings are done both in animals and in humans. In 1996, she and her co-author described designing a neuromime circuit to be used for modeling nerve networks from living organisms by using very large-scale integration (VLSI) technology.

Micheli-Tzanakou is a founding fellow of the American Institute for Medical and Biological Engineering. She is also a fellow of the Institute of Electrical and Electronics Engineers (IEEE) and has served on numerous boards and committees for that group. She is a member of the Society for Neuroscience, Association for Research in Ophthalmology, and Biophysical Society.

Further Resources

Rutgers University. Faculty website. http://cil.rutgers.edu/tzanakou/BriefCV.htm.

Mielczarek, Eugenie Vorburger

b. 1931
Solid-state Physicist, Biophysicist

Education: B.S., Queens College, 1953; M.S., Catholic University, 1957, Ph.D., physics, 1963

Professional Experience: physicist, U.S. National Bureau of Standards, 1953–1957; research assistant, Catholic University, 1957–1959, research associate, 1959–1962, assistant research professor, 1962–1965; professor, physics, George Mason University, 1965–, emerita

Concurrent Positions: visiting scientist, National Institutes of Health, 1965–

Eugenie Mielczarek is known for her work in biophysics, which is the conjunction between biology and physics. Her research includes solid-state low-temperature physics, semiconductors, Mossbauer spectroscopy of metal and biological compounds, biophysics, and Fermi surfaces of metals. Working with Mossbauer spectroscopy, she is applying the techniques of nuclear physics to biological materials in order to probe the molecular environment around iron atoms. She explains that our bodies contain iron. Persons who have sickle-cell anemia or Cooley's anemia suffer from damaged kidneys and spleens. The red blood cells break down more rapidly in these persons than in healthy individuals, and this breakdown dumps iron into those major organs. Iron chelators, or iron-grabbing compounds, are needed to clean up the excess iron, and we need to understand the atomic environment of iron in iron-chelating compounds in order to prevent the damage caused by the iron buildup.

Her early research was in solid-state metals physics, but she has moved into studies of metal in biological environments. Solid-state physicists increasingly are studying more complex biological systems, looking at hemoglobin, cell membranes, and brain waves, work that has application to living systems. For example, she has also studied the dangerously high noise levels from music played in aerobic exercise classes, making recommendations for how to protect the hearing of both instructors and participants in such situations.

She initially found it difficult to find employment as a female physicist, but finally obtained a position at the U.S. National Bureau of Standards, where she worked while completing her graduate degrees. Later, she was the founding chair of the physics department at George Mason University, a department that still maintains a higher-than-average number of female professors. Mielczarek is a member of the American Physical Society, Biophysical Society, American Association of Physics Teachers, and Association of Women in Science. She was co-editor (with Robert S. Knox) of *Biological Physics* and co-author (with Sharon Bertsch McGrayne) of *Iron, Life's Universal Element: Why People Need Iron and Animals Make Magnets* (2000).

Miller, Elizabeth Cavert

1920–1987
Biochemist

Education: B.S., biochemistry, University of Minnesota, 1941; M.S., biochemistry, University of Wisconsin, 1943, Ph.D., biochemistry, 1945

Professional Experience: postdoctoral fellow, McArdle Laboratory for Cancer Research, University of Wisconsin, 1945–1947, instructor, department of oncology, 1947–1949, assistant to associate professor, 1949–1969, professor, 1969–1987

Concurrent Positions: associate director, McArdle Laboratory for Cancer Research, University of Wisconsin, 1973–1987; senior research professor and emeritus professor, oncology, Wisconsin Alumni Research Foundation, 1980–1987

Elizabeth C. Miller was a biochemist recognized for her research on cancer and chemical carcinogens. She spent her entire career at the University of Wisconsin, where she collaborated with her husband, James A. Miller. The Millers were the first researchers to discover that an outside chemical could cause cancer in rats. They went on to research how carcinogens bind to DNA and, in the 1960s, worked on growing tumors in live tissue to understand how cancers spread. Their pathbreaking research provided insight into later studies and public awareness about potential cancer-causing toxins in the environment such as pollution, industrial chemicals, food additives, and drugs. Together, the Millers published more than 300 papers on chemical carcinogens.

Elizabeth Miller began her graduate research with a scholarship for joint work in biochemistry and home economics. After she received her doctorate at Wisconsin, she held a postdoctoral fellowship and then joined the faculty in 1947 as an instructor in oncology. She advanced through the tenure ranks, becoming a full professor in 1969. At the University of Wisconsin, she also served as associate director of the McArdle Laboratory for Cancer Research from 1973 until her retirement in 1987. She was the editor of *Cancer Research* (1954–1964) and president of the American Association for Cancer Research (1976 and 1978). Between 1978 and 1980, she served on President Carter's Panel of the National Cancer Institute.

Both Elizabeth Miller and James Miller were elected to the National Academy of Sciences in 1978. They received numerous awards and honors for their work, including the National Award in Basic Science of the American Cancer Society (1977), the first Founder's Award from the Chemical Industry Institute of Toxicology (1978), and a Mott Award from General Motors Cancer Research Foundation (1980). Elizabeth Miller was a member of the American Society of Biological Chemists, the American Association for Cancer Research, and the American Academy of Arts and Sciences. She died of kidney cancer in 1987.

Mintz, Beatrice

b. 1921
Biologist

Education: A.B., Hunter College, 1941; student, New York University, 1941–1942; M.S., University of Iowa, 1944, Ph.D., zoology, 1946

Professional Experience: assistant, Guggenheim Dental Clinic, 1941–1942; assistant, zoology, University of Iowa, 1942–1946, instructor, 1946; instructor, biological science, University of Chicago, 1946–1949, assistant to associate professor, 1949–1960; associate member, Institute for Cancer Research (now Fox Chase Cancer Center), Philadelphia, 1960–1965, senior member, 1965–

Beatrice Mintz has been recognized for her research on cellular biology and developmental genetics, and she particularly investigated inherited susceptibility to certain tumors. Her research has focused on gene control of differentiation and disease in mammals, including, most recently, the hereditary basis of melanoma or skin

Biologist Beatrice Mintz. (Courtesy of the Fox Chase Cancer Center)

cancer. Melanoma is a highly dangerous form of skin cancer, but is difficult to detect early and treat. She is renowned for her techniques in manipulating the genetic makeup of mouse embryos and for new methods for freezing cells. After receiving her undergraduate degree, she accepted a position at the University of Iowa as assistant and instructor while she completed her doctorate. She was hired at the University of Chicago as an instructor in 1946 and advanced to associate professor in 1955. During this time, she was awarded a Fulbright research fellowship to study in France. She left Chicago in 1960 to become an associate member of the Institute for Cancer Research, now the Fox Chase Cancer Center, where she is still a senior member and researcher, and holds an endowed chair.

Mintz was elected to the National Academy of Sciences in 1973 and was named an Outstanding Woman in Science by the New York Academy of Sciences in 1993. She has received numerous awards, including the Papanicolaou Award for Scientific Achievement (1979), the first medal of the Genetics Society of America (1981), Germany's first Ernst Jung Gold Medal for Medicine (1990), the first March of Dimes Prize in Developmental Biology (1996), and a National Medal of Honor for Basic Research from the American Cancer Society (1997). She is a fellow of the American Association for the Advancement of Science, the American Academy of Arts and Sciences, and the Pontifical Academy of Sciences. She is also a member of the Genetics Society of America, the Society for Developmental Biology, the International Society of Developmental Biology, the American Institute of Biological Sciences, and the American Philosophical Society.

Further Resources

Fox Chase Cancer Center. "Beatrice Mintz, PhD." http://www.fccc.edu/research/pid/mintz/.

Mitchell, Helen Swift

1895–1984
Nutritionist

Education: B.A., Mount Holyoke College, 1917; Ph.D., physiological chemistry, Yale University, 1921

Professional Experience: high school instructor, 1917–1918; director, nutrition research, Battle Creek Sanitarium, 1921–1932; professor, nutrition, Battle Creek College, 1924–1935; research professor, nutrition and home economics, Massachusetts State College, 1935–1941; principal nutritionist, Office of Defense, Health, and Welfare Services, Washington, D.C., 1941–1943; chief nutritionist, Office of Foreign Relief and Rehabilitation Operations, U.S. Department of State, 1943–1944; professor, nutrition, Carnegie Institute of Technology, 1946; dean, home economics, University of Massachusetts, 1946–1960

Concurrent Positions: exchange professor, Hokaido University, Japan, 1960–1962; research consultant, Harvard School of Public Health

Helen Mitchell was an authority on nutrition and vitamins who helped develop the idea of Recommended Dietary Allowances (or RDA), now required on all food labeling. After receiving her doctorate at Yale University, she became research director of nutrition at the Battle Creek Sanitarium in 1921 and then, beginning in 1924, served simultaneously as professor of physiology and nutrition at Battle Creek College. She was appointed a research professor at Massachusetts State College in 1935. During World War II, she took a leave from teaching to be chief nutritionist for the U.S. Office of Defense, Health, and Welfare, and then for the Department of State. It was during her tenure working for the government that Mitchell and other researchers prepared a report on nutrition and vitamin needs of enlisted men; their work led to the RDA recommendations for different groups, which eventually were applied to the population at large.

After the war, Mitchell returned to academia, first at the Carnegie Institute of Technology for a year, and then in an appointment as dean of home economics at the University of Massachusetts in 1946. She traveled widely to conduct research and attend international congresses, visiting Newfoundland, Russia, Scandinavia, Scotland, and the Middle East. She received an honorary degree from the University of Massachusetts at the time of her retirement in 1960. Even after her retirement, she was active as a consultant to the Harvard School of Public Health and as an exchange professor in Japan for two years. She was also a co-author of *Nutrition in Nursing* and of the fifteenth edition of *Nutrition in Health and Disease*, a standard textbook that has been regularly updated and reprinted for more than 70 years.

Mitchell was elected a fellow of the American Public Health Association, and was also a member of the American Dietetic Association, the American Home Economics Association, the American Institute of Nutrition, and the Institute of Food Technologists.

Mitchell, Joan L.

b. 1947
Physicist

Education: B.S., physics, Stanford University, 1969; M.S., University of Illinois, Urbana-Champaign, 1971, Ph.D., physics, 1974

Professional Experience: research staff member, International Business Machines (IBM) J. Watson Research Center, 1974–1994, research staff member, Image Applications, 1996–2007; fellow, Ricoh/IBM InfoPrint Solutions Company, 2007–

Concurrent Positions: visiting professor, University of Illinois

Joan L. Mitchell is a physicist whose work has had applications in computer science over the course of her long career in photographic image processing and technologies with IBM. She was a member and editor of the Joint Photographic Experts Group (JPEG) that developed and standardized the algorithm for color image compression, and she co-authored books in the mid-1990s on the JPEG and MPEG formats, both now standard international data compression formats. Mitchell received her doctorate in physics from the University of Illinois in 1974 and worked in various departments of the IBM T. J. Watson Research center for more than 30 years before joining the new Ricoh–IBM collaboration, InfoPrint

Solutions, as a fellow in 2007. She holds or shares more than 100 patents related to processes for photographic facsimile (fax) and image data compression.

Mitchell was elected to the National Academy of Engineering in 2004. She has received several IBM Outstanding Innovation Awards, including for Two-Dimensional Data Compression (1978), Teleconferencing (1982), Image View Facility (1985), Resistive Ribbon Thermal Transfer Printing Technology (1985), Speed-Optimized Software Implementations of Image Compression Algorithms (1991), and Q-Coder (1991), and an Outstanding Technical Achievement Award for Algorithms for Improved Printer Performance (2001). She was elected to the IBM Academy of Technology in 1997 and was named an IBM Fellow in 2001. Mitchell is a fellow of the Institute of Electrical and Electronics Engineers (IEEE). She also wrote a career-advice book, *Dr. Joan's Mentoring Book: Straight Talk about Taking Charge of Your Career* (2007).

Further Resources

"10 Minutes with Dr. Joan Mitchell, InfoPrint Fellow and Master Inventor." *InfoPrint Insights.* 8. (June 2009). Ricoh-IBM InfoPrint Solutions. http://www.infoprintsolutionscompany .com/internet/wwsites.nsf/vwWebPublished/ii_060109_us?OpenDocument#_18.

"Joan Mitchell." IBM Women in Technology. IBM Women Fellows. http://www-03 .ibm.com/ibm/history/witexhibit/wit_fellows_mitchell.html.

Mitchell, Mildred Bessie

1903–1983
Clinical Psychologist

Education: B.A., Rockford College, 1924; M.A., Radcliffe College, 1927; Ph.D., psychology, Yale University, 1931

Professional Experience: professor, education and mathematics, Lees College, 1927–1928; psychologist, George School, 1931–1933; chief psychologist, New Hampshire State Hospital, 1933–1936; vocational director, U.S. Employment Service in New Hampshire, 1936; psychologist, Bellevue Hospital, New York City, 1937; chief psychologist, Psychopathic Hospital, Iowa State University, 1938–1939; psychologist, Mt. Pleasant and Independence Street Hospitals, 1939–1941; clinical psychologist, State Bureau of Psychological Services, Minnesota, 1941–1942; member of Women Accepted for Voluntary Emergency Services (WAVES), 1942–1945; vocational appraiser, Veterans Guidance Center, City College New York, 1945–1946; psychologist, Domestic Relations Court,

New York City, 1946–1947; chief psychologist, Veterans Administration Center, Dayton, Ohio, 1951–1958; clinical psychologist, Aerospace Medical Laboratory, Wright-Patterson Air Force Base, 1958–1960, research psychologist, Bionics Section, Aeronautical Systems Division, 1960–1963; associate professor, psychology, University of Tampa, 1965–1967; lecturer, behavioral science, University of South Florida, 1967–1970

Mildred Mitchell had a distinguished career as a clinical psychologist, but she is best known for her early contributions to the development of the science of bionics. Bionics involves utilizing electronic devices and mechanical parts to assist humans in performing difficult, dangerous, or intricate tasks by supplementing or duplicating parts of the body. Tasks can range from the design of glove boxes to handling radioactive material in clean rooms to the design of artificial limbs to replace those lost to accident or disease. Bionics was a new science in the 1960s, and psychologists, biologists, physicians, chemists, physicists, mathematicians, and engineers teamed up to duplicate electronically the functions of people, animals, and plants.

Mitchell became involved in bionics in the late 1950s when she was asked by the U.S. Air Force to assist in the psychological evaluation of men competing for the astronaut training program. Initially, she was asked only to test the applicants' reaction to isolation, but later she was appointed to the selection team. The selection committee chose experienced pilots, and devised tests that simulated the pressures of high altitude and the resultant stresses on the body. The scientists knew that even experienced pilots had difficulty performing some actions such as manipulating the controls during takeoffs and landings because of high gravity (G) forces. When Mitchell was head of bionics at the Aerospace Medical Laboratory, she designed an artificial muscle that could take over such operations and could also assist if an astronaut who had experienced long periods of weightlessness found his muscles had become weak or impaired. She also designed a "nail bender" that can bend an iron nail with a puff of air. Her group designed a man-made "biological clock," which duplicates through machinery the natural mechanism that tells animals whether it is day or night, even if their environment has been artificially altered. There have been significant advances in materials, in computer simulation of muscle action, and in the need for specific bionic equipment since the beginnings of the space program. However, Mitchell and her teams early and made significant contributions to this new science.

After working with the Air Force, Mitchell accepted positions teaching at several academic institutions and, throughout her career, she was involved in improving the status of women psychologists. In 1951, she published a landmark report in the journal *American Psychologist* on the status of women psychologists who were members of the American Psychological Association. The data indicated that women had not been elected as fellows or officers, nor had they been appointed to committees in proportion to their numbers and qualifications. She also noted that women (such as herself) changed jobs frequently due to lack of opportunities for advancement. Her report garnered some criticism, but also resulted in reforms within the profession.

Mitchell was honored with distinguished technical achievement awards of the U.S. Air Force (1962 and 1964). She was a fellow of the American Association for the Advancement of Science, the American Psychological Association, and the International Council of Women Psychologists.

Moore, Emmeline

1872–1963
Aquatic Biologist

Education: A.B., Cornell University, 1905; A.M., Wellesley College, 1906; Ph.D., Cornell University, 1914

Professional Experience: teacher, public schools, 1895–1903; instructor, biology, normal school, 1906–1910; substitute professor, botany, Huguenot College, South Africa, 1911; instructor and assistant professor, Vassar College, 1914–1919; research biologist and director of biological survey, New York Conservation Commission, 1919–1944

Emmeline Moore was an aquatic biologist and one of the few women to be appointed the director of a state fisheries department. Her research focused on the effect of fishing, disease, and pollution on fish in freshwater lakes, ponds, and rivers. Her career followed the pattern of many women in that she taught public school for several years before receiving her undergraduate degree. She was appointed an instructor in biology at a normal school after receiving her master's degree from Wellesley and substituted as a botany professor in South Africa for a year before returning to Cornell to complete her doctorate in 1914. She was appointed instructor and then assistant professor at Vassar, but joined the New York State Conversation Department in 1919 as its first female research biologist. She became chief aquatic biologist and was eventually appointed director of the survey. While her main focus was on the waterways and lakes of New York, she also conducted research projects throughout the United States and Canada, as well as in Europe and Africa. In 1926, she published a study on *Problems in Fresh Water Fisheries*. Even after her formal retirement in 1944, Moore served as an honorary fellow at the University of Wisconsin and as a research assistant at the Yale University oceanography lab.

Moore received the Walker Prize of the Boston Society of Natural History in both 1909 and 1915. She was the first woman president of the American Fisheries Society in 1928 and was a member of the American Association for the Advancement of Science and the Ecological Society of America. In 1958, a state marine research ship, the *Emmeline M.*, was named after her.

Further Resources

Hennigan, Robert D. 2004. "Emmeline Moore: Pioneer Biologist and Fisheries Scientist." *Clearwaters*. 34(3). (Fall 2004). New York Water Environment Association, Inc. http://www.nywea.org/clearwaters/04-3-fall/EmmelineMoore.cfm.

Brown, Patricia Stocking. 1994. "Early Women Ichthyologists." *Environmental Biology of Fishes* 41: 9–30. (1994). http://swfsc.noaa.gov/uploadedFiles/Education/Women%20in%20Ichthyology.pdf.

Morawetz, Cathleen (Synge)

b. 1923
Applied Mathematician

Education: B.S., University of Toronto, 1944; M.S., Massachusetts Institute of Technology, 1946; Ph.D., mathematics, New York University, 1951

Professional Experience: research associate, Massachusetts Institute of Technology (MIT), 1951–1952; research associate, Courant Institute of Mathematical Science, New York University, 1952–1957, assistant to associate professor, mathematics, 1957–1966, professor, 1966–1993, associate director, Courant Institute, 1978–1984, director, 1984–1988, emerita

Cathleen Morawetz is renowned for her research in applied mathematics. She is the first woman in the United States to head a mathematical institution, the Courant Institute of Mathematical Science at NYU. Her early work involved the mathematical analysis of transonic flow, which has practical applications in the design of aircraft as it involves the study of flow past an airfoil, such as the wing of an airplane. At very fast speeds, shock waves will develop and will increase the drag on an aircraft, which has important implications for the design of supersonic aircraft. In the 1960s, her research indicated that the equations of transonic flow show that a shock wave must occur if a plane goes fast enough, no matter how the wings are designed; engineers now settle for designing airfoils with small

Mathematician Cathleen Morawetz is former director of the Courant Institute at NYU. (New York University Archives)

shocks. Later, she concentrated on the mathematics associated with the scattering of waves—electromagnetic, sound, or elastic—upon hitting a barrier. The problem was how to observe and analyze the interaction of the wave with the barrier, whether it was reflected, absorbed, or transmitted. Some applications of scattering theory are in x-ray diffraction, and mathematical analyses of high-frequency waves are the basis of techniques used in medicine to visualize internal organs as well as techniques used in geology to search for oil fields.

Morawetz's father was the mathematician John Synge, renowned for his work in tensor analysis. He did not push his daughter toward a career in mathematics, and she originally wanted to study engineering at the California Institute of Technology, but the school did not accept women at that time. She therefore concentrated on applied mathematics because she found it esthetically appealing to use mathematics to describe natural phenomena. She later obtained a temporary job at New York University in the Mathematics Department to edit mathematician Richard Courant's book *Supersonic Flow and Shock Waves* (1948). She never formally applied to the graduate school but began taking classes and eventually wrote a thesis on imploding shock waves. She gave birth to four children during her graduate and early career years and spent several years working as a part-time researcher supported by government contracts before joining the faculty at the Courant Institute. She eventually became assistant director and then, in 1984, director of the school.

Morawetz has received eight honorary degrees, including an honorary doctorate from her own institution, New York University, in 2007. She was elected to membership in the National Academy of Sciences in 1990, the first woman member of the Applied Mathematics Section. She was named Outstanding Woman of Science by the Association for Women in Science (1993), and is a recipient of a National Medal of Science (1998), the Leroy P. Steele Prize for Lifetime Achievement by the American Mathematical Society (2004), and the Birkhoff Prize in Applied Mathematics (2006), awarded jointly by the AMS and the Society for Industrial and Applied Mathematics. She is a fellow of the American Association for the Advancement of Science and the American Academy of Arts and Sciences, and a member of the American Mathematical Society (president, 1995–1997), Society for Industrial and Applied Mathematics, and Mathematical Association of America.

Further Resources

Wasserman, Elga. 2002. *The Door in the Dream: Conversations with Eminent Women in Science*. Washington, D.C.: Joseph Henry Press.

Murray, Margaret Anne Marie. 2000. *Women Becoming Mathematicians: Creating an Identity in Post–World War II America*. Cambridge, MA: MIT Press.

Morgan, Agnes Fay

1884–1968
Biochemist and Nutritionist

Education: B.S., University of Chicago, 1904, M.S., 1905, Ph.D., chemistry, 1914

Professional Experience: instructor, chemistry, Hardin College, 1905–1907; instructor, chemistry, University of Washington, 1910–1913; assistant to associate professor, nutrition, University of California, Berkeley, 1915–1923, professor, nutrition, 1923–1928, professor, home economics and biochemistry, 1938–1954

Concurrent Positions: biochemist, experiment station, University of California, Berkeley, 1938–1954

Agnes Fay Morgan was recognized as one of the pioneers in the development of home economics as a scientific discipline, and as one of the pioneers in nutrition research. The home economics department at Berkeley under Morgan had one of the outstanding programs in the country due to her emphasis on research and her insistence on chemistry as an integral part of the home economics curriculum. Between 1951 and 1954, she served as chair of departments at both Berkeley and Davis. She founded Iota Sigma Pi, a national society for women in chemistry. Although she had a fine record of research and teaching, she was proudest of her administrative skills in establishing a department of Household Science and Arts at Berkeley and in playing a major role in the growth of the science of home economics. Her research included the effect of heat on the biological value of proteins and the mechanism of action of vitamins. She was recognized for her pioneering work on the biochemistry of vitamins, which has had a lasting influence on research today. She was the first to produce graying of hair through vitamin deficiency and the first to note certain supplementary effects of vitamin D.

Morgan received the Garvan Medal of the American Chemical Society in 1949 for her work on vitamins, and she received the Borden Award in 1954. In 1961, the Berkeley campus named the home economics building in her honor. She received an honorary degree from the University of California in 1959. She published *Experimental Food Study* (1927, 1940). She was elected a fellow of the American Institute of Nutrition and was a member of the American Association for the Advancement of Science and the American Society of Biological Chemists.

Further Resources

King, Janet C. 2003. "Contributions of Women to Human Nutrition." *Journal of Nutrition.* 133: 3693–3697. (November 2003). http://jn.nutrition.org/cgi/content/full/133/11/3693.

Morgan, Ann Haven

1882–1966
Zoologist and Ecologist

Education: A.B., Cornell University, 1906, Ph.D., 1912

Professional Experience: assistant and instructor, zoology, Mount Holyoke College, 1906–1909; assistant and instructor, Cornell University, 1909–1911; associate professor, Mount Holyoke College, 1912–1913, professor, 1914–1947

Ann Morgan was a biologist and zoologist recognized for her pioneering research on ecology and conservation and wrote several popular books, including *Field Book of Ponds and Streams: An Introduction to the Life of Fresh Water* (1930), the source for information on collecting and preserving specimens for many amateur naturalists, and *Field Book of Animals in Winter* (1939). Her research included freshwater biology, respiration and ecology of aquatic insects, biology of mayflies, habits and conditions of hibernating animals, and conservation, and her students nicknamed her "Mayfly Morgan." Morgan studied at Wellesley before transferring to Cornell University, where she received her bachelor's degree in 1906 and her doctorate in 1912. She was a visiting scholar at numerous colleges and institutions, including the Marine Biological Laboratory at Woods Hole, Harvard University, Yale University, and the Tropical Laboratory at Kartabo, British Guiana. In the 1940s and 1950s, Morgan concentrated on reforming the science curriculum to include the topics of ecology and conservation in both schools and colleges. She gave lectures and workshops for teachers of geography, zoology, and sociology. Her last book, *Kinships of Animals and Man: A Textbook of Animal Biology* (1955), written for an introductory course in zoology, synthesized her work on this topic.

Morgan was a member of the American Association for the Advancement of Science, American Society of Naturalists, National Commission on Policies in Conservation Education, New York Herpetological Society, American Society of Zoologists, and Entomological Society of America.

Further Resources

Bonta, Marcia. 1991. *Women in the Field: America's Pioneering Women Naturalists.* College Station: Texas A&M University Press.

Moss, Cynthia Jane

b. 1940
Wildlife Biologist

Education: B.A., philosophy, Smith College, 1962

Professional Experience: reporter and researcher, *Newsweek*, 1964–1968; veterinarian research assistant, Nairobi, 1969; research assistant, Athi Plains and Tvavo National Park, 1970; freelance journalist, 1970–1971; editor, *Wildlife News*, 1971–1985; co-director, Amboseli Elephant Research Project, Kenya, 1972–

Concurrent Positions: senior associate, African Wildlife Foundation, 1985–

Cynthia Moss is one of the foremost experts on the African elephant in the world, and, for many years, she and her associate **Joyce Poole** led the fight to stop the world trade in ivory. The illegal killing of elephants for their ivory tusks has negative effects for the entire elephant community, since it is the older lead elephants or the strongest males that are the targets of poachers. During the 1980s, Moss and Poole temporarily set aside their research projects to work with Richard Leakey to protect the elephants in Kenya and to stop the worldwide ivory trade. The three worked together to have the African elephant designated an endangered species by the Convention on International Trade in Endangered Species in 1989. Moss and Poole created a worldwide movement to ban the ivory trade by inviting photographers and newspaper reporters to visit Amboseli to photograph the elephants and tell their stories.

Moss's unique research on animals has been compared to the work of English primatologist Jane Goodall. Moss developed a method of identifying elephants by their ears, and she and her researchers have identified more than 1,400 individual elephants. Like Goodall, Moss began naming the elephants according to their families. She also studied the elephants' family structure and social patterns and became an authority on the subject. She is famous for her research that shows the male African elephants experience *musth*, a condition of increased aggression and increased sexual activity that had previously been attributed only to male Indian elephants. Along with Poole, she has also conducted pioneer studies of elephant vocalizations and identified different calls and behaviors that signal what the elephants will do—either charge or move away. Another insight Moss discovered is that, in times of drought, the elephants do not breed and therefore reduce the number of babies that will require food.

Moss fell in love with Africa on a brief visit to the country in 1967, and after working as a journalist for a number of years, she moved to Africa permanently to work with several established researchers. In 1972, she helped found the Amboseli Elephant Research Project in Kenya. Her books, *Portraits in the Wild: Behaviour Studies of East African Mammals* (1975) and *Elephant Memories: Thirteen Years in the Life of an Elephant Family* (1988; rev. ed., 2000), describe her work in Amboseli National Park. She has also contributed to children's books and wildlife documentaries on the elephants. In 2000, Moss was named one of *Time* magazine's "Heroes for the Planet."

Further Resources

Amboseli Elephant Research Project. "Cynthia Moss." http://www.elephanttrust.org/node/41

Poole, Joyce. 1996. *Coming of Age with Elephants: A Memoir*. New York: Hyperion.

Murray, Sandra Ann

b. 1947
Molecular Biologist, Cell Biologist

Education: B.S., biology, University of Illinois, Chicago, 1970; M.S., biology, Texas Southern University, 1973; Ph.D., anatomy, University of Iowa, 1980

Professional Experience: instructor, biology, Texas Southern University, 1972–1973; National Institutes of Health (NIH) postdoctoral research fellow, University of California, Riverside, 1980–1982; assistant professor, anatomy, University of Pittsburgh Medical School, 1982–1989, associate professor, cell biology and physiology, 1989–

Concurrent Positions: researcher, Marine Biological Laboratory, Woods Hole Oceanographic Institution, 1986–1990; visiting scientist, Scripps Research Institute of Molecular Biology, 1991–1992; associate professor, Health Officers Institute, Office of Defense, Addis Ababa, Ethiopia, 1996–

Sandra Murray is known for her research in molecular and cell biology. She uses molecular biological, biochemical, and morphological methods to study how cells function, what brings about normal functions in a cell population, what controls the rate of cell population growth if a normal population has been injured, and how that compares with the daily process of aging and replenishing that population. She looks at what is different in cancer cell populations and examines the capacity of cells to send signals from one cell to an adjacent cell via structures called "connexins" that are associated with controlling the function of cells and the rate of cell population growth. She studies cells in culture and sometimes from human tissue taken from donors.

Murray became interested in science at a very early age. She did not feel any limitations on her career goals until she got to high school, when a counselor told her that "colored girls don't become scientists." While still in high school, however, she worked as a laboratory aide at the University of Illinois Medical School and was participating in Saturday science classes at the University of Illinois. After earning her B.S., she went on to graduate study at Texas Southern University and the University of Iowa, where a professor made racist comments about her ability to keep up in class. When she made good grades, he told her that her lighter skin probably indicated she had non-African blood that allowed her to do well. She transferred to a different department and received her doctorate in anatomy in 1980. Soon after, she became an assistant professor at the University of Pittsburgh Medical School, where she became the first African American to receive tenure.

Murray remains committed to encouraging women and minority students in the sciences. Recalling her own early interest in science, she also regularly serves as a mentor and judge for the National Technology Association of Science and the International Science and Engineers Fairs. She is a member of the American Society of Cell Biology (and served on the Minorities Affairs Committee), the American Society of Biological Chemists, American Association of Anatomists, Tissue Culture Association, and Endocrine Society.

Further Resources

University of Pittsburgh School of Medicine. Faculty website. http://www.cbp.pitt.edu/faculty/murray.html.

Ambrose, Susan A. 1997. *Journeys of Women in Science and Engineering: No Universal Constants*. Philadelphia, PA: Temple University Press.

N

Napadensky, Hyla Sarane (Siegel)

b. 1929
Combustion Engineer

Education: B.S. and M.S., mathematics, University of Chicago, ca. 1950

Professional Experience: design analysis engineer, International Harvester Company, 1952–1957; director of research, Illinois Institute of Technology (IIT) Research Institute, Chicago, 1957–1988; vice president, Napadensky Energetics, Inc., 1988–1994, engineering consultant, 1994–1998

Concurrent Positions: instructor, Mechanics Department, IIT, 1964–1966

Hyla Napadensky is a combustion engineer who spent her career as an expert in explosives and propellant safety. Her research included the study of accidental fires and explosions during the manufacture, transport, and storage of explosives, propellants, and pyrotechnics. She also studied explosive and initiation mechanisms, facility siting, and systems safety and risk analysis. After working for five years for the International Harvester Company, she began a career as director of research at the IIT Research Institute in Chicago, involved with research on a contract basis, some of it with federal agencies. Many of the studies she conducted for the government on materials used in explosive charges are probably classified as secret and therefore are not included in the standard databases. Napadensky prepared a 220-page book for the U.S. Army, *Development of Hazards Classification Data on Propellants and Explosives* (1978), and a similar book for the same agency, *Recommended Hazard Classification Procedures for In-Process Propellant and Explosive Material* (1980). As an internal publication, she prepared data on the TNT equivalency of black powder. She has also written about the risks of handling explosives on ships and in harbors.

Napadensky spent 30 years at the IIT Research Institute. She then established a consulting company, Napadensky Energetics, Inc., and formally retired in 1998. She was elected to membership in the National Academy of Engineering in 1984, and has been a National Associate of the National Academies since 2001.

Navrotsky, Alexandra A. S.

b. 1943
Geochemist, Geophysicist

Education: B.S., University of Chicago, 1963, M.S., 1964, Ph.D., chemistry, 1967

Professional Experience: research associate, theoretical metallurgy, Technische Hochschule, Clausthal, Germany, 1967–1968; research associate, geochemistry, Pennsylvania State University, 1968–1969; assistant professor, chemistry, Arizona State University, 1969–1974, associate professor, 1974–1978, professor, chemistry and geology, 1978–1985, director, Center for Solid State Science, Arizona State University, 1984–1985; professor, geological and geophysical science (affiliate in chemistry), Princeton University, 1985–1997; Interdisciplinary Professor, Ceramic, Earth, and Environmental Materials Chemistry, University of California, Davis, 1997–; director, Nanomaterials in the Environment, Agriculture, and Technology, Organized Research Unit (NEAT ORU), 2002–

Concurrent Positions: visiting research associate, James Franck Institute, University of Chicago, 1970–1971; visiting scientist, Technische Universitat, Clausthal, Germany, 1972; visiting scientist, Bell Telephone Laboratories, 1974; visiting lecturer, Massachusetts Institute of Technology, 1975; visiting associate professor, University of California, Berkeley, 1976; Program Director for Chemical Thermodynamics, National Science Foundation, 1976–1977; visiting professor, State University of New York, 1981; visiting summer faculty, IBM, T. J. Watson Research Center, 1988

Alexandra Navrotsky is recognized as one of the leaders in combining mineralogical and materials research. As new technological materials become increasingly complex in structure and bonding, they are beginning to resemble the materials that make up our planet; materials science is the study of the characteristics and uses of various materials such as glass, plastics, and metals. One of the areas she has investigated is the composition of the Earth, and she points out that although humans have explored the moon, a journey to the center of the Earth remains fictional and technologically unattainable. However, mineral physics can provide some information via laboratory and computational simulations of matter under high pressure and temperature. The Earth is composed of, in descending order, the crust, the upper mantle, the transition zone, the lower mantle, the outer core, and the inner core. Navrotsky has published on the topic of thermochemistry. In 2002, she became the director of a new research institute at the University of California, Davis called NEAT: Nanomaterials in the Environment, Agriculture, and Technology. NEAT is "a multidisciplinary research and education program which links the fundamental physics, chemistry, and engineering of small particles and nanomaterials to several challenging areas of investigation," making applications in agricultural and environmental technology and health sciences.

Navrotsky was elected to membership in the National Academy of Sciences in 1993. Her expertise has been recognized by invitations to lecture at universities around the world. She has served on visiting committees for several institutions and scientific organizations. She was a member of the Committee on Mineral Physics of the American Geophysical Union (1983–1993) and the Committee on High Temperature Chemistry of the National Academy of Sciences (1981–1985), a fellow of the Mineralogical Society of America (President, 1992–1993), and a fellow of the American Ceramic Society (2001). In 2002, she was awarded the prestigious Benjamin Franklin Medal in Earth Science and, in 2006, the Harry H. Hess Medal of the American Geophysical Union. She is the author of *Physics and Chemistry of Earth Materials* (1994), a textbook designed for advanced undergraduates and first-year graduate students. She holds a U.S. Patent (2005) for "Methods for Removing Organic Compounds from Nano-Compositic Materials."

Further Resources

University of California, Davis. Faculty website. http://navrotsky.engr.ucdavis.edu/.

Nelkin, Dorothy (Wolfers)

1933–2003
Sociologist

Education: B.A., sociology, Cornell University, 1954

Professional Experience: research associate, sociology, Cornell University, 1963–1969, senior research associate, 1970–1972, associate professor, 1972–1976, professor, Science, Technology and Society Policy Program, 1976–1990, professor, sociology, 1977–1989; professor, sociology and affiliate professor, law, New York University, 1989–2003

Dorothy Nelkin was a sociologist who wrote or co-authored more than 20 books as well as numerous papers on topics as diverse as migrant labor, nuclear power, housing innovation, university and military research, methadone maintenance, science, technological decisions, the atom, the creation controversy, animal rights, unsafe work conditions, genetics, and medical diagnosis. In her book *Workers at Risk: Voices from the Workplace* (1984), she reviewed the unsafe conditions that workers of all types encounter. Her research team interviewed workers in museums, beauty shops, research laboratories, and computer-assembly plants as well as steel mills, auto-assembly plants, and other obvious places for dangerous working conditions. The surveyors found there was no direct link between the actual hazards and people's perceptions of risk. For example, artists and research scientists often feel that the rewards of their job outweigh the risks of handling extremely toxic chemicals. However, many workers complained they lacked information about the chemicals with which they worked. Nelkin's hope was that workplace safety would improve as a result of the survey.

Nelkin was concerned about how scientific information and tests were used to control people's lives. Her earlier book on the workplace was revised and republished under the title of *Dangerous Diagnostics: The Social Power of Biological Information* (1989), with new research from Nelkin and her researchers on the myriad tests that pronounce people healthy or ill, or likely or unlikely to suffer any of hundreds of ailments. The authors focused on the social implications of the information that these tests provide and the power that accrued to employers who administer the tests. In her book *The DNA Mystique: The Gene as a Cultural Icon* (1995; co-authored with M. Susan Lindee), she weighed in on the increasing public and political interest in human genetics in relation to social questions of intelligence, homosexuality, or criminality. The authors concluded that a reliance on DNA testing obscures efforts to solve social problems through policy or sociological support. In 2001, she and co-author Lori B. Andrews continued the discussion about who controls genetic information in their book, *Body Bazaar: The Market for Human Tissue in the Biotechnology Age*. Always interested in the relationship between science and culture, at the time of her death in 2003 she was working on new projects on science and religion and, with Suzanne Anker, on the influence of genetic science on the arts. Their book, *The Molecular Gaze: Art in the Genetic Age*, was published after Nelkin's death.

Nelkin was a consultant to the Organization of Economic Cooperation and Development (OECD, 1975–1976) and the Institute of Environment, Berlin (1978–1979), and a member of the National Advisory Council to the Human Genome Project of the National Institutes of Health (1991–1995). She was a fellow of the American Association for the Advancement of Science and a member of the Society for Social Studies of Science (president, 1978–1979). In 1993, she was elected to the Institute of Medicine of the National Academies of Science.

Neufeld, Elizabeth (Fondal)

b. 1928
Geneticist, Biochemist

Education: B.S., Queens College, 1948; student, University of Rochester, 1949–1950; Ph.D., comparative biochemistry, University of California, Berkeley, 1956

Professional Experience: postdoctoral researcher, biochemistry, University of California, Berkeley, 1956–1963; research biochemist, National Institute of Arthritis, Metabolism and Digestive Diseases, National Institutes of Health, 1963–1973, chief, Section on Human Biochemical Genetics, 1973–1979, chief, Genetics and Biochemistry Branch, National Institute of Arthritis, Diabetes, and Digestive and Kidney Diseases, 1979–1984, deputy director, Division of Extramural Research, 1981–1983; professor and chair, biological chemistry, School of Medicine, University of California, Los Angeles, 1984–2005, emeritus

Concurrent Positions: U.S. Public Health Service fellow, University of California, Berkeley, 1956–1957, assistant research biochemist, 1957–1963

Biochemist Elizabeth Neufield has researched the genetic basis of metabolic diseases. (National Library of Medicine)

Elizabeth Neufeld is a leading international authority on human genetic diseases. Her research includes human biochemical genetics, mucopolysaccharidoses; Tay-Sachs disease; synthesis and transport of lysosomal enzymes; and inherited disorders of lysosomal functions. She provided new insights on the absence of certain enzymes that prevent the body from properly storing certain substances and has led to prenatal diagnosis of such life-threatening fetal disorders as Hurler syndrome. Her research on inherited disorders of the connective tissues focused on diseases in which cells lack certain enzymes needed to process complex sugars. The accumulation of sugars causes the cells to grow and put internal pressure on nerve tissues, which can die from too much pressure. Patients suffer from severe mental and motor deterioration, have vision and hearing problems, and die prematurely, usually before puberty. The diseases are known as the Hurler and Sanfilippo syndromes and are also related to Tay-Sachs and other diseases. After years of research, her team found that the problem was a defective gene that was causing the sugars to break down at an abnormally slow rate, and further study indicated that a series of enzymes were lacking in the patients. Her work has led to successful prenatal diagnosis and has contributed to the availability of genetic counseling for parents. Future treatments being considered are gene replacement therapy and bone marrow transplant.

Neufeld's parents were Russian refugees living in Paris after the Russian revolution when she was born; the family moved to New York City before the Germans occupied France in 1940. Her parents stressed the importance of education because education cannot be taken away, and she became interested in science while in high school through the influence of her biology instructor. She started her scientific studies at a time when few women were choosing science as a career and there were few positions open for women—partly because of the historical bias against women in science and partly because of the influx of men returning from World War II. Few women could be found on the science faculties of colleges and universities, but she persevered in her career because she enjoyed what she was doing.

Neufeld has received numerous honorary degrees and awards. She was elected to membership in the National Academy of Sciences in 1977. She received the Lasker Award in 1982, the highest honor in the United States for medical research and which often leads to the Nobel Prize. She also won the Wolf Prize in Medicine (1988) and was awarded the National Medal of Science (1994). She is a fellow of the American Association for the Advancement of Science and a member of the American Society of Human Genetics, American Chemical Society, American Society of Biological Chemists, American Society of Cell Biology, American Society of Biochemistry and Molecular Biology (president, 1992), and American Academy of Arts and Sciences. In 1990, she was named California Scientist of the Year.

Further Resources

University of California, Los Angeles. Faculty research page. http://dgsom.healthsciences.ucla.edu/research/institution/personnel?personnel_id=45290.

New, Maria (Iandolo)

b. 1928
Pediatrician

Education: B.A., Cornell University, 1950; M.D., University of Pennsylvania, 1954

Professional Experience: medical intern, Bellevue Hospital, New York, 1954–1955; resident, pediatrics, New York Hospital, 1955–1957, National Institutes of Health fellow, pediatrics, New York Hospital, Cornell Medical Center, 1957–1958, research pediatrician, Diabetic Study Group, Comprehensive Care Teaching Program, 1958–1961, instructor, pediatrics, 1958–1963, assistant to associate attending professor, pediatrics, 1963–1971, chief, pediatric endocrinology, Cornell University Medical College (now Joan and Sanford Weill Medical College of Cornell), 1964–2002, professor and attending pediatrician, 1971–2004, chair, pediatrics, 1980–2002, program director, Children's Clinical Research Center, 1996–2002; professor, pediatrics and human genetics, and attending pediatrician, Mount Sinai School of Medicine, New York, 2004–, and director, Adrenal Steroid Disorders Program, 2004–

Concurrent Positions: assistant pediatrician to outpatients, New York Hospital, 1957–1959, pediatrician, 1960–1963, director, Pediatric Metabolism Clinic, 1964–2003; attending pediatrician, New York-Presbyterian Hospital (formerly New York Hospital), 1971–, pediatrician-in-chief, 1980–2002; visiting physician, Rockefeller University Hospital, New York, 1973–; consultant, Albert Einstein College of Medicine, Bronx, New York, 1974–1976; consultant, pediatrics and endocrinology, New York United Hospital Medical Center, Port Chester, New York, 1977–; adjunct attending pediatrician, Memorial Sloan-Kettering Cancer Center, 1979–1993; consultant, pediatrics, North Shore University Hospital, Manhasset, New York, 1982–; consultant, pediatrics, Catholic Medical Center of Brooklyn and Queens, 1986–; honorary member, pediatrics, Blythedale Children's Hospital, Valhalla, New York, 1992–; consultant, Memorial Hospital for Cancer and Allied Diseases, 1993–; consultant, Memorial Sloan-Kettering Cancer Center, 1993–2006; director, pediatric endocrinology, New York-Presbyterian: University Hospital of Columbia and Cornell, 1998–2002

Maria New is an endocrinologist who specializes in pediatric endocrinology and renal diseases, juvenile hypertension, pediatric pharmacology, and growth and development from the biochemical viewpoint. She established the Maria I. New Children's Hormone Foundation in New York as a nonprofit organization to support medical research on pediatric endocrinology and provide services and support to patients and their families. Dr. New has been affiliated with several major hospitals in the New York area as a consulting physician and has trained hundreds of new pediatricians in her specialty. She had a long career at Cornell University Medical Center before becoming professor of pediatrics and director of the Adrenal Steroid Disorders Program at Mount Sinai School of Medicine in 2004. New has edited or co-edited numerous medical textbooks as well as more than 600 research papers. She helped edit a book for the general public, the two-volume Disney *Encyclopedia of Baby and Child Care* (1995), compiled by four pediatricians. In addition to her numerous publications, she served as editor-in-chief of the *Journal of Clinical Endocrinology and Metabolism* from 1994 to 1999.

New was elected to membership in the National Academy of Sciences in 1996. She has received multiple honors and awards, including the Robert H. Williams Distinguished Leadership Award (1988), medal of the New York Academy of Medicine (1991), Maurice R. Greenberg Distinguished Service Award (1994), Humanitarian Award of the Juvenile Diabetes Foundation (1994), Dale Medal of the British Endocrine Society (1995), MERIT Award of the National Institute of Child Health and Human Development (1998), Hall of Honor of the National Institute of Child Health and Human Development (2003), and Fred Conrad Koch Award, the highest honor of the Endocrine Society (2003).

She is a member of numerous associations, such as the American Association for the Advancement of Science, New York Academy of Sciences, American Society of Human Genetics, American Academy of Pediatrics, Society for Pediatric Research, Endocrine Society (president, 1991–1992), American Fertility Society, and American Academy of Arts and Sciences, and is an honorary member of the Italian Endocrine Society.

Further Resources

Maria New Children's Hormone Foundation. http://www.newchf.org/.

Endocrine Society. "Maria New." http://www.endo-society.org/about/Maria-New.cfm.

Nice, Margaret Morse

1883–1974
Ornithologist

Education: B.A., Mount Holyoke College, 1906; A.M., psychology, Clark University, 1915

Professional Experience: independent researcher, 1915–1974

Concurrent Positions: associate editor, *Bird-Banding*, 1935–1942 and 1946–1971

Margaret Nice was an internationally known ornithologist who adapted the techniques of scientific investigation from psychology to a new area of research, that of bird behavior. Her research interests included birds of Oklahoma; life history studies of birds, particularly mourning doves, warblers, and song sparrows; and speech development of children. After receiving a master's degree in psychology, she pursued an independent interest in ornithology. Her work was supported by occasional small grants, but she never held a faculty or museum appointment. Initially, she was interested in languages as a student at Mount Holyoke College, where she received her undergraduate degree in 1906. At that time, ornithology was taught in the zoology department and consisted of identifying dead species. Her interest shifted to psychology at Clark University, where she received her master's degree in 1915. She published 18 articles on child psychology from observations of her own children between 1915 and 1933. She began conducting field observations on birds and started corresponding with fellow ornithologists.

Nice was at the center of a network of women ornithologists whose scientific correspondence also served as a professional support system. She published approximately 250 papers and, due to her language skills, also contributed to the discipline by reviewing a large number of the leading European publications. She co-authored (with her husband, L. Blaine Nice) *The Birds of Oklahoma* (1924) and was the sole author of the two-part *Studies in the Life History of the Song Sparrow* (1937, 1943). These works established her reputation as one of the world's foremost ornithologists and bird behaviorists.

Nice was active in ornithological and conservation organizations and served as associate editor of the journal *Bird-Banding*. She published one bird book for the general public, *The Watcher at the Nest* (1939), which was reprinted in paperback. In later life, she increasingly turned her attention to educating the public about conservation and nature with lectures and talks on the radio. She often enlisted her entire family in her work; for example, her children would climb trees to observe nests for her. When the family lived in Columbus, Ohio, the local ornithology club was an all-male group and, even though by that time her work was known internationally, they invited her husband to join, but ignored her.

Nice was awarded the Brewster Medal of the American Ornithologists' Union in 1942. She was the first woman president of the Wilson Ornithological Society (1938–1939), and she was elected a fellow of the American Ornithologists' Union. She received an honorary degree from Mount Holyoke in 1955 and one from Elmira College in 1962. She published an autobiography, *Research Is a Passion with Me* (1979). She was listed in some sources as "Mrs. L. B. Nice."

Further Resources

Bonta, Marcia. 1991. *Women in the Field: America's Pioneering Women Naturalists.* College Station: Texas A&M University Press.

Nichols, Roberta J.

1931–2005
Environmental Engineer

Education: B.S., physics, University of California, Los Angeles, 1968; M.S., environmental engineering, University of Southern California, 1975, Ph.D., engineering, 1979

Professional Experience: mathematician, missile department, Douglas Aircraft Company, 1957; mathematician, propulsion department, TRW Space Technology Laboratory, 1958–1960; research associate, Aerospace Corporation, Aerodynamics and Propulsion Laboratory, 1960–1967, Chemical Kinetics Department, 1969–1978; consultant, Synthetic Fuels Office, State of California, 1978–1979; developer of synthetic fuels, Ford Motor Company, 1979–1989, manager, alternate fuels program, 1989–1995

Roberta Nichols was a research engineer who led the U.S. automobile manufacturers in developing alternate fuels and cars to use those fuels. She worked for many years for Ford Motor Company and acquired three patents related to the Flexible Fuel Vehicle (FFV). Nichols was one of the few people who had the foresight that future clean-air laws would alter the use of fuels used to power cars and trucks. She gave lectures worldwide and served as a consultant in industry and government on issues related to low-emission and alternative energies based on alcohol, methanol-gasoline blends, hydrogen power, and battery power. She joined the Ford Motor Company in 1979 and almost singlehandedly dragged the American automobile manufacturers into the alternative fuels age. She developed ethanol-fueled engines for Ford of Brazil; designed and developed 630 methanol-fueled Escorts, which were used primarily as government vehicles; designed and developed the power train for an alternate fuel vehicle exhibited in 1982; and oversaw the development of natural-gas trucks.

Nichols became interested in alternative fuels after her father, an aerospace engineer, introduced her to racing boats. She not only held the women's world water speed record for several years in the late 1960s, but she began learning about engines and fuel performance. She earned a degree in physics from the University of California, Los Angeles (UCLA) and worked for several aerospace and aircraft companies, including establishing the Air Pollution Laboratory at Aerospace Corporation. After she became a widow with two small children to rear, she returned to school to obtain graduate degrees in engineering and then got a job at Ford. She was also a longtime board member for the Center for Environmental Research and Technology at the University of California, Riverside.

Nichols was elected to the National Academy of Engineering in 1997 and was the first woman to be elected a fellow of the Society of Automotive Engineers. She received the Outstanding Engineer Merit Award of the Institute for the Advancement of Engineering, the Aerospace Corporation's Woman of the Year Award, the Society of Women Engineers National Achievement Award (1988), a Clean Air Award for Advancing Air Pollution Technology (1989), and the Gene Ecklund Award from the U.S. Department of Energy (1996).

Nickerson, Dorothy

1900–1985
Physicist

Education: unknown

Professional Experience: assistant and assistant manager, Munsell Research Laboratory, 1921–1926; color technologist, U.S. Department of Agriculture (USDA), 1927–1964; consultant, 1965–1974

Dorothy Nickerson was a physicist and color specialist who applied color-graded standards to agricultural and horticultural products and soil. She developed the Nickerson color fan of more than 300 color samples graded by light value, hue, and chroma. The color fan or chart is important in science and industry for grading the color of products such as new strains of vegetables or cotton for textiles. At the Munsell Color Research Laboratory, Nickerson specialized in color technology and rose to the level of assistant manager. She joined the USDA in 1927 as a color technologist in the bureau of agricultural economics. She authored more than 150 papers and articles on the Munsell color system. She left the USDA in 1964 and served as a U.S. expert on color rendering for the International Commission on Illumination from 1956 to 1967. After retiring from the USDA, she formed a private consulting firm.

Nickerson received several awards, such as the Superior Service Award from the USDA (1951), the Distinguished Achievement Award of the Instrument Society of America (1964), the Gold certificate of the American Horticultural Council (1957), the Godlove Award of the Inter-Society Color Council (ISCC) (1961), and the Gold Medal of the Illumination Society of England (1970). She was a founding member of the ISCC and served as secretary (1935–1952) and president (1954–1955). In 1980, the ISCC established a Nickerson Award in her honor. She was a member of the American Association for the Advancement of Science, the Optical Society of America, and the Illuminating Engineering Society.

Nielsen, Jerri Lin

1952–2009
Physician

Education: B.A., zoology, Ohio University; M.D., Medical College of Ohio, 1977

Professional Experience: physician, 1977–1998; physician, Amundsen-Scott station, Antarctica, 1998

Jerri Nielsen was the only physician working at the Amundsen-Scott South Pole Station in Antarctica in 1998 when she diagnosed herself with breast cancer. (AP/Wide World Photos)

Jerri Lin (Cahill) Nielsen was hired to spend a year as the only physician working at the Amundsen-Scott South Pole Station in Antarctica in 1998. She attracted media attention when, during the isolated winter, she discovered a lump in her breast and had to perform her own biopsy and administer her own chemotherapy before she could leave Antarctica. A longtime private physician and emergency room doctor, Nielsen, a divorced mother of three, joined the one-year expedition as the research station's sole physician, in charge of administering basic medical care to the scientists and staff. The station is completely isolated for nine long, dark months of the year, as it is too dangerous for supply planes to land or take off on the ice. After noticing the lump in her breast, Nielsen communicated via e-mail and videoconferencing with doctors back in the United States. She used a needle to extract samples of the tumor's cells, sending the images to other doctors via computer. After confirmation that the cells were cancerous, medical supplies and drugs for her treatment were airdropped into the station. As the only physician at the station, she had to rely upon assistance from the other nonmedical personnel to administer her chemotherapy. As soon as the weather permitted, she was airlifted back to the United States, where she underwent further treatment, including a mastectomy.

Because of the unique and dramatic nature of her ordeal, Nielsen became a popular media figure and an international motivational speaker. She wrote a book about her experience, *Icebound: A Doctor's Incredible Battle for Survival at the South Pole* (2001; with Maryanne Vollers), which was adapted as a television movie starring Susan Sarandon. She lived with the cancer for more than 10 years, but it eventually spread to her brain and she died in June 2009 in Massachusetts.

Nightingale, Dorothy Virginia

b. 1902
Organic Chemist

Education: A.B., University of Missouri, 1922, A.M., organic chemistry, 1923; Ph.D., organic chemistry, University of Chicago, 1928

Professional Experience: instructor, chemistry, University of Missouri, Columbia, 1923–1939, assistant to associate professor, 1939–1958, professor, 1958–1972

Concurrent Positions: consultant, Office of Scientific Research and Development, 1942–1945; research associate, University of California, Los Angeles (UCLA), 1946–1947

Dorothy Nightingale was a physical chemist who has been recognized for her work in organic synthetic reactions. Her research had important industry applications for the production of gasoline, synthetic rubber, cleaning products, and plastics. Nightingale was originally interested in studying history and languages, but was encouraged early on by a professor and changed her major to chemistry. She joined the faculty at the University of Missouri as one of only two women chemistry instructors after receiving her master's degree in 1923. She received her doctorate in organic chemistry from the University of Chicago in 1928, while still teaching at Missouri. She was not promoted to assistant professor until 1939 and full professor in 1958; during her tenure there, she directed the research of more than 50 graduate chemistry students. She drew upon this experience in writing *A History of the Department of Chemistry: University of Missouri-Columbia, 1843–1975*, published in 1975.

During World War II, Nightingale took a leave from the university to work as a civilian with the Office of Scientific Research and Development (OSRD). She worked with the Committee on Medical Research of the OSRD, contributing to compound studies important in the development of antimalarial drugs for the military. She spent a year conducting research at UCLA before returning to Missouri. She retired from the University of Missouri in 1972 after nearly 50 years at that institution. Nightingale was a member of the American Chemical Society (ACS) and received the Garvan Medal of the ACS in 1959.

Northrup, Christiane

b. 1949
Physician

Education: M.D., Dartmouth Medical School, New Hampshire, 1975; diplomate, American Board of Obstetrics and Gynecology, 1981; diplomate, American Board of Holistic Medicine, 2005

Professional Experience: director, Resident's Outpatient Obstetrics and Gynecology Clinic, St. Margaret's Hospital Boston, 1979–1980; associate clinical professor, obstetrics and gynecology, Tufts University School of Medicine, 1979–1980; clinical instructor, obstetrics and gynecology, University of Vermont College of Medicine, 1979–1982, assistant clinical professor, 1982–2001; co-founder, Women to Women, Yarmouth, Maine, 1985–1997; physician, private practice, obstetrics and gynecology, Portland and Yarmouth, Maine, 1979–2005

Christiane Northrup is a physician and women's health advocate who has built an international following as a proponent of holistic healthcare and wellness through combining Western medicine, vitamins and herbal supplements, and mind–body healing. Northrup specializes in obstetrics, gynecology, and women's general health, and has written on childbirth, menopause, and breast cancer, among other topics. Her first book, *Women's Bodies, Women's Wisdom*, was published in 1994 (rev. 2006), and sold more than 1 million copies worldwide and was translated into 15 different languages. The book launched her career as a popular media figure and women's health expert, and she began to make guest appearances on shows such as the *Oprah Winfrey Show, Today, The View, Good Morning America, 20/20,* and numerous other news programs. Her second bestselling book, *The Wisdom of Menopause: Creating Physical and Emotional Health and Healing during the Change* (2001; rev. 2006) was radical in taking an empowering and positive view of the changes women undergo as a new phase of life rather than focusing on only the negatives and losses. She followed with the publication of *The Secret Pleasures of Menopause* (2008). In 2005, Northrup published *Mother-Daughter Wisdom: Understanding the Crucial Link between Mothers,*

Daughters, and Health, which explores the physical and mental connection between mothers and daughters, and the effect on our health over the course of a lifetime. Her books also inspired her own public-television specials in the late 1990s and early 2000s. She has fans and followers around the world and publishes a monthly newsletter on "Women's Health Wisdom," and has organized a "Women's Wisdom Community" through her popular website.

Dr. Northrup sits on a number of medical advisory boards related to women's issues and holistic health strategies, including for *Natural Health Magazine*, Alternative Therapies in Health and Medicine, American Holistic Health Association, Pilates Health, Heal Breast Cancer Foundation, and A Woman's Nation, a research and policy center founded by California First Lady Maria Shriver. Northrup is a member of the American Holistic Medical Association (AHMA) (president, 1986–1988) and the American College of Obstetrics and Gynecology. In addition to the awards and acknowledgements for her books, she has been named a Pioneer of Holistic Medicine by the AHMA (2003), and has received a Maine Media Women's President's Award (2003), Campaign for Better Health Celebrating Excellence Award (2003), American Heart Association's Learn & Live Gold Heart Award (2004), Lamaze International Irwin Chabon Award (2006), and Excellence in Integrative Medicine Award from the Heal Breast Cancer Foundation (2007), among other awards. She received an honorary doctorate from the University of Maine, Farmington (2002).

Further Resources

"Christiane Northrup, M.D." http://www.drnorthrup.com.

Novello, Antonia (Coello)

b. 1944
Pediatrician

Education: B.S., University of Puerto Rico, 1965; M.D., University of Puerto Rico, San Juan, 1970; M.S., public health, Johns Hopkins University School of Hygiene, 1982; diplomate, American Board of Pediatrics

Professional Experience: intern, pediatrics, University of Michigan Medical Center, 1970–1971, resident, pediatrics, 1971–1973, pediatric nephrology fellow, 1973–1974; pediatric nephrology fellow, Georgetown University Hospital, 1974–1975; physician, private practice, 1976–1978; project officer, National Institute of Arthritis, Metabolism and Digestive Diseases, National Institutes of Health (NIH),

Pediatrician and former U.S. Surgeon General, Antonia Novello. (Getty Images)

1978–1979, staff physician, 1979–1980, executive secretary, General Medicine Study Section, Division of Research Grants, 1981–1986, deputy director, National Institute of Child Health and Human Development, 1986–1990; Surgeon General, U.S. Department of Health and Human Services, 1990–1993; Special Representative for Health and Nutrition, UNICEF, 1993–1996; visiting professor, Health Policy and Management, Johns Hopkins School of Hygiene and Public Health, 1996–1999; Health Commissioner, State of New York, 1999–2006; vice president, Women's and Children's Health and Policy Affairs, Disney Children's Hospital, Orlando, 2008–

Concurrent Positions: clinical professor, pediatrics, Georgetown University Hospital, 1986, 1989; adjunct professor, pediatrics and communicable diseases, University of Michigan Medical School, 1993; adjunct professor, international health, Johns Hopkins University School of Hygiene and Public Health

Antonia Novello was the first woman to be selected Surgeon General of the United States, and also the first Hispanic person to hold that post. The Surgeon General is the nation's chief advisor on matters of public health, is a spokesperson for the president in such areas, and oversees a corps of public-health research and policy teams. Novello used the position to attract national media attention to issues such as the healthcare of minorities, women, and children; injury prevention; and the problems of domestic violence, alcohol abuse among the nation's youth, and smoking among women and young people. Although she opposed abortion, she seldom discussed the issue while Surgeon General, feeling that women should not view abortion as the only issue to tackle. Novello made headlines in 1992 when she and the executive vice president of the American Medical Association held a news conference to urge R. J. Reynolds Tobacco Company to withdraw its ads featuring the cartoon character Joe Camel because of its appeal to young people. She also attacked the practice of using sports heroes in alcohol advertising, targeting young people and thus encouraging underage drinking. She was also concerned about the number of children who are not vaccinated against common infectious diseases and the widespread lack of proper prenatal care.

After receiving her M.D. in Puerto Rico, Novello and her husband moved to the University of Michigan to continue their education. She then had additional training at Georgetown University before she joined the NIH. While with the NIH, she received a master's degree in public health from Johns Hopkins University and rose rapidly through the ranks of government service and policymaking. She helped draft the Organ Transplantation Procurement Act of 1984 and served on the Senate Committee on Labor and Human Resources before being appointed Surgeon General by President George H.W. Bush. After she left the Surgeon General's office, she accepted a position with UNICEF and then returned to Johns Hopkins as a visiting professor. She served as Commissioner of Health for the State of New York for seven years and in 2008 was appointed Vice President for Women's and Children's Health and Policy Affairs at Disney Children's Hospital at Florida Hospital in Orlando.

Novello has received numerous honorary degrees and awards, including the Public Health Service Outstanding Medal (1988), Surgeon General Medallion Award (1990), Alumni Award of the University of Michigan Medical School (1991), and Distinguished Public Service Award (1993). Novello was also presented with the Legion of Merit Medal by U.S. Secretary of State Colin Powell. She was inducted into the National Women's Hall of Fame in 1994. She is a member of the American Medical Association, International Society of Nephrology, and American Society of Nephrology.

Further Resources

"Antonia Novello, M.D." Academy of Achievement. http://www.achievement.org/autodoc/page/nov0bio-1.

Ocampo, Adriana C.

b. 1955
Planetary Geologist

Education: student, aerospace engineering, Pasadena City College, ca. 1972–1975; B.S., geology, California State University, 1983

Professional Experience: planetary geologist, Jet Propulsion Laboratory, NASA, National Aeronautics and Space Administration (NASA), 1983–1998; program executive, Office of Space Science and Office of External Relations, 1998–2002; senior research scientist, European Space Agency, Noordwijk, Netherlands, 2002–2004

Adriana Ocampo is a planetary geologist with expertise in remote sensing. She is primarily involved in applying traditional geological principles to other celestial bodies, such as stars, moons, comets, and asteroids, and to objects on Earth that are of extraterrestrial origin, such as meteorite remnants. At the Jet Propulsion Laboratory in Pasadena, California, she was involved in the *Viking* space mission to explore Mars and the outer planets and in the *Hermes* mission to explore Mercury. In 1984, she produced the only available photo atlas of Phobos, one of the moons of Mars. For the *Mars Observer* mission, she was responsible for the thermal emission spectrometer, an instrument that was supposed to measure the heat produced by the planet, thus enabling cartographers to create accurate maps. Unfortunately, the mission failed in 1993, and the instrument remained untested. As a science coordinator for the *Galileo* mission to Jupiter, she was responsible for operation of one of the spacecraft's four remote sensing instruments, the Near-Infrared Mapping Spectrometer (NIMS), which measured reflected sunlight and heat from Jupiter's atmosphere to help scientists determine the planet's composition, cloud structure, and temperature.

In the early 1990s, Ocampo and her husband, Kevin O. Pope, were part of a team sent to the Yucatan to locate the crater made by an asteroid when it impacted the Earth at the time of the Cretaceous-Tertiary (KT) boundary (65 billion years ago). The theory was that the sulfurous cloud that rose from that impact circled the Earth, blocked the sun, and killed the vegetation on which the dinosaurs and large mammals fed, causing the extinction of both. Ocampo and Pope helped verify this theory, and their work was cited in Walter Alvarez's book, *T-Rex and the Crater of Doom* (1997).

Ocampo was born in Colombia and lived in Argentina until her family moved to California when she was a teenager. While still in high school, she obtained a summer job at the Jet Propulsion Laboratory and continued to work there during her last two years of high school and while she was in college. When she joined the lab as a full-time employee in 1983, she had already worked there 10 years. It was through her work there that she decided on a career in planetary geology. In recent years, she has held a number of high-profile research positions with NASA and other international space exploration agencies. In particular, she has worked in the recent Mars Program Science Division and has been active in educational outreach on programs related to science education for children and promoting women's careers in the sciences. In 2002, she was featured in a National Science Foundation program on "Women in Science." She is the recipient of the Woman of the Year Award in Science from the Comision Feminil (1992), Advisory Council for Women Award from the Jet Propulsion Laboratory (1996), and Science and Technology Award from the Chicano Federation (1997). In 2002, *Discover* magazine named her one of the "Top 50 Women in Science." In some sources, she appears under the name Adriana Ocampo Uria.

Further Resources

National Aeronautics and Space Administration. "Women of NASA: Adriana C. Ocampo." http://quest.arc.nasa.gov/people/bios/women/ao.html.

Ochoa, Ellen

b. 1958
Electrical Engineer, Astronaut

Education: B.S., San Diego State University, 1980; M.S., Stanford University, 1981, Ph.D., electrical engineering, 1985

Professional Experience: researcher, Imaging Technology Division, Sandia National Laboratory, 1985–1988; Group Leader to Chief, Intelligent Systems Branch, Ames Research Center, National Aeronautics and Space Administration (NASA), 1988–1990; astronaut, missions STS-56 (1993), STS-66 (1994), STS-96 (1999), and STS-110 (2002); deputy director, Flight Crew Operations, Johnson Space Center, 2002–2006; director, Flight Crew Operations, 2006–2007; deputy director, Johnson Space Center, 2007–

Ellen Ochoa is an electrical engineer and astronaut specializing in optics and optical recognition in robotics. While working at Sandia National Laboratory, she developed a process that implements optics for image processing that is normally done by computer. For example, one method she devised removes noise from an image through an optical system rather than using a standard digital computer to do the work. She was chosen for the astronaut program in 1990. Her first flight was in 1993 on the orbiter *Discovery* mission STS-56, which carried the Atmospheric Laboratory for Applications and Science, known as Atlas-2. She deployed instruments in space to enable scientists to look at the sun's corona, and she operated the robotic arm to deploy and retrieve the Spartan 201 satellite. Her second mission in 1994 continued the Spacelab

Astronaut Ellen Ochoa during test activities at the Kennedy Space Center in Florida, 2002. (NASA)

flight series to study the sun's energy during an 11-year solar cycle in order to learn how changes in the irradiance of the sun affect the Earth's environment and climate. For the 1999 *Discovery* mission, she was part of the team who made the first docking to the International Space Station. Her fourth flight was *Atlantis* in 2002, which again visited the International Space Station, and Ochoa was in charge of operating the robotic arm to move supplies and crewmembers.

While still a graduate student, Ochoa developed and patented a real-time optical inspection technique for defect detection, and she considers it her most important scientific achievement to date. She joined the technical staff in the Imaging Technology Division of Sandia after receiving her doctorate, and there her research centered on developing optical filters for noise removal and optical methods for distortion-invariant object recognition. She was co-author of two additional patents, one for an optical system for the nonlinear median filtering of images and another for a distortion invariant optical pattern recognition system. Since her flights as an astronaut, Ochoa has been a director of flight crew operations at NASA and, in 2007, was named Deputy Director of Johnson Space Center.

As the first female Hispanic astronaut, Ochoa quickly became a role model for young girls and Hispanics, and frequently speaks before school groups. She has received several awards, including the NASA Group Achievement Award for Photonics Technology (1991), NASA Space Flight Medal (1993), Women in Science and Engineering (WISE) Engineering Achievement Award (1994), National Hispanic Quincentennial Commission Pride Award (1990), *Hispanic* magazine's Hispanic Achievement Science Award (1991), and Congressional Hispanic Caucus Medallion of Excellence Role Model Award (1993). She is a member of the Optical Society of America and the American Institute of Aeronautics and Astronautics.

Further Resources

Kevles, Bettyann H. 2003. *Almost Heaven: The Story of Women in Space*. New York: Basic Books.

National Aeronautics and Space Administration. "Ellen Ochoa (Ph.D.)." http://www.jsc .nasa.gov/Bios/htmlbios/ochoa.html.

Ogilvie, Ida Helen

1874–1963
Geologist

Education: A.B., Bryn Mawr College, 1900; student, University of Chicago, 1900–1901; Ph.D., geology, Columbia University, 1903

Professional Experience: lecturer, geology, Barnard College, 1903–1905, tutor and instructor, 1905–1912, assistant to associate professor, 1912–1938, professor, 1938–1941; farm owner and operator

Concurrent Positions: director, Women's Agricultural Camp, 1917–1920

Ida Ogilvie helped expand science education for women as the founder and first chair of Barnard College's department of geology in 1903, one of the first such programs in a women's college. Her own research focused on glacial geology and petrology (rock origins), and she conducted research and mapping expeditions in Maine, New Mexico, California, New York, and Mexico.

Ogilvie attended schools in Europe before enrolling at Bryn Mawr, where she worked with **Florence Bascom** in the new geology department. After she received her doctorate from Columbia University, she was appointed the first lecturer in geology at Barnard (Columbia's women's college) in 1903, and then advanced through the faculty ranks over the next 35 years from tutor to associate professor. Throughout her entire tenure at Barnard, she was chair of the geology department, but did not become a full professor until just a few years before her retirement. She had an interest in farming and established a Women's Agricultural Camp in Bedford, New York, recruiting female students to work there during World War I. She later purchased a 660-acre farm in Germantown, New York, where she bred cattle and horses.

Ogilvie was only the second woman elected a fellow of the Geological Society of America. She also was a member of the American Association for the Advancement of Science, the Ecological Society of America, and the New York Academy of Sciences.

Osborn, Mary Jane (Merten)

b. 1927
Molecular Biologist, Biochemist

Education: B.A., physiology, University of California, Berkeley, 1948; Ph.D., biochemistry, University of Washington, 1958

Professional Experience: postdoctoral fellow, microbiology, New York University School of Medicine, 1959–1961, instructor, 1961–1962, assistant professor, 1962–1963; assistant to associate professor, molecular biology, Albert Einstein College of Medicine, 1963–1968; professor, microbiology, University of Connecticut Health Center, 1968–

Mary Osborn was the first person to demonstrate the mode of action of a major cancer chemotherapeutic agent called *methotrexate*, an agent that also opposes the physiological effects of folic acid. She is best known for her research into the biosynthesis of a complex polysaccharide known as *lipopolysaccharide*, which is a molecule that is essential to bacterial cells and is responsible for major immunological reactions and for the bacteria's characteristic toxicity. She thus helped to identify a potential target for the development of new antibiotics and chemotherapeutic agents, especially for leukemia. She entered college as a pre-med student, but by her senior year she realized she was more interested in research than in treating patients. Her thesis research examined the functions of the vitamins and enzymes whose action depended on folic acid. As a postdoctoral student, she moved into the biosynthesis of lipopolysaccharide.

Molecular biologist Mary Osborn is one of the pioneers of immunofluorescence microscopy, a method for the observation of cell structure. (Micheline Pelletier/Sygma/Corbis)

Osborn was elected to membership in the National Academy of Sciences in 1978. She has served on numerous commissions of the National Institutes of Health, the American Heart Association, and the National Academy of Sciences, and from 1980 to 1986, she was a member of the prestigious National Science Board, the board that advises the National Science Foundation. She is a fellow of the American Academy of Arts and Sciences and a member of the American Association for the Advancement of Science, American Society of Biological Chemists (president, 1981), American Chemical Society, American Society for Biochemistry and Molecular Biology (president 1981), Federation of American Societies for Experimental Biology (president, 1982), and American Society for Microbiology.

Further Resources

University of Connecticut Health Center. Faculty website. http://grad.uchc.edu/faculty/bios/osborn.html.

Ostrom, Elinor

b. 1933
Economist

Education: B.A., political science, University of California, Los Angeles, 1954, M.A., 1962, Ph.D., political science, 1965

Professional Experience: visiting assistant professor, government, Indiana University, Bloomington, 1965–1966, assistant professor and graduate advisor, political science, 1966–1969, associate professor, 1969–1974, professor, 1974–1991, Arthur F. Bentley Professor of Political Science, 1991–

Concurrent Positions: co-director of the Center for the Study of Institutions, Population, and Environmental Change (CIPEC), Indiana University, 1996–2006; co-director, Workshop in Political Theory and Policy Analysis, Indiana University, Bloomington, 1973–2009, senior research director, 2009–; professor (part-time), School of Public and Environmental Affairs, Indiana University; founding director and research professor, Center for the Study of Institutional Diversity, Arizona State University, Tempe

Elinor (Lin) Ostrom is a social scientist who won the Nobel Prize in Economics in 2009 for her research on the development, self-governance, and collective action of small communities. Ostrom, the first woman to win in Economic Sciences since the prize was added in 1968, shared

Elinor Ostrom was awarded the 2009 Nobel Prize in Economic Sciences, the first woman to win the prize in this category. (The Nobel Foundation. Photo: Ulla Montan)

the award with Oliver Williamson of the University of California, Berkeley. Trained as a political scientist, Ostrom's research has focused on integrating political and economic concerns to understand how communities come together to manage resources (both natural and political) and to understand the relationship between these practices and political, economic, and ecological sustainability. Her goal is to understand what kinds of policy initiatives and institutions best support local needs. She has authored, co-authored, or edited numerous books, including, *Governing the Commons: The Evolution of Institutions for Collective Action* (1990); *Institutional Incentives and Sustainable Development: Infrastructure Policies in Perspective* (1993); *The Samaritans' Dilemma: The Political Economy of Development Aid* (2005); and *Seeing the Forest and the Trees: Human-Environment Interactions in Forest Ecosystems* (2005).

Ostrom earned a bachelor's degree in political science from UCLA in 1954, but then moved to Boston to work in a law firm for three years. She returned to Los Angeles to continue her education, earning a master's degree and then doctorate in political science in 1965 with a thesis on water management. At UCLA, she also met her future husband, political scientist Vincent Ostrom. The couple spent time in Washington, D.C., before Vincent joined the faculty at Indiana University, where Elinor taught introductory courses in American government before also being offered a tenure-track position. The Ostroms researched police forces for what was eventually a comparative study of 80 major U.S. urban centers. Their analysis focused on the importance of information and coordination at the local and even neighborhood levels, as opposed to management from above of a larger decentralized force. Frustrated with the difficulty of conducting research across several disciplines (political science, economics, sociology), the Ostroms founded the Workshop in Political Theory and Policy Analysis in 1973, which now brings together researchers and projects across the social and natural sciences. Elinor Ostrom went on to study other types of community initiatives and aid efforts, such as in farming, forestry, and fishing.

Elinor Ostrom was elected to the National Academy of Sciences in 2001, and is a fellow of the American Academy of Arts and Sciences, American Philosophical Society, and American Academy of Political and Social Science. She has been a member of the American Political Science Association (vice president, 1975–1976; president, 1996–1997), Public Choice Society (president, 1982–1984), Midwest Political Science Association (president, 1984–1985), Association for Politics and Life Sciences, and International Association for the Study of Common Property. Before being awarded the Nobel Prize in 2009, she received numerous other awards, honors, and recognitions, including the Thomas R. Dye Service Award of the Policy Studies Organization (1997), Frank E. Seidman Distinguished Award in Political Economy (1997), Lifetime Achievement Award of Atlas Economic Research Foundation (2003), John J. Carty Award for the Advancement of Science from the National Academy of Sciences (2004), James Madison Award of the American Political Science Association (2005), Sustainability Science Award of the Ecological Society of America (2005), Cozzarelli Prize of the National Academy of Sciences (2006), William Riker Award for Understanding Institutional Diversity from the American Political Science Association (2006), Galbraith Award of the American Agricultural Economics Association (2008), and Fellowship from the Beijer Institute of Ecological Economics, Stockholm, Sweden (2007). She has received honorary doctorates from universities in Sweden, Norway, Germany, Canada, and the United States.

Further Resources

Indiana University. Faculty website. http://www.cogs.indiana.edu/people/homepages/ostrom.html.

Indiana University. "Elinor Ostrom: 2009 Nobel Laureate in Economic Sciences." http://www.iu.edu/nobel/.

Owens, Joan Murrell

b. 1933
Marine Geologist, Paleontologist

Education: B.A., art, Fisk University, 1954; M.S., counseling, University of Michigan, 1956; B.S., geology, George Washington University, 1972, M.A., 1976, Ph.D., geology, 1984

Professional Experience: reading therapist, Children's Psychiatric Hospital, University of Michigan, 1955–1957; reading specialist, English department, Howard University, 1957–1964; curriculum specialist, Education Services, Inc., 1964–1971; museum technician, Smithsonian Institute, 1972–1973; instructor to associate professor, geology and geography, Howard University, 1976–1995, associate professor, biology, 1991–1995

Joan Owens is a marine scientist who spent 20 years as an educator before returning to college to pursue a different career. She is considered the first African American woman to earn a doctorate in geology. Owens was fascinated with water animals as a child. Growing up in Florida, she had opportunities to see unusual species, such as manatees, alligators, and otters, and in high school, she dreamed of a career in marine science. However, when she entered Fisk University, she found that neither women nor African Americans were welcome in that field.

She majored instead in art, with a double minor in psychology and mathematics, and took education courses as well. She combined her interests in art and science by working as an illustrator for medical school students and then a hospital.

She was admitted to the graduate commercial art program in the School of Architecture at the University of Michigan, but she did not enjoy the program. A fellow graduate student suggested she transfer to the Bureau of Psychological Services, which is part of the School of Education, and she enjoyed her work there because she turned out to have a special talent for working with brain-damaged and emotionally disturbed children. She received a master's degree in counseling and joined the English Department at Howard University, where she taught remedial reading. When her husband's job took them to Massachusetts, she obtained a position with Education Services, Inc., where she developed new procedures and programs for teaching English to educationally disadvantaged high school students and designing college remedial programs, later transferring to the company's Washington, D.C., offices.

At the age of 37, Owens decided to change careers and returned to college to study her original passion, marine sciences. She earned another bachelor's at George Washington University and went on to receive her master's and Ph.D. in geology and zoology. For her thesis, she studied the Smithsonian Institution's collection of button deep-sea corals, and also worked at the Smithsonian as a museum technician. After completing her doctorate, she accepted a position at Howard University, where she taught geology, paleontology, and oceanography, and continued her research on the classification of corals with support from major oil companies.

Further Resources

Warren, Wini. 1999. *Black Women Scientists in the United States*. Bloomington: Indiana University Press.

P

Palmer, Katherine Hilton Van Winkle

1895–1982
Paleontologist

Education: B.S., University of Washington, Seattle, 1918; Ph.D., paleontology, Cornell University, 1925

Professional Experience: assistant geologist, University of Oregon, 1918–1922; fellow, geology, Cornell University, 1918–1920, assistant, paleontology and historical geology, 1921–1925, postdoctoral fellow, 1925–1927; curator, paleontology, Oberlin College, 1928; special lecturer, paleontology, Cornell, 1942–1945; technical expert, zoology, New York State Museum, 1945–1946; special technical expert, Redpath Museum, McGill University, 1950–1951; director, Paleontological Research Institute, Ithaca, New York, 1951–1978

Concurrent Positions: assistant professor, history of geology and paleontology, University of Washington, Seattle, 1922; special technical assistant, Provincial Museum, Quebec, 1951

Katherine Palmer was a notable paleontologist whose research interests were paleontology, stratigraphy, and conchology, in particular the study of mollusk fossils. For more than two decades, she was director of the Paleontological Research Institute in Ithaca, New York. After receiving her undergraduate degree from the University of Washington, she was appointed an assistant in geology at the University of Oregon in 1918. The same year, she became affiliated with the geology department of Cornell University, where her husband was a professor. She received her Ph.D. from Cornell in 1925 and continued to teach there until 1946. During these years, she held interim appointments at other colleges and museums, including the University of Washington, Oberlin College, McGill University, and the Provincial Museum of Quebec. In 1951, she became director of the Paleontological Research Institute, a position she retained until her retirement at the age of 83. Even after formally retiring, she continued her research into mollusk fossils until the time of her death in 1982.

Palmer received grants from several sources, including the National Science Foundation, and received numerous honors and awards. She received an honorary degree from Tulane University and was a fellow of the Paleontology Society, the Geological Society of America, and the American Association for the Advancement of Science. She was elected president of the American Malacological Union (1960), and for many years served as secretary-treasurer, then vice president (1958), then president (1960) of the Cushman Foundation, a foraminiferal research group. She was also a member of the American Association for the Advancement of Science and the American Association of Petroleum Geologists, and an honorary member of the Society of Economic Paleontologists and Mineralogists.

Pardue, Mary Lou

b. 1933
Cell Biologist, Geneticist

Education: B.S., College of William and Mary, 1955; M.S., radiation biology, University of Tennessee, 1959; Ph.D., biology, Yale University, 1970

Professional Experience: postdoctoral fellow, Institute of Animal Genetics, University of Edinburgh, 1970–1972; associate professor, biology, Massachusetts Institute of Technology, 1972–1980, professor, 1980–, Boris Magasnik Professor of Biology, 1995–

Concurrent Positions: instructor, molecular cytogenetics, Cold Spring Harbor Laboratory, 1971–

Mary Lou Pardue is a cell biologist who is known for her work in insect genetics. Her area of specialization is the structure and function of chromosomes in eukaryotic organisms (organisms whose DNA, or deoxyribonucleic acid, which provides the information for reproduction, is contained in their cells' nuclei, or centers). Her work excludes lower organisms such as bacteria and viruses, which are prokaryotic organisms (these have their genetic material located in the cell area surrounding the nucleus, the cytoplasm). Her studies have primarily centered on the breed of fruit fly known as *Drosophila melanogaster*. Because fruit flies have very short lifetimes, the rapid succession of fruit fly generations facilitates a time-saving study of genetic developments. An added benefit is that the flies' gene activity is similar, and therefore applicable, to that of higher organisms.

In the late 1960s, while a graduate student at Yale, she and her major professor developed a technique called "*in situ* hybridization" for localizing, with intact chromosomes, specific nucleotic sequences, which determine traits imparted during reproduction. These experiments were carried out using the chromosomes for the *Drosophila*'s salivary glands. The technique, which was designed to locate genes on the chromosomes, is used to identify the chromosomal regions of DNA that are complementary to specific nucleic acid molecules, or RNAs. Pardue later concentrated on heat-shock response, which refers to the effects of temperature on genetic activity. Studies of the fruit fly indicated that increases in its environmental temperature exceeding 10 degrees result in the suspension of some genetic activity. Her studies attempted to determine what genes are affected by the heat increase. In related research on stress response in insect muscle cells, she found that stress also resulted in suspending some genetic activity and the associated synthesis of proteins. This research is significant for its potential application in cancer treatment, for an understanding of how to turn genetic activity on and off carries potential benefits in establishing new forms of cancer therapy as well as other scientific/medical treatments.

Pardue was elected to membership in the National Academy of Sciences in 1983. She has received numerous awards, including the Esther Langer Award for Cancer Research (1977) and the Lucius Wilbur Cross Medal of Yale Graduate School (1989). She was a member of the Science Advisory Council of Abbott Laboratories, the American Cancer Society Advisory Committee on Nucleic Acids and Protein Synthesis, the Howard Hughes Medical Institute Science Review Board, and the National Research Council Board of Biology. She is a fellow of the American Association for the Advancement of Science and a member of the American Society for Cell Biology (president, 1985–1986), Genetics Society of America (president, 1982–1983), and American Academy of Arts and Sciences.

Further Resources

Wasserman, Elga. 2002. *The Door in the Dream: Conversations with Eminent Women in Science*. Washington, D.C.: Joseph Henry Press.

Massachusetts Institute of Technology. Faculty website. http://mit.edu/biology/www/facultyareas/facresearch/pardue.html.

Parsons, Elsie Worthington Clews

1875–1941
Anthropologist and Sociologist

Education: A.B., Barnard College, 1896, A.M., 1897, Ph.D., sociology, Columbia University, 1899

Professional Experience: high school teacher, 1897; fellow, Barnard College, 1899–1902, lecturer, sociology, 1902–1905; independent researcher and author, 1900–1941

Elsie Clews Parsons was recognized as one of the leading women anthropologists of the twentieth century, but began her career in sociology. Based on her early lectures at Barnard, she published her first book, *The Family* (1906), in which she used sociological arguments to make the case for equal opportunities for women. Her next work was a study of sexual practices associated with various religions, *Religious Chastity* (1913), which she wrote under a pseudonym. Her other major books of this period were *The Old Fashioned Woman* (1913), *Fear and Conventionality* (1914), *Social Freedom* (1915), and *Social Rule* (1916). Although she had thus published widely on sociological topics, the direction of Parsons's research changed around 1915 when, on a trip to the Southwest with her husband, she first encountered Native Americans. She then shifted from sociology to anthropology and began making annual extended field trips to the pueblos to interview and collect stores from native peoples.

It was considered scandalous at that time for a woman, especially a mother, to spend time in the field and live among the native peoples, as Parsons did. Her studies resulted in numerous papers and books, including her major work, the encyclopedic *Pueblo Indian Religion* (1939). She then extended her study of folklore to other groups, such as the Gullahs of the Carolina coastal islands. Parsons was interested in both original stories and cross-cultural influences. One of her last research projects was investigating the degree of Spanish influence on twentieth-century Native American cultures. Parsons's later ethnographic publications included *The Social Organization of the Tewa of New Mexico* (1929) and *Pueblo Indian Religion* (1939).

Parsons was born into a wealthy family and used her resources to pursue a life of independence and commitment to education and scholarship. She supported the founding of the Free School of Political Science (later the New School for Social Research) in New York City and was politically active as a feminist and as a pacifist during World War I. She was elected president of the American Folklore Society (1918–1920) and the American Ethnological Association (1923–1925), and was the first female president of the American Anthropological Association (1940–1941).

Further Resources

Deacon, Desley. 1997. *Elsie Clews Parsons: Inventing Modern Life*. Chicago: University of Chicago Press.

Jacobs, Margaret D. 1999. *Engendered Encounters: Feminism and Pueblo Cultures, 1879–1934*. Lincoln: University of Nebraska Press.

Lavender, Catherine J. 2006. *Scientists and Storytellers: Feminist Anthropologists and the Construction of the American Southwest*. Albuquerque: University of New Mexico Press.

Partee, Barbara (Hall)

b. 1940
Anthropologist, Linguist

Education: B.A., mathematics, Swarthmore College, 1961; Ph.D., linguistics, Massachusetts Institute of Technology, 1965

Professional Experience: assistant to associate professor, linguistics, University of California, Los Angeles, 1965–1971, associate professor, linguistics and philosophy, 1971–1972; associate professor, linguistics and philosophy, University of Massachusetts, 1972–1973, professor, 1973–1990, Distinguished University Professor, 1990–2003, emerita

Concurrent Positions: visiting professor, El Colegio de Mexico, Charles University, Prague, Moscow State University, Russian State Humanities University, University of Leipzig, University of Canterbury; fellow, Center for Advanced Study in Behavior Sciences, 1976–1977; member, Board of Managers, Swarthmore College, 1990–2002; honorary permanent guest professor, Charles University, Prague, 1995–

Anthropologist and linguist, Barbara Partee. (Courtesy of the University of Massachusetts)

Barbara Partee is known for her philosophical approach to linguistics, the science of language that includes phonetics, phonology, syntax, semantics, pragmatics, and historical linguistics. Her research combines mathematical and psychological or cognitive approaches to understanding the development of language and speech. She published *Fundamentals of Mathematics for Linguistics* (1978) and is the co-author of *Mathematical Methods in Linguistics* (1990). The updated volume included many of the new theories in linguistics, such as phonology and syntax, that had emerged since her first book, and included information on formal languages, grammars, and linguistic trees. Partee also co-edited *Properties, Types and Meaning* (1989), a two-volume set of essays on foundational and semantic issues in linguistics. A later book, *Quantification in Natural Languages* (1995), which she co-edited with her husband, Emmon Bach, and others, consists of 20 papers on the subject of semantics, which is the study of meaning, or the study of linguistics developed by classifying and examining change in meaning and form.

Partee's most recent book is *Compositionality in Formal Semantics: Selected Papers of Barbara Partee* (2004). She has been an invited guest lecturer at several international universities, and spends a significant amount of time conducting research in Russia, where she continues (post-retirement) to teach theoretical and applied linguistics at Russian State Humanities University and Moscow State University.

Partee was elected to membership in the National Academy of Sciences in 1989. She has received numerous grants for work, both individual and collaborative, including National Science Foundation grants, a National Endowment for the Humanities (NEH) fellowship, and the Max Planck Research Award. She has received honorary doctorates from colleges in the United States, Europe, and Russia. She has been a member or fellow of the Linguistics Society of America (president, 1986), American Philosophical Association, Association for Computational Linguistics, American Academy of Arts and Sciences, American Association for the Advancement of Science, and Massachusetts Academy of Sciences, and in 2002 was elected a Foreign Member of the Royal Netherlands Academy of Arts and Sciences.

Further Resources

University of Massachusetts. Faculty website. http://people.umass.edu/partee/.

Patch, Edith Marion

1876–1954
Entomologist

Education: B.S., University of Minnesota, 1901; M.S., University of Maine, 1910; Ph.D., Cornell University, 1911

Professional Experience: high school instructor, 1901–1903; instructor, entomology and English, University of Maine, 1903–1904, head, Department of Entomology, Maine Agricultural Experiment Station, 1904–1937

Edith Patch was an entomologist known as an international authority on the life histories and ecology of migratory aphids. She was one of the earliest critics of chemical pesticides. Patch grew up in Worcester, Massachusetts, and then on a 10-acre farm in Minnesota, where she spent her early years exploring nature and studying local wildlife and insects. As a high school student, she wrote a prize-winning report on the monarch butterfly. She went on to college at the University of Minnesota, where she became interested in aphids and their effect on agriculture. Like many college-educated women of her generation, the primary job available to her after graduation was as a school teacher, and she taught high school for two years while pursuing work as an entomologist. She secured a position at the University of Maine, where she remained affiliated for the remainder of her career, teaching English and entomology before founding and becoming head of the new Department of Entomology at the Maine Agricultural Experiment Station in Orono (which was affiliated with the University). She was named the director of the station in 1924. During this same time, she began her graduate education, earning a master's degree in 1910 and a doctorate from Cornell in 1911. Her scientific publications included 15 books and nearly 100 papers, and two new genera and several species of insects were named in her honor. Her most important publication was her 1938 *Food-Plant Catalogue of the Aphids of the World*, still an important reference book.

After Patch's formal retirement, she wrote a number of nature books for children. She was also committed to the science education of women and wrote papers on entomology as a career for women. She received an honorary doctorate from the University of Maine (1937), and was a member of the American Association for the Advancement of Science and the American Society of Naturalists, and served as president of both the Entomological Society of America (1936) and the American Nature Study Society (1937).

Further Resources

Bonta, Marcia. 1991. *Women in the Field: America's Pioneering Women Naturalists*. College Station: Texas A & M University Press.

Paté-Cornell, (Marie) Elisabeth Lucienne

b. 1948
Industrial Engineer

Education: B.S., mathematics and physics, University of Marseilles, 1968; M.S. and engineer degree, computer science and applied mathematics, Polytechnic

Institute of Grenoble, 1970 and 1971; M.S., operations research, Stanford University, 1972, Ph.D., engineering-economic systems, 1978

Professional Experience: engineer-economist, Régie Autonome des Transports de Paris, France (Transportation Planning), 1972–1973; assistant professor, civil engineering, Massachusetts Institute of Technology (MIT), 1978–1981; assistant professor, industrial engineering and engineering management, Stanford University, 1981–1984, associate professor, 1984–1991, professor, management science and engineering, 1991–1999, Burt and Deedee McMurtry Professor of Engineering, 1999–, professor and chair, management science and engineering, Stanford University, 2000–

Concurrent Positions: senior fellow, Institute of International Studies, Stanford University, 2000–

Elisabeth Paté-Cornell is known for her research in engineering systems analysis that is combined with economic analysis to assess risk and find realistic solutions to real-world problems. In pulling together what had been thought to be separate disciplines to offer a unique approach to problems, she has drawn on her studies in mathematics and physics, computer engineering with an electrical engineering component, economics, and operations research. Operations research (OR), which was developed around 1940 to 1945, during World War II, for military operations, is the analysis,

Industrial engineer, Elisabeth Paté-Cornell. (Courtesy of the Stanford University News Service Library)

usually involving mathematical treatment, of a process, problem, or operation to determine its purpose and effectiveness and to gain maximum efficiency. For her doctoral dissertation, she studied seismic risk from a public-policy viewpoint, looking at the costs and benefits of reducing earthquake risks. Her more recent research has had applications in industrial, medical, and government programs, including assessing National Aeronautics and Space Administration (NASA) shuttle missions and government intelligence regarding terrorist attacks. In 1998, she was a member of the Marine Board of the National Research Council (NRC), a committee on risk assessment and management of marine systems, such as offshore platforms.

Born in Senegal, she attended high schools in both Senegal and France, where she was influenced toward studying science by her engineer father. She earned degrees in mathematics, physics, and computer science before coming to Stanford University in California in 1971 to study in the interdisciplinary program of engineering and economic systems. She became a U.S. citizen in 1986, by which time she was an assistant professor at MIT, before returning again to Stanford as a faculty member in 1981. She has led the department of Management Science and Engineering at Stanford since 2000.

Paté-Cornell was elected to the National Academy of Engineering in 1995. She has served on the President's Foreign Intelligence Advisory Board (2001–2004) and has been a member of the Advisory Council of NASA's Jet Propulsion Laboratory and the Board of Trustees of the Aerospace Corporation since December 2004. She has also served as a member of the Army Science Board, the NASA Advisory Council, and the Air Force Scientific Advisory Board, and is a member of the Society for Risk Analysis (president, 1995) and the Institute for Operations Research and Management Science (INFORMS). She is the recipient of a Distinguished Achievement Award from the Society for Risk Analysis (2002) and was elected to the French Académie des Technologies in 2003.

Further Resources

Stanford University. Faculty website. http://www.stanford.edu/dept/MSandE/people/faculty/mep/index.html.

Patrick, Jennie R.

b. 1949
Chemical Engineer

Education: student, Tuskegee Institute, 1969–1970; B.S., chemical engineering, University of California, Berkeley, 1973; Ph.D., chemical engineering, Massachusetts Institute of Technology, 1979

Professional Experience: research engineer, General Electric Research and Development Center, 1979–1983; project manager, Phillip Morris Company, 1983–1985; department manager, fundamental chemical engineering research, Rohm and Haas Company, 1985–1990; assistant to executive vice president, Southern Company Services, 1990–1993; 3M Eminent Scholar and Professor of Chemical Engineering, Tuskegee Institute, 1993–1997; senior consultant, Raytheon Engineers and Constructors (Washington Group International), Alabama, 1997–

Concurrent Positions: assistant engineer, Dow Chemical Company, 1972; Stauffer Chemical Company, 1973; Chevron Research, 1974; Arthur D. Little, 1975; adjunct professor, Rensselaer Polytechnic Institute, 1982–1985, and Georgia Institute of Technology, 1983–1987

Jennie Patrick is a chemical engineer, manager, and educator who has worked in a variety of research, industry, and academic settings. She was the first African American woman to earn a doctorate in chemical engineering, which she received at the Massachusetts Institute of Technology (MIT) in 1979. Her working-class parents emphasized to their five children that knowledge was an escape from poverty. Jennie attended segregated elementary and middle schools, but in high school, she was one of the first participants in an integrated school in her hometown in Georgia. She wanted to attend the integrated school because it had all the scientific equipment she needed for her studies, while the school for blacks had none. She entered Tuskegee Institute as a chemistry major but transferred to the University of California, Berkeley to complete her undergraduate degree. She began working for chemical companies to support herself while still in school. She then went to MIT to obtain her doctorate in chemical engineering with research on superheating, in which a liquid is raised above its boiling temperature but does not become a vapor. She investigated the temperature to which pure liquids and mixtures of two liquids could be superheated.

After receiving her Ph.D., Patrick joined the General Electric Research and Development Center, where her work involved research on energy-efficient processes for chemical separation and purification, particularly the use of supercritical extraction. She worked for several other corporations, as well as taking positions as an adjunct professor, before returning to academia full-time as an endowed chair and professor chemical engineering back at the Tuskegee Institute. At Tuskegee, she was committed to helping minority students find success, particularly in the fields of science and engineering. In 1997, she returned to industry as an engineering consultant at Raytheon.

Patrick received the Outstanding Women in Science and Engineering Award (1980) and the Black Achievers in Chemical Engineering Award of the American

Institute of Chemical Engineers (2008). She is identified in some sources as Jennie Patrick-Yeboah.

Further Resources

Williams, Clarence G. 2003. *Technology and the Dream: Reflections on the Black Experience at MIT, 1941–1999*. Cambridge, MA: MIT Press.

Patrick, Ruth

b. 1907
Botanist, Limnologist

Education: B.S., Coker College, South Carolina, 1929; M.S., University of Virginia, 1931, Ph.D., botany, 1934

Professional Experience: assistant, Coker College, 1929; assistant, research, Temple University, 1934; phycology researcher and volunteer curator, Academy of Natural Sciences, Philadelphia, 1933–1937, associate to assistant curator, Leidy Microscopical Collection, 1939–1947, chair (and founder), department of limnology, 1947–1973, curator, 1947–, Francis Boyer research chair, 1973–

Concurrent Positions: lecturer, Marine Biological Laboratory, Woods Hole, Massachusetts, 1951–1955; lecturer, botany, University of Pennsylvania, 1952–1970, adjunct professor, 1970–

Ruth Patrick is a botanist and limnologist, or hydrobiologist, a multidisciplinary scientist who studies freshwater ecosystems. Patrick's specific expertise has been on the biodynamic cycle of rivers, and on the taxonomy, ecology, and physiology of diatoms, a family of microscopic one-celled algae that is the basic food for many organisms in the freshwater ecology. She was employed as an assistant at Coker College and Temple University before receiving her doctorate in botany from the University of Virginia in 1934. Soon after, she began her long career with the Academy of Natural Sciences in Philadelphia, leading expeditions to build the world-renowned collection of the Diatom Herbarium and becoming founding chair and curator of a new Department of Limnology there in 1947 (a department now known as the Patrick Center for Environmental Research). Although she celebrated her one-hundredth birthday in November 2007, she has never formally retired and still maintains an affiliation with the Academy.

Patrick's invention of a device called the *diatometer* made it possible for the first time to determine accurately the presence of pollution in fresh water. For many years, she was a consultant for government and corporate projects, assessing the ecological impact of nuclear power plants, groundwater pollution, and acid rain. In 1975, she became the first woman to sit on the board of directors of the Du Pont company. Along with **Rachel Carson**, Patrick was among the scientists largely responsible for calling attention to such ecological concerns in the mid-twentieth century; she published a book on the topic, *Groundwater Contamination in the United States*, in 1983.

Patrick was elected to membership in the National Academy of Sciences in 1970 and received the National Medal of Science in 1996. She has received more than 25 honorary degrees and an astonishing list of awards and honors from government, industry, and citizen's groups. The most prestigious of these include a $150,000 John and Alice Tyler Ecology Award (1975), Public Service Award from the U.S. Department of the Interior (1975), Golden Medal of the Royal Zoological Society of Antwerp, Belgium (1978), Founders Award of Society of Environmental Toxicology and Chemistry (1982), Commonwealth of Pennsylvania Governor's Award for Excellence in the Sciences (1988), Benjamin Franklin Award for Outstanding Scientific Achievement from American Philosophical Society (1993), Lifetime Achievement Award from American Society of Limnology and Oceanography (1996), Mendel Medal from Villanova University (2002), Chairman's Medal of the Heinz Family Foundation (2002), and Lifetime Achievement Award from the National Council for Science and the Environment (2004). She has been a member of the Phycological Society of America (president, 1954–1957), American Society of Naturalists (president, 1975–1977), American Philosophical Society, Botanical Society of America, South Carolina Academy of Sciences, American Academy of Arts and Sciences, American Society of Limnology and Oceanography, American Institute of Biological Sciences, Ecological Society of America, and American Society of Plant Taxonomists.

Further Resources

Wasserman, Elga. 2002. *The Door in the Dream: Conversations with Eminent Women in Science*. Washington, D.C.: Joseph Henry Press.

Patrick Center for Environmental Research, Academy of Natural Sciences. "Dr. Ruth Patrick." http://www.ansp.org/research/pcer/rp/index.php.

Patterson, Flora Wambaugh

1847–1928
Plant Pathologist

Education: A.B. Antioch College, 1860; M.L.A., Cincinnati Wesleyan College, 1865, A.M., 1883; A.M., University of Iowa, 1895

Professional Experience: assistant, Gray Herbarium, Harvard University, 1895; private school instructor, 1896; assistant pathologist, herbarium, U.S. Department of Agriculture (USDA), 1896–1901, mycologist, pathological collections, Bureau of Plant Industry, 1901–1923

Flora Patterson was a plant pathologist whose research included fungal diseases of plants and insects and systemic mycology. She was only the second woman scientist employed by the USDA; the first was Effie (Southworth) Spalding. Patterson worked as an assistant at the Gray Herbarium at Harvard University and as a private school teacher before obtaining a position at the USDA in 1896, where she remained until retiring in 1923. One benefit for women scientists working at the USDA in the early twentieth century was that, unlike in many academic research labs, they were able to publish their research under their own names. Patterson published numerous papers on her mycological research in addition to the pamphlets she prepared for the USDA series. She was co-author of *Mushrooms and Other Common Fungi* (1915), with fellow mycologist **Vera Charles**, and she wrote a chapter on "The Plant Pathologist" for a 1920 guide to *Careers for Women* (edited by Catherine Filene).

After college, Patterson married and had two children. When her husband became debilitated and then died, Patterson was forced to find a way to support herself and her children. She returned to college and received another master's degree from Cincinnati Wesleyan, then on to continue her studies at the University of Iowa, where she became interested in botany. She moved to Massachusetts with her brother, and studied botany at Radcliffe for three years and became an assistant at the Gray Herbarium at Harvard. During this time, she also became interested in mycology and served as assistant editor of *Economic Fungi*. She received another master's degree from the University of Iowa in 1895 and began teaching biology at a private school in Boston. Soon after, she began working for the USDA as a vegetable pathologist and then as a mycologist overseeing collections for the new Bureau of Plant Industry.

Patterson was a member of the American Association for the Advancement of Science, the American Phytopathological Society, and the Botanical Society of America.

Further Resources

Rossman, Amy Y. 2002. "Flora W. Patterson: The First Woman Mycologist at the USDA." *The Plant Health Instructor.* APSnet Education Center. http://www.apsnet.org/education/feature/patterson/.

Payne, Nellie Maria de Cottrell

1900–1990
Entomologist and Agricultural Chemist

Education: B.S., Kansas State Agricultural College, 1920, M.S., 1921; Ph.D., zoology, University of Minnesota, 1925

Professional Experience: assistant zoologist and entomologist, Kansas State Agricultural College, 1918–1921; instructor, science and math, Lindenwood College, 1921–1922; assistant and librarian, entomology, University of Minnesota, 1925–1930, lecturer, 1933–1937; National Research Foundation fellow, University of Pennsylvania, 1925–1927; scientific staff, *Biological Abstracts*, 1927–1933; assistant research entomologist, American Cyanamid Company, 1937–1943, entomologist, 1943–1944, zoologist, 1944–1957; literature chemist, Velsicol Chemical Corporation, 1957–1971; consultant

Concurrent Positions: National Research Council fellow, zoology, University of Pennsylvania, 1925–1927; research investigator, University of Vienna and University of Berlin, 1930–1931

Nellie Payne was an entomologist and agricultural chemist whose research interests included hydroid pigments, hibernation and low-temperature effects in insects, and the mathematics of population growth. She had a varied career, involving both academic and corporate appointments. She was employed as an assistant zoologist and entomologist while she was working toward both her bachelor's and master's degrees at Kansas State. She taught for one year in chemistry and mathematics at Lindenwood College, then received an appointment as assistant entomologist while she completed her doctorate in invertebrate zoology at the University of Minnesota. After positions as a fellow at the University of Pennsylvania and a member of the scientific staff of the major index *Biological Abstracts*, she returned to Minnesota as a lecturer for five years. She was appointed entomologist and zoologist in research at American Cyanamid in 1937. In 1957, she accepted a position as a literature chemist at Velsicol Chemical, then became a consultant starting in 1971. Payne also worked for the Entomological Society of America. Prior to the 1960s, many women scientists were employed as indexers and abstracters rather than in research positions in industry. Today, corporations hire both men and women scientists in their information centers to keep abreast of both the internal and external research data.

Payne was elected a fellow of the American Association for the Advancement of Science, the Entomological Society of America, and the American Institute of Chemists. She also was a member of the American Chemical Society, the Biometric Society, the Zoological Society of America, and the New York Academy of Sciences.

Payne-Gaposchkin, Cecilia Helena

1900–1979
Astronomer

Education: A.B., natural sciences, Newnham College, Cambridge University, 1923; Ph.D., astronomy, Radcliffe College, 1925

Professional Experience: National Research Fellow, Harvard University, 1925–1927, astronomer, Harvard College Observatory, 1927–1938, Phillips Astronomer, 1938–1967, Phillips Professor and Chair, astronomy, Harvard University, 1956–1967; staff member, Smithsonian Astrophysical Observatory, 1967–1979

Cecilia Payne-Gaposchkin, an authority on variable stars and galactic structure, was the first woman to achieve the rank of full professor at Harvard. Early in her career, she developed new techniques for ascertaining stellar magnitudes from photographic plates. She applied these techniques to a large collection of photographic plates dating back to 1890 that were stored at the observatory. In the mid-1930s, she concentrated on the study of variable stars. Her research team made several million observations over the entire sky. She often collaborated with her husband, Sergei I. Gaposchkin, and other staff members, and published more than 300 papers on galactic structure and novae. In addition to her scientific publications, she was the author of several books, including *Variable Stars* (1938), *Stars in the Making* (1952), *Variable Stars and Galactic Structure* (1954), and *Galactic Novae* (1957).

After receiving her undergraduate degree from Cambridge in 1923, she won a National Research Fellowship to study at Radcliffe and to work at the Harvard College Observatory, where she spent her entire career. In 1925, she was the first scholar at Radcliffe to receive a doctorate in astronomy, changing the career pattern for women astronomers (many of whom received degrees in physics) and broadening their research and employment opportunities. She continued working at the observatory and was appointed a permanent member of the staff in 1927. At the time, there were numerous other prominent women astronomers working at Harvard, including **Annie Jump Cannon**, **Antonia Maury**, and others. Payne-Gaposchkin was eventually promoted to full professor of astronomy and chaired the department at Harvard. After she retired in 1967, she became a staff member at the Smithsonian Astrophysical Observatory.

Payne-Gaposchkin received the first Annie J. Cannon Prize of the American Astronomical Society (AAS) (1935) and was the first woman to give the Henry Norris Russell Prize Lecture of the AAS (1976), the Society's highest honor for lifetime achievement in astronomy. She received honorary doctorates from Wilson College (1942), Smith College (1943), Western College (1951), Cambridge University (1952), Colby College (1958), and Women's Medical College of Philadelphia (1961). She was a member of the American Astronomical Society, the American Philosophical Society, the American Academy of Arts and Sciences, and the Royal Astronomical Society.

Further Resources

Byers, Nina and Gary A. Williams. 2006. *Out of the Shadows: Contributions of Twentieth-Century Women to Physics*. New York: Cambridge University Press.

Payton, Carolyn (Robertson)

1925–2001
Psychologist

Education: B.S, home economics, Bennett College, 1945; M.S., clinical psychology, University of Wisconsin, Madison, 1948; Ed.D., counseling and administration, Teachers College, Columbia University, 1962

Professional Experience: instructor, psychology, Livingstone College, North Carolina, 1948–1953; dean of women and instructor, psychology, Elizabeth City State Teachers College, North Carolina, 1953–1956; associate professor, psychology, Virginia State College, 1956–1959; assistant professor, psychology, Howard University, 1959–1964; Chief Field Selection Officer, U.S. Peace Corps, 1964–1966; deputy director, Peace Corps Eastern Caribbean Section, 1966–1971; assistant professor and director, Counseling Services, Howard University, 1971–1977; director, U.S. Peace Corps, 1977–1978; dean, counseling and career development, Howard University, 1978–1995

Carolyn Payton was a psychologist known for her work in counseling and career development, and served for one year as the first black and the first female director of the U.S. Peace Corps. When the Peace Corps was formed in 1961, it was charged with sharing technical skills with requesting countries. Trained volunteers spent two years in host countries working primarily in the areas of agriculture, rural development, health, and education. At first, the corps sent volunteers to Latin America, Africa, and the Middle East, but after 1990 and the end of the Cold War, Eastern Bloc countries also began requesting volunteers. Payton joined the Peace Corps in 1964 as a field selection officer and progressed in rank until she was deputy director of the Eastern Caribbean Section in 1966. She returned to Howard University to teach until 1977, when she was named by President Carter director of the Peace Corps. At that time, most recruits were experienced, highly skilled persons who could fill the specialized needs of developing countries; however, they tended to "teach down" to the people they were sent to help. Payton planned a program to train the volunteers to be better teachers and planned to recruit more blacks, women, and college graduates from varied backgrounds for the program. The Peace Corps was no longer an autonomous organization, however, and it was being administered by the American Council to Improve Our Neighborhoods (ACTION), whose head did not agree with her plans. Payton was forced to resign. However, her resignation had a positive impact in that President Carter restored the Peace Corps to an independent agency in 1981.

Payton worked to promote world understanding through cross-cultural interactions in both public and private forums. She was convinced that the inequalities in America were related to worldwide problems of poverty, hunger, and illiteracy, and was committed to the idea that professional scientists had an ethical imperative to work for social justice. She published a 1984 article in *American Psychologist* entitled "Who Must Do the Hard Things?," in which she argued that the discipline of psychology must have application to social problems and policy. She urged psychologists to "place our talents, our expertise, and our energy in the service of our conscience as well as our discipline." She was involved in the Public Policy Committee of the American Psychological Association (APA) and supported psychological research and education through the establishment of a scholarship fund at her alma mater, Bennett College.

Payton was a fellow of the APA and was awarded the APA's Distinguished Professional Contributions Award (1982) and the APA Committee on Women in Psychology Leadership Citation Award (1985). The APA honored her again in 1997 with the Award for Outstanding Lifetime Contribution to the field of psychology.

Further Resources

Keita, Gwendolyn P. 2001. "Carolyn Robertson Payton (1925–2001)." *The Feminist Psychologist.* 28(3). Newsletter of the Society for the Psychology of Women, Division 35

of the American Psychological Association. (Summer 2001). http://www.psych.yorku
.ca/femhop/Carolyn%20Robertson%20Payton.htm.

O'Connell, Agnes N. and Nancy Felipe Russo, eds. 1988. *Models of Achievement: Reflections of Eminent Women in Psychology*. Vol. 2. Hillsdale, NJ: Lawrence Erlbaum Associates.

Pearce, Louise

1885–1959
Pathologist

Education: A.B., physiology, Stanford University, 1907; student, Boston University School of Medicine, 1907–1909; M.D., Johns Hopkins University School of Medicine, 1912

Professional Experience: intern, Johns Hopkins Hospital, 1912; fellow, Rockefeller Institute for Medical Research, 1913–1923, associate member, 1923–1951

Concurrent Positions: visiting professor, syphilology, Peiping Union Medical College, China, 1931–1932; president, Women's Medical College of Philadelphia, 1946–1951

Louise Pearce was one of the foremost American women scientists of the early twentieth century and one of the principal figures in developing the drug tryparsamide to control African sleeping sickness. Her results, in collaboration with pathologist Wade Hampton Brown, were published in the *Journal of Experimental Medicine* in 1919, and she went to Africa in 1920 to supervise tests of the drug on humans. She spent her entire career at the Rockefeller Institute for Medical Research after receiving her medical degree from the Johns Hopkins University School of Medicine. Her other work included the biology of infectious and inherited diseases, such as syphilis and smallpox. In her study of syphilis in rabbits, she found that it closely resembled the human variety. The observations were therefore valuable to students of immunity and to physicians engaged in treating syphilitic patients. She and her collaborators found a tumor in rabbits that was capable of being grown in a laboratory and transplanted. The Brown-Pearce tumor was subsequently studied in cancer laboratories throughout the world. The breeding program and studies led the research team to isolate a virus similar to human smallpox when an epidemic of rabbit pox nearly destroyed the carefully developed rabbit colony. In the 1930s, the team enlarged its breeding program for rabbits, and by 1940, more than two dozen hereditary diseases and deformities were represented in the rabbit colony. Unfortunately, many of Pearce's files were destroyed after her death, and she had not completed writing up the results of all of her research.

Pearce also worked to advance the cause of women in medicine and science, and served as a member of the board of the Women's Medical College of Philadelphia from 1941 to 1946, and as president from 1946 to 1951. She also served on the scientific advisory council of the American Social Hygiene Association. She received several honors from the Belgian government for her work on sleeping sickness in the Belgian Congo (now Zaire), including the Ancient Order of the Crown, membership in the Belgian Society of Tropical Medicine, and the King Leopold II Prize in 1953.

Further Resources

National Institutes of Health. "Dr. Louise Pearce." Changing the Face of Medicine: Celebrating America's Women Physicians. National Library of Medicine, National Institutes of Health. http://www.nlm.nih.gov/changingthefaceofmedicine/physicians/biography_248.html.

Peckham, Elizabeth Gifford

1854–1940
Arachnologist and Entomologist

Education: B.A., Vassar College, 1876, A.M., 1889; Ph.D., Cornell University, 1916

Professional Experience: independent researcher

Elizabeth Peckham was an early entomologist and taxonomist recognized by her contemporaries for her research on spiders and wasps, fields known as *arachnology* and *hymenoptera*. Before receiving her doctorate, she collaborated and co-authored numerous papers and articles with her husband, entomologist George Williams Peckham, a high school biology teacher and public-library director with a medical degree. George Peckham was an innovator in emphasizing scientific research in secondary school, and the couple met when she came to work in his high school laboratory. They lived and worked in Wisconsin, and many of their publications were issued by organizations such as the Natural History Society of Wisconsin, the Wisconsin Geological Survey, and the Wisconsin Academy of Sciences, Arts, and Letters. Elizabeth Peckham had a solid educational background, with undergraduate and master's degrees from Vassar. After George Peckham's death in 1914, she returned to New York to pursue a doctorate from Cornell University, which housed one of the preeminent programs in entomology in the nation. She earned her Ph.D. from Cornell in 1916, at the age of 62.

In 1898, the Peckhams published a book, *On the Instincts and Habits of Solitary Wasps*. Influenced by the new theories of Charles Darwin on adaptability and variability within species, they also published research in the new field of insect psychology, emphasizing insect behavior and not just physical characteristics in some of the first papers on the mental powers of spiders and courtship and sexual selection among insects. Elizabeth Peckham was listed as the primary author of their 1905 book, *Wasps Social and Solitary*, which details their firsthand observations of wasp communities and the working habits of wasps. Among their important discoveries detailed in this book was the use of tools by one species of wasps.

Elizabeth and George Peckham have a distinguished legacy as early arachnologists. A genus of jumping spiders, *Peckhamia*, is named in their honor, as well as 20 individual species and subspecies. The Peckham Society was founded in 1977 to honor their work and to bring together both amateur and professional scientists interested in studies of salticid, or jumping spiders.

Further Resources

Bonta, Marcia. 1991. *Women in the Field: America's Pioneering Women Naturalists*. College Station: Texas A & M University Press.

The Peckham Society. http://peckhamia.110mb.com/.

Peden, Irene (Carswell)

b. 1925
Electrical Engineer, Radio Scientist

Education: B.S., University of Colorado, 1947; M.S., Stanford University, 1957, Ph.D., electrical engineering, 1962

Professional Experience: junior engineer, Delaware Power and Light Company, 1947–1949; junior engineer, Aircraft Radio Systems Laboratory, Stanford Research Institute, 1949–1950, research engineer, 1950–1952, antenna research group, 1954–1957; research engineer, Midwest Research Institute, 1953–1954; research assistant, Hansen Laboratory, Stanford University, 1958–1961, acting instructor in electrical engineering, 1959–1961; assistant to associate professor, University of Washington, Seattle, 1961–1971, professor, 1971–, associate dean of engineering, 1973–1977, associate chair, Electrical Engineering Department, 1983–1986

Irene Peden is a specialist in radio science and electromagnetic waves who conducted geophysical studies of radio wave propagation through the Antarctic ice pack, and she was the first American woman scientist to live and work in the interior of that continent. At the Byrd Antarctic Research Station in the 1970s, she developed new methods to analyze the deep glacial ice by studying the effect it has on radio waves directed through it, and she has continued this line of research by studying certain properties in the lower ionosphere over Antarctica. She developed the methodology for her own experiments and invented the mathematical models needed to study and interpret the data the team collected. She and her students were the first researchers to measure many of the electrical properties of Antarctic

Electrical engineer and radio scientist, Irene Peden. (Courtesy of University of Washington/ UnivPhoto)

ice and to describe important aspects of very low frequency (VLF) propagation over long paths in the polar region. Later, she turned her attention to subsurface exploration technologies, using very high frequency (VHF) radio waves to detect and locate subsurface structures and other targets.

Although women scientists from other countries had been conducting research at their countries' research stations in Antarctica for a number of years, American women were excluded from the U.S. station before Peden applied to go in 1970. The U.S. Navy was in charge of the research station and was responsible for transportation to and from the area, plus any travel within Antarctica, and the Navy argued that the weather was too harsh and the living quarters inadequate for women. Even when Peden received a grant from the National Science Foundation (NSF) for research on Antarctic ice, only her male graduate students could visit the site. Finally, under pressure from the NSF, the Navy approved Peden to make the trip in 1970 with the requirement that she have another female scientist accompany her. Peden described her experiences in Barbara Land's 1981 book, *The New Explorers*.

Peden later served as a Division Director at NSF and served on the Polar Research Board of the National Academy of Sciences. She was also a council member for the International Arctic Research Center in Fairbanks, Alaska.

She was elected to the National Academy of Engineering in 1993. She has received numerous awards for her research, including the Society of Women Engineers Achievement Award (1973), U.S. Army's Outstanding Civilian Service Medal (1987), and Centennial Medals from the Institute of Electrical and Electronics Engineers (1984) and the University of Colorado (1988), and was named to the Hall of Fame of the American Society for Engineering Education. She is a fellow of the Institute of Electrical and Electronics Engineers (IEEE), from whom she has received numerous honors, including the 2000 Third Millennium Medal and the 2000 Distinguished Achievement Award of the IEEE Education Society. She is a member of the American Association for the Advancement of Science, the Explorers' Club, the American Geophysical Union, the New York Academy of Science, and the Society of Women Engineers.

Further Resources

University of Washington. Faculty website. http://www.ee.washington.edu/faculty/peden/.

Shoemaker, Brian. 2005. "Dr. Irene Peden, 8 May 2002." Interview. Polar Oral History Program. Byrd Polar Research Center Archival Program. The Ohio State University Libraries. http://hdl.handle.net/1811/6058.

Peebles, Florence

1874–1956
Zoologist

Education: A.B., Goucher College, 1895; Ph.D., Bryn Mawr, 1900

Professional Experience: assistant, biology, Bryn Mawr College, 1897–1898; instructor, Goucher College, 1899–1902, associate professor, 1902–1906; lecturer, Bryn Mawr, 1913; professor, biology, Newcomb College, Tulane University, 1915–1917; associate professor, physiology, Bryn Mawr, 1917–1919; professor, biology, California Christian (Chapman) College, 1928–1942

Florence Peebles was recognized for her work on tissue regeneration in both plants and animals. Her research included the morphology of regeneration, growth and development, and the embryology of chicks. She conducted early research at Woods Hole Marine Biological Laboratory and was appointed an assistant in biology at Bryn Mawr while completing her doctorate, which she received in 1900. She also completed coursework at the Universities of Halle and Munich and postdoctoral work at the Naples Zoological Station and several European universities. She received her undergraduate degree at the Women's College of Baltimore

(Goucher College) and in 1899 was appointed as an instructor. She was promoted to associate professor in 1902. Between 1898 and 1927, she worked five times at the Naples Zoological Station, and 10 times at the Woods Hole Marine Laboratory between 1895 and 1924. Since she was recognized as an important contributor to scientific literature, she received support from fellowships for many of these research sessions. She held a position at Newcomb College in 1915, returning to Bryn Mawr in 1917. She moved to California and established a bacteriology department at California Christian College (now known as Chapman College) in 1928 and a biology department in 1935. Peebles continued teaching and research even after her formal retirement, establishing a biology laboratory at Lewis and Clark College in Portland, Oregon, in 1942. Both the lab and a science scholarship fund at Lewis and Clark are named in her honor.

Peebles received an honorary LL.D. from her alma mater, Goucher College, in 1954. She was a member of the American Association for the Advancement of Science and the American Society of Naturalists.

Pennington, Mary Engle

1872–1952
Chemist, Food Scientist

Education: certificate of proficiency, University of Pennsylvania, 1892, Ph.D., 1895

Professional Experience: fellow, botany, University of Pennsylvania, 1895–1897; fellow, physiological chemistry, Yale University, 1897–1898; researcher, University of Pennsylvania, 1898–1901; director, chemical laboratory, Women's Medical College of Pennsylvania, 1898–1906, lecturer, 1898–1906; owner, Philadelphia Chemical Laboratory, 1901–1905; director, bacteriological laboratory, Philadelphia Health Department, 1904–1907; bacteriological chemist, Bureau of Chemistry, U.S. Department of Agriculture (USDA), 1907–1908, chief, Food Research Laboratory, 1908–1919; director, research and development, American Balsa Company, 1919–1922; consultant, 1922–1952

Mary Pennington was a chemist and authority on food refrigeration, chemico-bacteriology of milk, and the chemistry, bacteriology, and histology of fresh and frozen foods. She developed methods for preserving dairy products and standards for milk inspection that were later employed throughout the country. She conducted a series of studies that led to methods of processing, storing, and shipping food that greatly increased its quality and availability. During World War I, she devised standards for railroad refrigerator cars that were used nationally. After several years working in various academic and government positions, she established her own consulting firm in 1922, where she specialized in food handling, storage, and transportation for the next 30 years. She did original research on frozen foods, earning her the nickname in one article of the "Ice Lady."

Pennington faced various hurdles as a woman scientist. Although she completed the requirements for a B.S. at the University of Pennsylvania, as a woman she was given only a certificate of proficiency instead of a degree. She received her doctorate and went on to Yale for another year of study in physiological chemistry. Unable to find a regular position, she briefly operated her own laboratory for chemical analysis, the Philadelphia Chemical Laboratory. She later secured a position as a bacteriological chemist with the USDA by taking the civil service exam under the name "M. E. Pennington" and accepting the job before the officials knew she was a woman. She used the same strategy when she was made chief of the Food Research Laboratory of the USDA in 1908.

Pennington was awarded the Garvan Medal of the American Chemical Society in 1940. She was the first woman member of the American Society of Refrigerating Engineers and was the first woman elected to the American Poultry Historical Society's Hall of Fame. She was a member of the American Association for the Advancement of Science, American Chemical Society, American Society of Biological Chemists, American Institute of Refrigeration, and Society of American Bacteriologists.

Pert, Candace Dorinda (Bebe)

b. 1946
Neurophysiologist, Pharmacologist

Education: B.A., biology, Bryn Mawr, 1970; Ph.D., pharmacology, School of Medicine, Johns Hopkins University, 1974

Professional Experience: postdoctoral research fellow, National Institutes of Health, Johns Hopkins University, 1974–1975; staff fellow, National Institute of Mental Health, 1975–1977, senior staff fellow, 1977–1978, research pharmacologist, 1978–1982, chief, Section on Brain Chemistry, 1982–1988; founder and scientific director, Peptide Design, 1987–1990; Chief Scientific Officer and Director, RAPID Pharmaceuticals, 2007–

Concurrent Positions: research professor, physiology and biophysics, Georgetown University School of Medicine, Washington, D.C.

Candace Pert is a neuroscientist and pharmacologist who is one of the world's foremost researchers on the chemistry of the brain and chemical receptors, which are the places in the body where molecules of a drug or natural chemical can be inserted, thus stimulating or inhibiting various physiological or emotional effects. As a graduate student, she was the co-discoverer, with her professor, of the brain's opiate receptors, the areas in which painkilling substances such as morphine can be inserted. Her work led to the discovery of endorphins, the naturally occurring substances manufactured in the brain that relieve pain and produce sensations of pleasure, by two Scottish scientists, who were awarded the Lasker Award in 1978. Pert's major professor at Johns Hopkins University medical school, neuroscientist Solomon Snyder, also shared the Award, but her name was omitted although she had conducted the early research and had already received her doctorate. This oversight created a controversy in the scientific world, because the Lasker Award is often an early step toward receiving the Nobel Prize.

Pert continued her work on neurotransmitters at the National Institute of Mental Health for a number of years. She examined Valium receptors in the brain and the receptors where the street drug PCP, or "angel dust," takes hold, and she also led the team that discovered peptide-T. She left the government laboratory to form her own company, Peptide Design, to encourage research on peptides, and worked there from 1987 to 1990. Pert's work on peptides and their receptors has led to a new area of research, the use of a chemical called peptide-T as a potential treatment for AIDS. She has evidence that the purified peptide-T prevents viruses from getting into cells by blocking the receptor sites on the cells, and there is also evidence that peptide-T reverses the symptoms of the disease. The first work was done in 1985, and clinical trials started in the early 1990s. She continues to investigate immune systems and the nature of HIV/AIDS as an adjunct professor of physiology at Georgetown University, and, in 2007, co-founded RAPID Pharmaceuticals.

Pert has explored the mind–body connection and the effect of brain chemicals on emotional and spiritual well-being. She wrote a book entitled *Molecules of Emotion: Why You Feel the Way You Feel* (1999), and co-authored *Everything You Need to Know to Feel Go(o)d* (2006). She also produced a guided imagery and music CD, *Psychosomatic Wellness: Healing Your Body-Mind*. She won the Arthur S. Fleming Award in 1979 for her research. She is a member of the American Society of Pharmacology and Experimental Therapeutics, American Society of Biological Chemists, and Society for Neuroscience. In 1980, Pert was the primary founder of Women in Neuroscience (WIN), a professional organization and committee of the Society for Neuroscience dedicated to assessing the status of women in the field.

Further Resources

Candace Pert, PhD. http://www.candacepert.com/.

Petermann, Mary Locke

1908–1975
Biochemist

Education: A.B., Smith College, 1929; Ph.D., physiological chemistry, University of Wisconsin, 1939

Professional Experience: technician, Yale University, 1929–1930; researcher, Boston Psychopathic Hospital, 1930–1934; postdoctoral researcher, physical chemistry, University of Wisconsin, 1939–1945; research chemist, Memorial Hospital, New York, 1945–1946; associate professor, biochemistry, medical school, Cornell University, 1952–1966, professor, 1966–1973

Concurrent Positions: professional assistant, Committee on Medical Research, 1942–1944; associate, Sloan-Kettering Institute for Cancer Research, 1946–1960, associate member, 1960–1963, member, 1963–1973

Mary Petermann was the first person to isolate and characterize animal ribosomes, which are the site of protein synthesis in cells. Her research included the physical chemistry of proteins, electrophoresis, plasma proteins, and ribosomes. Petermann showed an early interest in science, but was deterred from science as a career path. Undaunted, she became a chemistry major at Smith College, receiving her degree in 1929. She graduated from Smith with high honors and went on to work and conduct research at Yale and then at the Boston Psychopathic Hospital for four years investigating the acid–base balance of mentally unstable patients. In 1936, she entered the University of Wisconsin and received her doctorate in physiological chemistry in 1939. She remained at Wisconsin as a postdoctoral researcher and was the recipient of several prestigious fellowships, including a Rockefeller Foundation fellowship. In 1952, she was appointed associate professor of biochemistry at Cornell University medical school, and was the first woman promoted to full professor there in 1966.

Petermann was also a longtime member and researcher at Sloan-Kettering Institute for Cancer Research in New York City, and in 1963 became the first woman appointed a full member of the Institute. Because of her research at Sloan-Kettering, the ribosomes (previously known as "particles") were referred to by her colleagues as "Petermann's particles." However, simultaneous research was being conducted at the Rockefeller Institute by another researcher, George Palade, who received public credit as the "father of the particles." Palade, at least, acknowledged Petermann's work and privately gave her credit as "mother of the particles." In addition to nearly 100 scientific papers, Petermann was the author of a book, *The Physical and Chemical Properties of Ribosomes* (1964).

Petermann received the Sloan Award in cancer research (1963) and used the money to conduct research and give lectures in Europe. She also received the Garvan Medal of the American Chemical Society (1966) and a Distinguished Service Award from the American Academy of Achievement. In 1974, she organized the Memorial Sloan-Kettering Cancer Center Association for Professional Women and served as its first president. She was elected a fellow of the New York Academy of Sciences, and was a member of the American Society of Biological Chemists, the Harvey Society, and the Biophysical Society.

American biochemist Mary Locke Petermann. She was the first person to isolate and characterize animal ribosomes, which are the site of protein synthesis in cells. (Bettmann/ Corbis)

Phillips, Melba Newell

1907–2004
Physicist

Education: A.B., mathematics, Oakland City College, 1926; A.M., physics, Battle Creek College, 1928; Ph.D., physics, University of California, Berkeley, 1933

Professional Experience: high school teacher, 1926–1927; instructor, Battle Creek College, 1928–1930; research associate, University of California, Berkeley, 1933–1934, instructor, 1934–1935; research fellow, Bryn Mawr College, 1935–1936; fellow, Institute for Advanced Study, 1936–1937; instructor, physics, Connecticut College for Women, 1937–1938; instructor, Brooklyn College, 1938–1944, assistant professor, 1944–1952; lecturer, physics, Washington University, St. Louis, 1957–1962; professor, physics, University of Chicago, 1962–1972

Concurrent Positions: lecturer, University of Minnesota, 1941–1944; member, theoretical group, radio research laboratory, Harvard University, 1944; visiting professor, State University of New York, Stony Brook, 1972–1975; visiting lecturer, University of Science and Technology, Chinese Academy of Science, Beijing, 1980

Melba Phillips was a physicist whose research included theory of complex spectra and theory of light nuclei. She began her career at Oakland City College and retired as a professor of physics at the University of Chicago. She then was an instructor at Battle Creek College for three years after receiving her master's degree from that institution. She worked as an instructor at Berkeley after receiving her doctorate in 1933. Jobs were difficult to find during the Depression, but she was appointed an instructor at Connecticut College for Women for two years. She then moved to Brooklyn College in 1938, was promoted to assistant professor in 1944, and helped found the Federation of American Scientists in 1945.

At Berkeley, Phillips had worked under the direction of J. Robert Oppenheimer, who later became head of the Manhattan Project on development of the atomic bomb. In the 1930s, they had identified "the Oppenheimer-Phillips effect" to explain the behavior of the nuclei of radioactive hydrogen atoms. Despite her accomplishments of nearly 15 years at Brooklyn College, she was fired in 1952 for refusing to testify about the Manhattan Project before the McCarthy-era U.S. Senate subcommittee on internal security. Brooklyn College later publicly apologized to Phillips, but by then she had retired from the University of Chicago, where she had spent 10 years as a professor. She co-authored two textbooks: *Principles of Physical Science* (1957) and *Classical Electricity and Magnetism* (1955; rev. ed., 2005). After her formal retirement in 1972, she continued to teach for several years as a visiting lecturer at the State University of New York, Stony Brook, and at the Chinese Academy of Science in Beijing.

Phillips was especially active with the American Association of Physics Teachers (AAPT) throughout her career, serving as the first female president of the AAPT (1966–1967) and later acting executive officer (1975–1977). She received numerous awards from the AAPT, including a Distinguished Service Citation (1963) and the Oersted Medal (1974), and she was the first recipient of the Melba Newell Phillips Award (1982), established in her honor. She also received the Compton Award of the American Institute of Physics (1981), an Outstanding Teaching Award in Undergraduate Physics from Vanderbilt University (1988), and the Joseph Burton Forum Award of the American Physical Society (2003). She was a fellow of the American Physical Society and a member of the American Association for the Advancement of Science.

Further Resources

University of Chicago News Office. "Melba Phillips, Physicist, 1907–2004." http://www-news.uchicago.edu/releases/04/041116.phillips.shtml.

Pitelka, Dorothy Riggs

1920–1994
Zoologist

Education: B.A., zoology, University of Colorado, Boulder, 1941; Ph.D., zoology, University of California, Berkeley, 1948

Professional Experience: assistant, zoology, University of California, Berkeley, 1941–1943, 1945–1946, lecturer, 1949–1952, assistant research zoologist, 1953–1960, associate research zoologist, 1960–1966, research zoologist, 1966–1984, adjunct professor of zoology, 1971–1984

Concurrent Positions: fellow, University of Paris, 1957–1958

Dorothy Pitelka conducted research on protozoa, single-cell organisms, in order to understand other simple organisms, such as cancer-causing viruses. Her research interests included ultrastructure, function, and carcinogenesis in mammary glands; epithelial cell differentiation in cell culture; interactions of epithelium and stroma; and the ultrastructure and morphogenesis of protozoa. She was one of the early biologists (and one of the first at Berkeley) to use the new electron microscope, and in addition to her scientific papers, she published an early book, *Electron-Microscopic Structure of Protozoa*, in 1963. She isolated and studied mammary-gland cells at the University of California, Berkeley's Cancer Research Laboratory and was one of the first researchers to identify congenitally transmitted tumor viruses. She served on the editorial boards of the *Journal of Protozoology, Journal of Morphology*, and *Transactions of the American Microscopical Society.*

Pitelka was born in Turkey. Her family moved to the United States when she was a young child and settled in Colorado, where she completed her undergraduate degree in zoology at the University of Colorado. She went on to receive a Ph.D. in zoology from the University of California, Berkeley in 1948, and spent the remainder of her career at Berkeley. At Berkeley, she met and married her husband, also working on a Ph.D. in zoology. She spent her entire career at Berkeley, first as a research fellow and lecturer in zoology before being promoted through the ranks as a research scientist. She was supervisor of the electron microscope and also taught as an adjunct professor before retiring in 1984. In the 1950s, she spent a year conducting research in Paris as a fellow of the U.S. Public Health Service's National Cancer Institute.

Pitelka was elected the first woman president of the Society of Protozoologists (1964–1967). She was a member of the American Association for the Advancement of Science, the American Society for Cell Biology, the American Association of Cancer Research, and the Tissue Culture Association. She was also elected an honorary member of the Societe Francaise des Protistologues.

Further Resources

University of California. "Dorothy Riggs Pitelka, Zoology, Berkeley: 1920–1994." http://content.cdlib.org/xtf/view?docId=hb5g50061q&doc.view=frames&chunk.id=div00079&toc.depth=1&toc.id=.

Pittman, Margaret

1901–1995
Bacteriologist

Education: A.B., Hendrix College, Arkansas, 1923; M.S., University of Chicago, 1926, Ph.D., bacteriology, 1929

Professional Experience: principal and instructor, Galloway Woman's College, 1923–1925; fellow, Influenza Commission, Metropolitan Life Insurance Company, 1926–1928; research assistant, Rockefeller Institute for Medical Research, 1928–1934; assistant bacteriologist, New York State Department of Health, 1934–1936; associate bacteriologist, National Institutes of Health (NIH), U.S. Public Health Service, 1936–1941, bacteriologist, 1941–1947, senior bacteriologist, 1948–1954, principal bacteriologist, 1954–1958, chief, Laboratory of Bacterial Products, Division of Biological Standards, 1958–1971, guest scientist, 1971–1972; guest scientist and consultant, Center for Biological Evaluation and Research, Food and Drug Administration, 1972–1975

Concurrent Positions: consultant, World Health Organization, 1958–1959, 1962, 1969, 1971–1973; U.S. Pharmacopeia Panels, 1966–1975; guest lecturer, Howard University, 1967–1970

Margaret Pittman was known for her work standardizing the pertussis vaccine for whooping cough and for her international involvement in standardizing other vaccines, such as cholera and typhoid. Her work led to a dramatic decrease in whooping cough mortality by the 1950s. After several teaching and research positions that earned her renown as a bacteriologist, she joined the NIH/U.S. Public Health Service in 1936, where she had a long career, advancing quickly through the ranks to chief of the laboratory of bacterial products in 1958. After her official retirement from the NIH in 1971, she continued to consult and work for the Food and Drug Administration. She was a consultant for the World Health Organization numerous times and was active on the U.S. Pharmacopeia Panels.

Pittman grew up in rural Arkansas, where she and her sister assisted their father, a doctor, in his practice. She went on to study biology and mathematics at Hendrix College. She taught science and Spanish at Galloway Women's College in Searcy, Arkansas, and became principal of the school as well. She was saving her money to attend medical school, but decided to pursue graduate study in bacteriology at the University of Chicago, where she received a research fellowship to pay for her studies. She earned both a master's and a doctorate in bacteriology at Chicago, focusing on the bacterium responsible for pneumonia. She moved to New York, where she spent several years as a research scientist at the Rockefeller Institute for Medical Research studying bacterium responsible for childhood meningitis. During the Depression, she was lucky to continue her work with the New York State Department of Health before joining the NIH, where in 1958 she was named the first female laboratory chief. The NIH later named the Margaret Pittman Lectureship series in her honor.

Pittman received numerous awards and honors, such as the Superior Service Award (1963) and Distinguished Service Award from the U.S. Department of Health, Education, and Welfare (1968), the Federal Woman's Award (1970), and the Alice Evans Award from the American Society for Microbiology (1990). She also received an honorary doctorate from her alma mater, Hendrix College. She has been a member of the American Association for the Advancement of Science, the American Academy of Microbiology, the Society for Experimental Biology and Medicine, and the International Association of Biological Standardization.

Pool, Judith Graham

1919–1975
Physiologist

Education: B.S., biochemistry, University of Chicago, 1939, Ph.D., physiology, 1946

Professional Experience: assistant, physiology, University of Chicago, 1940–1942; instructor, physics, Hobart and William Smith Colleges, 1943–1945; assistant, physiology and pharmacology, toxicity laboratory, University of Chicago, 1946; research associate, Stanford Research Institute, 1950–1953; research fellow, Stanford University Medical Center, 1953–1956, research associate, 1957–1960, senior research associate, 1960–1970, senior scientist, 1970–1972, professor, medicine, 1972–1975

Concurrent Positions: Fulbright research scholar, Norway, 1958–1959

Judith Pool was renowned for her work in blood coagulation, which resulted in major contributions to the treatment of hemophilia. She developed the method of isolating the anti-hemophilic factor (AHF) in blood plasma that can be removed and frozen for later use, a method that is used for transfusions to correct bleeding in hemophiliac patients and improve their quality of life. This process, called *cryoprecipitation*, has since become the standard. She did not receive credit, however, for her participation as a graduate student in the development of a microelectrode to determine the electrical potential of a muscle fiber, later referred to as the *Ling-Gerard electrode*. (Another woman medical researcher, **Ida Hyde**, had also made early discoveries in this area.)

Pool became interested in science in high school and studied biochemistry as an undergraduate at the University of Chicago. She worked as a research assistant before following her husband, a political science professor, to Hobart and William Smith Colleges in New York, where she taught physics. She returned to Chicago to complete requirements for her doctorate in physiology, which she received in 1946. She held temporary teaching and research positions before moving to the Stanford Research Institute as a research associate in 1950. She then became a fellow in the school of medicine at Stanford University, where she switched from muscle physiology to research on blood. She was senior scientist before being promoted to full professor of medicine in 1972, just three years prior to her death.

Among her numerous honors, the National Hemophilia Foundation established a Judith Graham Pool Postdoctoral Research Fellowship in her name. She received the Murray Thelin Award of the National Hemophilia Foundation (1968), the Elizabeth Blackwell Award of Hobart and William Smith Colleges (1973), and a Professional Achievement Award from the University of Chicago (1975). She was president of the Association for Women in Science in 1971, and was a member of the American Association for the Advancement of Science, American Physiological Society, and Society for Experimental Biology and Medicine, and chair of Professional Women of Stanford University Medical Center.

Poole, Joyce

b. 1956
Wildlife Biologist

Education: B.A., biological sciences, Smith College, 1979; Ph.D., animal behavior, Cambridge University, 1983

Professional Experience: researcher, Amboseli Elephant Research Project, 1974–1990; coordinator, elephant conservation and management, Kenya Wildlife Service, 1990–1994; consultant and independent researcher, 1994–2000; founder and director, ElephantVoices, 2000–

Concurrent Positions: research director, Amboseli Elephant Research Project, 2002–2007

Joyce Poole is one of the world's authorities on the African elephant. Along with her colleague, **Cynthia Moss**, she has made several significant contributions to our knowledge of elephants. In particular, she and Moss were the first to recognize that male African elephants experience *musth*—an aggressive period of increased sexual activity—just as Asian elephants do. Poole is also credited for her research on vocalization among elephants and the discovery that elephants communicate in sound ranges that are below what the human ear is able to detect. She spearheaded the campaign against ivory poaching by providing counts and identification of individual elephants to the African Wildlife Fund and World Wildlife Fund, which led to African elephants being placed on the endangered species list in 1989.

Poole has lived in Africa most of her life. Her family first moved there in 1962 when her father was appointed director of the Peace Corps program in Malawi when she was six years old. After a brief return to the United States, the family moved in 1965 to Kenya for four years. Poole decided on biology as a career path after hearing primatologist Jane Goodall speak at the National Museum of Kenya about her research. Poole took a year off from her studies at Smith College when her father accepted a job in Nairobi with the African Wildlife Leadership Foundation. During this time she held an unpaid position with Cynthia Moss at the Amboseli Elephant Research Project. Poole helped compile vast records on all of the individual elephants in the preserve, identifying them through photographs of their ears and tusks. Poole returned to Smith the following year, but spent each summer and some of the Christmas holidays at Amboseli. Since she was concentrating on identifying the male elephants, she took note of their aggressive behavior during mating and identified it as musth, previously thought to be found only in Asian elephants. Poole used some of the early data for an undergraduate thesis at Smith and later expanded the data for her doctoral work at Cambridge University. During a postdoctoral fellowship at Princeton, Poole gained access to infrared sound equipment used to study whale vocalizations, and applied the technology to the study of elephant sounds.

Once she earned her doctorate, Poole decided to leave Moss's group and became elephant coordinator of the Kenya Wildlife Service with Richard Leakey; she resigned in protest when Leakey was fired in 1994, but continued her work as an independent elephant researcher. In 2000, she co-founded (with husband Petter Granli) the Savanna Elephant Vocalization Project, now known as ElephantVoices. In 2004, Poole left Africa after more than 30 years and set up headquarters for ElephantVoices in Norway. Her work has been profiled in documentaries and in wildlife and conservation magazines, such as *National Geographic* and *Smithsonian*, and she published an autobiography, *Coming of Age with Elephants* (1996). She has published numerous scientific papers and book chapters on the African elephants, and is a member of various advisory boards, including the Captive Elephant Management Coalition, Species Survival Network, and Amboseli Trust for Elephants, still run by her colleague, Cynthia Moss.

Further Resources

ElephantVoices. http://www.elephantvoices.org/.

Pour-El, Marian Boykan

1928–2009
Mathematician, Computer Scientist

Education: B.A., physics, Hunter College, 1949; M.A., mathematics, Harvard University, 1951, Ph.D., mathematical logic, 1958

Professional Experience: assistant professor, mathematics, Pennsylvania State University, 1958–1962, associate professor, 1962–1964; associate professor, mathematics, University of Minnesota, 1964–1968, professor, 1968–2000

Concurrent Positions: visiting faculty member, Institute for Advanced Study, Princeton, New Jersey, 1962–1964

Marian Pour-El was a mathematician who pioneered investigations on the interface among mathematical logic, mathematical analysis, computer science, and physics. Among the topics she studied in her research are the computability or noncomputability of the propagation of waves, the diffusion of heat, eigenvalues, and eigenvectors. She studied physics as an undergraduate at Hunter College in

New York and went on to graduate study at Harvard University. At Harvard in the 1950s, it was still unusual for a woman to prepare for a career as a mathematician or scientist. She recalled her first day in class at Harvard, when she was surrounded by empty chairs, as none of the other students, all men, would sit within two or three places of her, but she was soon accepted as a fellow student. After receiving a master's degree and then doctorate in mathematical logic in 1958, she joined the faculty of Pennsylvania State University. She received tenure a few years later, and then moved to the University of Minnesota, where she spent the remainder of her career.

Her husband, a biochemist, took a position in Illinois at the time she moved to the University of Minnesota (at that time, the University of Minnesota had a strong anti-nepotism rule, so it was not possible for both husband and wife to hold faculty positions there). Pour-El commented publicly on the dynamics of a long-distance marriage as a choice in order to pursue careers, and she has been committed to encouraging women to achieve satisfying careers in mathematics and science.

Pour-El was an invited lecturer on numerous occasions at colloquia, conferences, seminars, and symposia throughout Europe and the United States, and in Japan and China. She has also co-authored, with Ian Richards, *Computability in Analysis and Physics* (1989). She was a fellow of the American Association for the Advancement of Science, and a member of the American Mathematical Society, the Mathematical Association of America, and the Association for Symbolic Logic.

Further Resources

Henrion, Claudia. 1997. *Women in Mathematics: The Addition of Difference*. Bloomington: Indiana University Press.

Pressman, Ada Irene

1927–2003
Control Systems Engineer

Education: B.S., mechanical engineering, Ohio State University, 1950; M.B.A., Golden Gate University, 1974

Professional Experience: project engineer, Bailey Meter Company, 1950–1955; project engineer, Bechtel Power Corporation, 1955–1974, chief control engineer, 1974–1979, engineering manager, 1979–1987

Ada Pressman was an authority in power-plant controls and process instrumentation, and an expert in both fossil-fuel (coal, oil, and diesel) and nuclear power plants. She was especially known for the measures she devised to safeguard people working on the sites of nuclear power plants from the danger of radiation and to protect people living in the vicinity of the plant. She specialized in the area of shut-down systems for these plants and worked to find ways to ensure that a nuclear power plant's turbine, steam engine, and reactor work together properly and safely to generate electrical power. She contributed to the technology of emergency systems, including developing a secondary cooling system that operates from a diesel generator in the event of a primary power source loss. After working for Bailey Meter Company for a few years, Pressman accepted a position as a project engineer with Bechtel Corporation in Los Angeles, a company that manages nuclear power plants throughout the world. She advanced in responsibilities to the position of engineering manager in 1979. Before she retired in 1987, she managed 18 design teams for more than 20 power-generating plants scattered around the world.

In the 1970s, Pressman successfully campaigned to have control-systems engineering classified as a separate field with the state engineering board of California, and she was the first person to be registered in the new discipline; she was also a registered mechanical engineer in California and Arizona. She received several honors and awards, including a Distinguished Alumni Award of Ohio State University (1974), Society of Women Engineers Annual Achievement Award (1976), and E. G. Bailey Award of the Instrument Society of America (1985). She was a member of the American Nuclear Society, Instrument Society of America, and Society of Women Engineers (president, 1979–1980).

Further Resources

Hatch, Sybil E. 2006. *Changing Our World: True Stories of Women Engineers*. Reston, VA: American Society of Civil Engineers.

Prichard, Diana (Garcia)

b. 1949
Chemical Physicist

Education: L.V.N. (nursing) degree, College of San Mateo, 1969; B.S., chemistry and physics, California State University, Hayward, 1983; M.S., University of Rochester, 1985, Ph.D., chemical physics, 1988

Professional Experience: research scientist, Photo Science Research Division, Eastman Kodak Company, 1983–

Diana Prichard is a research scientist who conducts research on fundamental photographic materials for Eastman Kodak Company. She received praise for her graduate work on the behavior of gas phases at the University of Rochester, and the inventiveness of her project brought unusual attention and recognition by the scientific community. Her graduate work involved the high-resolution infrared absorption spectrum, which basically tells how much or what type of atoms or molecules are present, and she was able to construct the first instrument ever to be able to measure van der Waals clusters, which allows scientists to predict the behavior of gases. Van der Waals clusters are weakly bound complexes that exist in a natural state but are low in number, and Prichard's work allows scientists to produce these rare clusters by experimental methods in order to study them. Her graduate publications on the subject, such as a 1988 article in the *Journal of Chemical Physics*, have been cited in more than 100 subsequent publications.

In her position at Eastman Kodak, Prichard conducts basic studies in silver halide materials for photographic systems, and such work is in stark contrast to her early education. Although her parents had themselves received little education, they knew the value of education and supported her interest in learning. She received a degree in nursing and spent several years working and raising her children, but she had always been intrigued by the creativity required to do scientific research. She enrolled in California State University, Hayward, for her undergraduate degree, and then moved to the University of Rochester for her master's and doctorate degrees.

In 1992, she served on President Clinton's Transition Cluster for Space, Science, and Technology. She is active in encouraging students to undertake science and engineering careers, and founded a program in Rochester called Partnership in Education that provides Hispanic role models in the classroom to teach science and mathematics to students with only limited English proficiency. She also co-founded the Hispanic Organization for Leadership and Advancement (HOLA) at Eastman Kodak, and she is an active member of the Society of Hispanic Professional Engineers.

Prince, Helen Walter Dodson

1905–2002
Astronomer

Education: A.B., Goucher College, 1927; A.M., University of Michigan, 1932, Ph.D., astronomy, 1934

Professional Experience: assistant statistician, State Department of Education, Maryland, 1927–1931; assistant, astronomy, University of Michigan, 1932–1933;

instructor, astronomy, Wellesley College, 1933–1937, assistant professor, 1937–1945; associate professor to professor, astronomy and mathematics, Goucher College, 1945–1950; astronomer, McMath-Hulbert Observatory, University of Michigan, 1949–1957, associate director and professor of astronomy, 1957–1976; emerita professor and researcher, 1976–1979; consultant, Applied Physics Laboratory, Johns Hopkins University, 1979–2002

Concurrent Positions: summer observer, Maria Mitchell Observatory, Nantucket, 1934 and 1935; summer research assistant, Observatoire de Paris, 1938 and 1939; staff member, Radiation Laboratory, Massachusetts Institute of Technology, 1943–1945

Helen Dodson Prince spent 50 years observing solar activity, particularly the outbreak of solar flares and their effect on space, on light, and on the Earth's magnetic field. Prince was quoted in a 1963 *Time* magazine article, advising that the United States postpone missions to the moon until after 1972, when the then-current period of solar flare activity had passed and space travel would be safer and more effective. Prince worked as a statistician for a Maryland state agency before returning to graduate school to pursue a master's degree and then doctorate in astronomy. She joined the astronomy faculty at Wellesley after receiving her master's degree.

In 1934, while on the faculty at Wellesley, she completed her Ph.D. with a thesis entitled "A Study of the Spectrum of 25 Orionis." She spent several summers conducting research on solar flares and on the sun in residence at the Maria Mitchell Observatory in New England and at the Paris Observatory. Prince (then under the name Dodson) published the results of several years of her observations in the *Astrophysical Journal* in 1940. During World War II, she worked at the MIT Radiation Laboratory on the mathematical development of radar. She also taught astronomy and math at Goucher College before joining the staff of the McMath-Hulbert Observatory at the University of Michigan, where she was appointed full professor and associate director of the observatory. She published later articles on solar flares jointly with a colleague, Ruth Hedeman, and with the founder of the McMath-Hulbert Observatory, Robert McMath. Prince co-authored a biographical memoir of McMath for the National Academy of Sciences after his death in 1962.

Prince was also a revered teacher and received an honorary degree from Goucher College in 1952. Among her awards was the Annie Jump Cannon Prize of the American Astronomical Society (1955) and a Distinguished Achievement Award from the University of Michigan (1974). After retiring from the University of Michigan in 1979, Prince remained active as an independent consultant, working with the Applied Physics Laboratory at Johns Hopkins University. She was elected a fellow of the American Astronomical Society and held memberships in the American Association for the Advancement of Science and the American Geophysical Union. She married later in life, in her fifties, and many of her publications appeared under the name Helen Dodson

Prinz, Dianne Kasnic

1938–2002
Solar Physicist

Education: B.S., University of Pittsburgh, 1960; Ph.D., physics, Johns Hopkins University, 1967

Professional Experience: E. O. Hulbert fellow in physics and astronomy, University of Maryland, 1968–1971; research physicist, Space Science Division, U.S. Naval Research Laboratory, Washington, D.C., 1967–1968 and 1971–2001

Concurrent Positions: payload specialist, National Aeronautics and Space Administration (NASA), 1985

Dianne Prinz was known for her expertise in solar-terrestrial physics, and was a specialist in designing optical instrumentation. Her research includes infrared spectroscopy of atmospheric gases and ultraviolet spectroscopy of solar and atmospheric gases. She conducted research for 30 years at the U.S. Naval Research Laboratory, beginning in 1967, taking time in the 1980s for a special assignment with NASA as a payload specialist on Spacelab-2. She began NASA training in 1978 and was finally called up on the Spacelab-2 mission in 1985 as a liaison between the experimenters and NASA, defining page displays as they evolved, developing the mission timeline, and working up detailed ground command paths. As a specialist in optical instrumentation, she designed the optics and the flight software for instruments aboard Spacelab-2. The *Challenger* accident of 1986 delayed subsequent shuttle missions and cut short any further opportunities for Prinz to participate in space flight.

At the Naval Research Laboratory, she headed a research team on solar radiation and developed new instruments and data analysis software for measuring ultraviolet radiation in the Earth's upper atmosphere, a field of study known as "space weather." Her team took high-resolution images of the sun, and their Solar Ultraviolet Spectral Irradiance Monitor (SUSIM) has been used on space shuttle flights as well as other NASA and government research missions.

Prinz was a member of the American Geophysical Union, American Astronomical Society, Washington Academy of Science, and National Capital Section of the Optical Society of America (vice president, 1976). She received the Navy Award of Merit for Group Achievement (1985), the NASA Public Service Group Achievement Award (1987), and the Navy Meritorious Civilian Service Award (2001).

Further Resources

Cook, John William and Russell Alfred Howard. "Obituary: Dianne K. Prinz, 1938–2002." *Bulletin of the American Astronomical Society*. 35(5). (December 2003). http://adsabs.harvard.edu/abs/2003BAAS...35.1469C.

Profet, Margie

b. 1958
Biomedical Researcher, Evolutionary Biologist

Education: B.A., political philosophy, Harvard University, 1980; B.S., physics, University of California, Berkeley, 1985

Professional Experience: independent researcher and author

Margie Profet is an evolutionary biologist who has presented new theories relating to how humans adapt to their environment and, in particular, has challenged accepted theories on allergies, pregnancy sickness, and menstruation. Her own allergies to various foods and chemicals inspired her inquiries into an explanation for allergies. She published her early findings in a 1991 article entitled "The Function of Allergy: Immunological Defense against Toxins," in which she proposed that humans develop allergic reactions as a means of protecting the body from harmful toxins. She even noted that people with allergies are less likely to develop cancer than individuals without allergies, and believes that allergies are an internal warning device for the body. Another area of research was the cause of morning sickness during pregnancy; again, she theorized that the brain's ability to discern what is toxic becomes recalibrated during pregnancy so that almost any food or odor can cause an aversion. Her hypothesis is that all plants contain toxins and that pregnancy sickness is a natural defense mechanism that reduces the amount of toxins one ingests during the first trimester, the period when the embryo is particularly vulnerable to toxins that could cause birth defects. She presented her research in two books: *Protecting Your Baby-to-Be: Preventing Birth Defects in the First Trimester* (1995) and *Pregnancy Sickness: Using Your Body's Natural Defenses to Protect Your Baby-to-Be* (1997).

Profet next turned to an investigation into why women menstruate, and she presented the theory that sperm carry pathogens into the uterus, and that the menstrual flow allows the uterus to rid itself of bacteria and infection. Rather than being merely a monthly waste of blood and energy, Profet theorized that the myriad bacteria found in and around the genitals of both men and women hitch rides on sperm, thus gaining access to the uterus and fallopian tubes, and that menstruation in fact washes away the contaminants that could cause infection or infertility. She published her controversial theory in the September 1993 issue of *Quarterly Review of Biology* as, "Menstruation as a Defense against Pathogens Transported by Sperm."

Profet received two undergraduate degrees (in political philosophy and physics), but was not interested in the constraints of university research. Without an advanced degree or faculty position, she embarked upon a career as an independent researcher and evolutionary biologist, supporting herself with grants and various laboratory affiliations. Her article on menstruation led to a prestigious five-year MacArthur "genius" fellowship in 1993. In 2005, Profet disappeared while working at Harvard University and has not been seen since.

Further Resources

Martin, Mike. 2009. "Margie Profet's Unfinished Symphony: A Promising Scientist Vanishes Without a Trace." *Weekly Scientist*. (29 June 2009). http://weeklyscientist.blogspot.com/2009/07/margie-profets-unfinished-symphony.html.

Q

Quimby, Edith Hinkley

1891–1982
Radiological Physicist

Education: B.S., Whitman College, 1912; M.A., physics, University of California, Berkeley, 1916

Professional Experience: high school teacher, 1912–1914; assistant, physics, University of California, 1914–1915; assistant to associate physicist, New York City Memorial Hospital for Cancer and Allied Diseases, 1919–1942; assistant professor, radiology, Medical College, Cornell University, 1941–1942; associate professor, radiological physics, College of Physicians and Surgeons, Columbia University, 1942–1954, professor, 1954–1960

Edith Quimby was a pioneer in the new fields of radiology and nuclear medicine in the first half of the twentieth century. Her research helped physicians in the use of x-rays for diagnostic purposes and determining safe levels of radiation therapy for the treatment of cancer and other tumors. When she started working at Memorial Hospital in 1919, commercial radium had been in production in the United States for only six years. She was one of the scientists who brought the field to maturity; between 1920 and 1940, she published more than 50 papers describing the results of her research. She not only prepared data on radiation hazards and radiation safety, but also developed training courses in medical physics. She attended Whitman College in Walla Walla, Washington, on a full scholarship to study physics and mathematics. She taught high school for two years after graduating from college and then returned to the University of California, Berkeley on a physics scholarship, receiving her master's degree in 1916. At Berkeley, she met and married fellow physics student Shirley L. Quimby, who went on to teach at Columbia University. Edith followed her husband to New York, accepting a position as an assistant physicist at the new Memorial Hospital for Cancer and Allied Diseases, where she began her career in the medical use of x-rays and radiation. She was promoted to associate physicist in 1932, but she accepted a position as associate professor of radiological physics at Columbia's medical college in 1942. She was promoted to full professor in 1954 and retired in 1960.

At Columbia, Quimby helped found the Radiological Research Laboratory, where she researched radiation therapy for thyroid disease, brain tumors, and other diseases. Not surprisingly, her research into radioactive isotopes had implications for the U.S. government's World War II–era interest in the development of a nuclear bomb, and Quimby was involved in the Manhattan Project and worked as a consultant for the Atomic Energy Commission. She was also head of the National Council on Radiation Protection and Measurements. Over the course of her long career, Quimby published her findings in numerous scientific journals and was the author or co-author of three books: *Radioactive Isotopes in Clinical Practice* (1958), *Safe Handling of Radioactive Isotopes in Medical Practice* (1960), and *Physical Foundations of Radiology* (1970).

Quimby received honorary science doctorates from her alma mater, Whitman College (1940), and from Rutgers University (1957). She was the first woman (and still one of the few) to receive the Janeway Medal of the American Radium Society (1940), and was also the recipient of the Gold Medal of the Radiological Society of North America (1941), an Achievement Medal from the International Women's Exposition of Arts and Industries (1947), the Medal of the American Cancer Society (1957), the Gold Medal of the Inter-American College of Radiology (1958), and the Gold Medal of the American College of Radiology (1963). She was a fellow of the American Physical Society and of the American College of Radiology, and was a member of the American Roentgen Ray Society and the American Radium Society (vice president, 1929; president, 1954).

R

Ramaley, Judith (Aitken)

b. 1941
Endocrinologist, Reproductive Biologist

Education: B.A., Swarthmore College, 1963; Ph.D., anatomy, University of California, Los Angeles, 1966

Professional Experience: postdoctoral fellow, anatomy and physiology, Indiana University, 1967–1968, assistant professor, 1969–1972; assistant to associate professor, physiology and biophysics, University of Nebraska Medical Center, 1972–1978, professor, 1978–1982, assistant vice president for academic affairs, 1981–1982; vice president of academic affairs, State University of New York, Albany, 1982–1984, acting president, 1984–1985, executive vice president of academic affairs, 1985–1987; executive vice chancellor, University of Kansas, 1987–1990; acting president, State University of New York, Albany, 1990; president and professor, biology, Portland State University, Oregon, 1990–1997; president and professor, biology, University of Vermont, 1997–2001; assistant director, Education and Human Resources Directorate, National Science Foundation, 2001–2004; president, Winona State University, Minnesota, 2005–

Judith Ramaley is an endocrinologist whose specialty is the physiology of puberty and the control of male and female fertility. She has been prominent both in academic research and in administration, having now served as president or acting president of four major state universities. In addition to numerous scientific publications, she published two early books, *Progesterone Function: Molecular and Biochemical Aspects* (1972) and *Essentials of Histology* (1974, rev. ed., 1978), and edited a volume of papers from the American Association for the Advancement of Science on *Covert Discrimination of Women in the Sciences* (1978). Ramaley has remained committed to educational opportunity and science education. In addition to her administrative roles within academia, as assistant director of the Education and Human Resources Directorate of the National Science, she worked on initiatives for leadership education in science, engineering, technology, and mathematics. She has written dozens of papers and articles on higher education reform, responsibility, and opportunity.

Ramaley has been an active participant in the communities in which she has lived, involving herself in sports leagues, historical and cultural societies, Girl Scouts, Planned Parenthood, and other women's and family resources. She has been a member of the American Association of Colleges and Universities (AACU) board of directors in 1995, board member of the American Association of Higher Education, member of the National School-to-Work Advisory Board, member of the Advisory Council for the National Institute on Alcohol Abuse and Alcoholism, and fellow of the Margaret Chase Smith Center for Public Policy. In 2005, she was a visiting senior scientist at the National Academy of Sciences.

Ramaley is a fellow of the American Association for the Advancement of Science and a member of the American Association of Anatomists, Endocrine Society, Society for the Study of Reproduction, Society for Neuroscience, and American Physiological Society.

Further Resources

Winona State University. "Office of the President." http://www.winona.edu/president/.

Ramey, Estelle Rosemary White

1917–2006
Endocrinologist

Education: B.S., mathematics and biology, Brooklyn College, 1937; M.S., physical chemistry, Columbia University, 1940; Ph.D., physiology, University of Chicago, 1950

Professional Experience: teaching fellow, chemistry, Queens College, New York, 1938–1941; lecturer, biochemistry, University of Tennessee, 1942–1947; postdoctoral fellow and instructor, endocrinology, University of Chicago, 1950–1954, assistant professor, physiology, 1954–1958; assistant to associate professor, school of medicine, Georgetown University, 1956–1966, professor, physiology, 1966–1987, professor, biophysics, 1980–1987

Concurrent Positions: visiting professor, Stanford University, Harvard University, Yale University

Estelle Ramey researched endocrinology metabolism chiefly in the field of adrenal function, sex hormones, and insulin action. She began her teaching career at Queens College while completing work for her master's degree, which she received from Columbia University in 1940. She followed her husband's career to Knoxville, Tennessee, but when she first applied for a teaching job at the local university, she was told by the chairman that "he had never hired a woman, would never hire a woman, and I ought to go home and take care of my husband." After many male faculty members were called to duty in World War II, however, the same chairman called to offer her a teaching job. She stayed on at the University of Tennessee for five years. She went on to obtain her doctorate from the University of Chicago in 1950 and continued teaching there for several years as a U.S. Health Service postdoctoral fellow and then as the first female faculty member at the medical school. She accepted a position at Georgetown University medical school as assistant professor in 1956, and was promoted to associate professor in 1960 and professor in 1966. She was named professor of biophysics in 1980 and emeritus professor upon her retirement in 1987.

Ramey was committed to women's equality, in science and in society at large. Even after formally retiring, she continued to lecture, often donating her fees to women's organizations. She was a longtime member of the Association for Women in Science (AWIS) and founder of the AWIS Educational Foundation. As president of the AWIS (1972–1974), Ramey pressured the publisher of a standard medical school textbook to remove unnecessary photos of nude women from a new edition of the book. Her own research on sex hormones even had feminist implications in the 1970s, as she spoke out against people who would use "hormones" as a basis of sexism, rejecting the idea "that ovarian hormones are toxic to brain cells." She published more than 150 scientific papers or articles and was the co-author of *Electrical Studies on the Unanesthetized Brain* (1960).

Ramey was awarded numerous honorary doctorates, including one from her employer, Georgetown University (1977). Her other awards and honors include an Outstanding Alumna Award from the University of Chicago (1973), the Public Broadcasting Company Woman of Achievement Award (1984), and the National Women's Democratic Club Woman of Achievement Award (1993). In 1989, *Newsweek* magazine named her "one of 25 Americans who have made a difference." Ramey's expertise was widely sought, and she sat on the advisory boards of numerous government and medical institutions, including Planned Parenthood, the National Institutes of Health, the National Academy of Science, the Veteran's Administration for Women Veterans, and President Carter's Committee on the Status of Women. She was a member of several professional societies, including the American Physiological Society, the American Chemical Society, the Endocrine Society, the American Diabetes Association, and the American Academy of Neurology.

Further Resources

Fox, Margalit. 2006. "Estelle R. Ramey, 89, Who Used Medical Training to Rebut Sexism, Is Dead." *New York Times*. (12 September 2006). http://www.nytimes.com/2006/09/12/obituaries/12ramey.html.

Rand, (Marie) Gertrude

1886–1970
Psychologist

Education: A.B., experimental psychology, Cornell University, 1908; M.A. and Ph.D., psychology, Bryn Mawr College, 1911

Professional Experience: postdoctoral research fellow, Bryn Mawr College, 1911–1913, associate, experimental and applied psychology, 1913–1927; associate professor, research in ophthalmology, Wilmer Ophthalmological Institute, Johns Hopkins University School of Medicine, 1928–1932, physiological optics, 1932–1936, associate director, Research Laboratory of Physiological Optics, 1936–1942; research associate, ophthalmology, Knapp Foundation, Columbia University College of Physicians and Surgeons, 1943–1957

Gertrude Rand was an experimental psychologist and leading researcher in the field of physiological optics. In collaboration with her husband, Clarence E. Ferree (also her dissertation director), she developed numerous ophthalmological tools, including a way to map the retina for its perceptual abilities and sensitivity to color. The Ferree-Rand perimeter became an important tool for diagnosing vision problems. She and Ferree moved to Johns Hopkins in 1928, where she taught first in the area of research ophthalmology, then physiological optics, before becoming associate director of the research laboratory of physiological optics in 1936. Besides their academic work, the couple served as consultants on a variety of industrial lighting projects, including consulting for New York City on plans for glare-free illumination of the Holland Tunnel and for the U.S. government on night vision for the military. After her husband's death in 1942, she moved to Columbia University, where she resumed her earlier work on color perception. It was at Columbia that she and two colleagues developed plates for testing color vision and color blindness, a test known as the H-H-R (or Hardy-Rand-Rittler, for the collaborators) test.

Rand was the first woman elected a fellow of the Illuminating Engineering Society (1952) and she was the recipient of a Gold Medal from the Society (1963). She was also the first woman to win the Edgar Y. Tillyer Medal of the

Optical Society of America (1959), and in 1971, one of her students, Louise Sloan, became the second woman to receive the Tillyer Medal. Rand was a member of the American Association for the Advancement of Science and the American Psychological Association.

Ranney, Helen Margaret

1920–2010
Hematologist

Education: B.A., Barnard College, 1941; M.D., Columbia University, 1947

Professional Experience: assistant professor, clinical medicine, Columbia University, 1958–1960; associate professor, medicine, Albert Einstein College of Medicine, 1960–1965, professor, 1965–1970; professor, State University of New York at Buffalo, 1970–1973; chair, Department of Medicine, University of California, San Diego, 1973–1986, professor of medicine, 1973–1990, professor emeritus

Concurrent Positions: board member, Squibb Corporation, 1975–1989; distinguished physician, Veterans Administration Medical Center, San Diego, 1986–1991; staff member and consultant, Alliance Pharmaceutical Corporation, San Diego, 1991–2010

Helen Ranney was known for her research in abnormal hematology, the study of blood. Her research involved the relationship of hemoglobin and red cell membrane in sickle-cell disease and red cell survival. For many years, she was a major force in medical education, clinical hematology, and blood-related research and training, and for more than 40 years, her work extended into disciplines and directions as diverse as biochemistry, physical chemistry, immunology, metabolism, genetics, rheology, pharmacology, and analytical technologies. She received early renown for identifying the hereditary or genetic aspect of sickle-cell anemia, a disease that affects primarily African Americans.

Ranney began her college studies at Barnard, the women's annex of Columbia University. When she applied for graduate study at Columbia's College of Physicians and Surgeons in 1941, she was denied acceptance. It was not until after World War II that more women were needed and therefore admitted to such programs, and she was able to complete her M.D. at Columbia by 1947. She had a distinguished early teaching and research career at the Albert Einstein College of Medicine, where she founded a heredity clinic and trained important hematologists, and then at the State University of New York at Buffalo. In 1973, she became chair of the Department of Medicine at the University of California, San Diego, where there is now an endowed chair in her name. She authored a textbook, *Genetics in Hematology* (1990).

Ranney was elected to membership in the National Academy of Sciences in 1973. Among her awards are the J. M. Smith Prize of Columbia University (1955), the Dr. Martin Luther King, Jr., Medical Achievement Award (1972), Gold Medal of the College of Physicians and Surgeons (1978), and May H. Soley Research Award of the Western Society of Clinical Investigation (1987). She was a fellow of the American Association for the Advancement of Science and a member of the American Academy of Arts and Sciences, American College of Physicians, American Society of Clinical Investigation, and American Physiological Society, and was the first female president of both the American Society of Hematology (1974) and the Association of American Physicians (1984–1985). In 1979, she received an honorary doctorate from the University of Southern California.

Further Resources

Bunn, H. Franklin. "Helen Margaret Ranney: A Woman of Many Firsts." *The Hematologist*. American Society of Hematology. (1 March 2008). http://www.hematology.org/Publications/Hematologist/2008/1296.aspx.

National Institutes of Health. "Dr. Helen M. Ranney." Changing the Face of Medicine: Celebrating America's Women Physicians. National Library of Medicine, National Institutes of Health. http://www.nlm.nih.gov/changingthefaceofmedicine/physicians/biography_260.html.

Ratner, Sarah

1903–1999
Biochemist

Education: A.B., Cornell University, 1924; A.M., Columbia University, 1927, Ph.D., biochemistry, 1937

Professional Experience: assistant, pediatrics, Long Island College of Medicine, 1926–1930; assistant biochemist, College of Physicians and Surgeons, Columbia University, 1930–1931, teaching assistant, 1932–1934, Macy research fellow, department of biochemistry, 1937–1939, instructor, 1939–1943, associate, 1943–1946, assistant professor, 1946; assistant to associate professor, pharmacology, college of medicine, New York University, 1946–1954; associate member, division of nutrition and physiology, Public Health Research Institute of the City of New York, Inc., 1954–1957, member, division of biochemistry, 1957–1992

Concurrent Positions: Fogarty scholar in residence, National Institutes of Health, 1978–1979

Sarah Ratner was one of the leading researchers in the biochemistry of amino acids and protein metabolism. She did early work on acids and hormones excreted in urine and blood, but one of her most important discoveries was argininosuccinic acid, an indicator of a genetic defect related to neurological damage and even mental retardation. She was employed at Columbia University starting in 1930 as an assistant biochemist and advancing to assistant professor in 1946. A portion of this time was spent in completing her doctorate, which she received in 1937. Her slow advancement could be due to working during the years of the Depression, when faculty positions of any type were scarce, especially for women. She later taught pharmacology at New York University before accepting a position at the Public Health Research Institute of the City of New York in 1954, where she worked until her retirement in 1992 at the age of almost 90 years old.

Ratner served on the editorial boards of *Journal of Biological Chemistry* and *Analytical Biochemistry*. She received numerous awards, such as the Neuberg Medal (1959), the Garvan Medal of the American Chemical Society (1961), and the Freedman Award of the New York Academy of Sciences (1975). She was elected to membership in the National Academy of Sciences in 1974 and was awarded an honorary doctorate from the State University of New York, Stony Brook (1984). She was elected a fellow of the Harvey Society and of the New York Academy of Sciences. She was a member of the American Academy of Arts and Sciences, the American Society of Biological Chemists, and the American Chemical Society.

Further Resources

Bentley, Ronald. "Sarah Ratner. June 9, 1903–July 28, 1999." *Biographical Memoirs*. National Academies Press. http://www.nap.edu/readingroom.php?book=biomems &page=sratner.html.

Ray, (Marguerite) Dixy Lee

1914–1994
Zoologist

Education: B.A., zoology, Mills College, 1937, M.A., 1938; Ph.D., biological sciences, Stanford University, 1945

Professional Experience: public school teacher, 1939–1942; instructor, zoology, University of Washington, Seattle, 1945–1947, assistant to associate professor, zoology, 1947–1976; Assistant Secretary of State, International Environmental and Scientific Affairs' Bureau of Oceans, U.S. Department of State, 1975; governor, Washington State, 1977–1981

Concurrent Positions: director, Pacific Science Center, Seattle, Washington, 1963–1972; visiting professor, Stanford University, 1964; chief scientist, International Indian Ocean Expedition's *Te Vega*, 1964; consultant, Argonne National Laboratory and Livermore National Laboratory, 1987–1994

Dixy Lee Ray was trained as a marine biologist and later received recognition as the first female governor of the state of Washington. Her scientific research focused on crustacean and other invertebrates. She spent 30 years teaching zoology at the University of Washington, Seattle, during which time she also served as director of the Pacific Science Center, an institution committed to encouraging public interest in and awareness of science. In the early 1960s, Ray was a chief scientist for the International Indian Ocean Expedition, a multinational exploration of that ocean's marine environment. Ray was involved in many national and international projects in environmental science and policy issues. She was a consultant for the National Science Foundation, U.S. representative to the Organization for Economic Cooperation and Development for Science, member of the President's Task Force on Oceanography, and member and last chairperson of the U.S. Atomic Energy Commission under President Nixon. In this capacity, Ray was concerned about environmentally sound alternatives to fossil fuel, and she promoted the safety of nuclear power plants, a position that brought her into conflict with environmentalist groups. She published several articles and co-authored two books on environmentalism (both with Louis R. Guzzo): *Trashing the Planet* (1990) and *Environmental Overkill* (1994).

Marine biologist and environmental scientist Dixy Lee Ray was the only woman to chair the Atomic Energy Commission (AEC), appointed by President Richard Nixon in 1972, and, in 1976, became the first woman governor of Washington. (Washington State Archives)

Because of the nature of her work, spanning academic research to government policy to community development, she has been honored by various groups and was the recipient of several honorary degrees. A small selection of Ray's impressive awards includes a Guggenheim fellowship (1952), Foreign Fellow Award of the Danish Royal Society of Natural History (1965), Axel-Axelson Johnson Award from the Swedish Royal Academy of Science and Engineering (1974), Achievement Award of the American Association of University Women (1975), Abram Sacher Award from Brandeis University (1976), Walter H. Zinn Award of the American Nuclear Society (1977), Washington Award of the Western Society of Engineers (1978), Centennial Medallion Award of the American Society of Mechanical Engineers (1980), Outstanding Woman in Energy Award from Nuclear Energy Women (1981), Centennial Medal from the Institute of Electrical Engineers (1984), Woman of Achievement in Energy Award (1988), and being named among the One Hundred Honored Citizens at the State of Washington Centennial (1989). She appeared on the cover of *Time* magazine (December 12, 1977) and, in 1998, the American Society of Mechanical Engineers (ASME) established an annual Dixy Lee Ray Award for contributions to the field of environmental protection.

Further Resources

Pace, Eric. "Dixy Lee Ray, 79, Ex-Governor; Led Atomic Energy Commission." *New York Times*. (3 January 1994). http://www.nytimes.com/1994/01/03/obituaries/dixy-lee-ray-79-ex-governor-led-atomic-energy-commission.html.

Rees, Mina Spiegel

1902–1997
Mathematician

Education: A.B., Hunter College, 1923; A.M., Columbia University, 1925; Ph.D., mathematics, University of Chicago, 1931

Professional Experience: teacher, Hunter College High School, 1923–1926; instructor, mathematics, Hunter College, 1926–1932, assistant to associate professor, 1932–1943; principal technical aide, Applied Mathematics Panel, National Defense Research Committee, Office of Scientific Research and Development, 1943–1946; head, mathematics division, Office of Naval Research, 1946–1949, director, mathematics science division, 1950–1952, deputy science director, 1952–1953; professor, mathematics, and dean of faculty, Hunter College, 1953–1961; dean of graduate studies, City University of New York

Mathematician Mina Rees was the first woman president of the American Association for the Advancement of Science in 1971. (Bettmann/ Corbis)

(CUNY), 1961–1968, provost, graduate studies, president, CUNY Graduate School and University Center, 1969–1972

Mina Rees was a researcher of linear algebra, numerical analysis, and the history of computers, and helped set up programs for government support of mathematical research. She was employed by Hunter College for 35 years, starting as an instructor in mathematics in 1926 and rising through the ranks to full professor and then dean of faculty in the 1950s. During World War II, she took a leave from Hunter to work for the Applied Mathematics Panel of the Office of Scientific Research and Development (OSRD). She worked on military applications for jet rocket propulsion and high-speed computers, receiving certificates and medals of service from both the U.S. and British governments. After the war, she established the program in mathematics at the Office of Naval Research (ONR) and was the deputy science director there from 1952 to 1953. When the National Science Foundation was established in 1950, her ONR program for connecting government with academia was used as the model for government funding of mathematical and computer research. In 1953, she returned to Hunter College as professor and dean of faculty, then moved to CUNY as dean of graduate studies in 1961. She became founding president of the CUNY graduate school, where, in 1985, the Mina Rees Library was named in her honor.

Rees received many honorary degrees, honors, and awards. Among the latter were the President's Certificate of Merit (1958) and the first Award for Distinguished Service to Mathematics of the Mathematical Association of America (1962). She was the first female president of the American Association for the Advancement of Science (1971) and was a fellow of both the American Association for the Advancement of Science and the New York Academy of Sciences. She received honorary membership in the National Academy Sciences (NAS) when she was awarded the NAS Public Welfare Medal (1983). She was a member of the American Mathematical Society, the Mathematical Association of America, and the Society for Industrial and Applied Mathematics.

Further Resources

Williams, Kathleen Broome. 2001. *Improbable Warriors: Women Scientists and the US Navy in World War II*. Annapolis, MD: Naval Institute Press.

Reichard, Gladys Amanda

1893–1955
Anthropologist

Education: A.B., Swarthmore College, 1919; A.M., Columbia University, 1920, Ph.D., anthropology, 1925

Professional Experience: school teacher, 1909–1915; instructor, anthropology, Barnard College, 1923–1928, assistant to associate professor, 1928–1951, professor, 1951–1955

Concurrent Positions: Guggenheim fellow, Hamburg, Germany, 1926–1927

Gladys Reichard was an anthropologist known for her expertise in Navajo language and culture, but she studied other tribes. She spent her entire career at Barnard College, which for many years was the only anthropology department in a women's college in the United States. Starting about 1923, Reichard spent summers each year on Southwestern reservations learning languages, learning to weave, and observing daily life by living with families from time to time. Much of her work was financially supported by another female anthropologist, **Elsie Clews Parsons**. In 1934, Reichard made the first attempt to teach native speakers to write the Navajo language. Since Navajo society traditionally is matriarchal, women anthropologists were more successful than men in working with these tribes, and much of Reichard's work was focused on women's roles and contributions to native society. In addition to scientific articles, she published a number of books, including *Social Life of the Navajo Indians* (1928), which traced Navajo genealogy back several generations. She also published on textile production and designs, as well as books on *Navajo Religion: A Study of Symbolism* (1950), and *Navajo Grammar* (1951). The latter book on Navajo language was controversial in that she did not accept a method of transcription that was newer than the one she developed.

Reichard was a member of the American Ethnological Society (secretary, 1924–1926), the American Folklore Society (secretary, 1924–1935), and the American Association for the Advancement of Science.

Further Resources

University of South Florida. "Gladys Amanda Reichard (1893–1955)." Celebrating Women Anthropologists. http://anthropology.usf.edu/women/reichard/reichard.html.

Lavender, Catherine J. 2006. *Scientists and Storytellers: Feminist Anthropologists and the Construction of the American Southwest.* Albuquerque: University of New Mexico Press.

Reichmanis, Elsa

b. 1953
Computer Scientist, Organic Chemist

Education: B.S., chemistry, Syracuse University, 1972, Ph.D., organic chemistry, 1975

Professional Experience: intern, organic chemistry, Syracuse University, 1975–1976, Chaim Weizmann fellow of scientific research, 1976–1978; technical staff, organic chemistry, AT&T Bell Laboratories, 1978–1984, technical manager, Radiation Sensitive Material and Applications, 1984–1994, director, Polymer and Organic Materials Research, 1994–

Elsa Reichmanis is known for her contributions to the science of manufacturing integrated circuits, or computer chips, specifically her research centers on developing sophisticated chemical processes and materials for computer chips. She holds 11 patents, some of which are for the design and development of organic polymers, called *resists*, which are used in microlithography (the principal process by which circuits, or electrical pathways, are imprinted upon the tiny silicon chips used in computers). During the multistage process of chip manufacture, layers of resist material are applied to a silicon base and exposed to patterns of ultraviolet light. As portions of the resists harden, they become templates for the application of subsequent layers of positively and negatively charged semiconductors that serve as the channel through which electric current travels. As computer products have become smaller and smaller, it has become more and more of a challenge to develop materials and processes to manufacture them.

In addition to publishing more than 100 scientific papers, Reichmanis has edited four volumes for the American Chemical Society (ACS): *The Effects of*

Radiation on High-Technology Polymers (1989), *Polymers in Microlithography: Materials and Processes* (1989), *Irradiation of Polymeric Materials: Processes, Mechanisms, and Applications* (1993), and *Microelectronics Technology: Polymers for Advanced Imaging and Packaging* (1995). She also edited a volume of the proceedings of an International Society for Optical Engineering symposium, *Advances in Resist Technology and Processing VI* (1989).

Reichmanis was elected to membership in the National Academy of Engineering in 1995. She is a fellow of the Society of Women Engineers (SWE) and a member of the ACS (president, 2003), American Association for the Advancement of Science, Materials Research Society, and Society of Photo-Optical Instrumentation. She has received several awards, including *Research and Development Magazine's* R&D 100 Award for one of the 100 most significant inventions of 1992, SWE Annual Achievement Award (1993), American Society for Metals (ASM) Engineering Materials Achievement Award (1996), Photopolymer Science and Technology Award (1998), ACS Award in Applied Polymer Science (1999), and Perkin Medal (2001). She was a member of the Committee to Survey Materials Research Opportunities and Needs for the Electronics Industry of the National Research Council and the Air Force Science Advisory Board.

Further Resources

Bell Laboratories, Physical Sciences Research. "Elsa Reichmanis." http://www.bell -labs.com/org/physicalsciences/profiles/reichmanis.html.

Reinisch, June Machover

b. 1943
Psychologist

Education: B.S., New York University, 1966, M.A., Columbia University Teachers College, 1970, Ph.D., psychology, Columbia University, 1976; diplomate, American Board of Sexology, 1989

Professional Experience: instructor, psychology, Columbia University Teachers College, 1972, 1974–1975; staff research associate, psychiatry, University of California, Los Angeles School of Medicine, 1973–1974; assistant to associate professor, psychology, Rutgers University, 1975–1982; professor, psychology and psychiatry, Indiana University, 1982–1993; director and professor, Kinsey Institute for Research in Sex, Gender, and Reproduction, 1982–1993, director and professor emeritus, senior research fellow, and trustee, 1993–

Psychologist June Reinisch was the director of the Kinsey Institute for Research in Sex, Gender, and Reproduction from 1982–1993. (Douglas Kirkland/Corbis)

Concurrent Positions: adjunct assistant professor, psychiatry, College of Medicine and Dentistry of New Jersey, Rutgers University Medical School, 1976–1981, adjunct associate professor, psychiatry, 1981–1982

June Reinisch is a developmental psychobiologist who served as director of one of the most controversial social science institutes in the United States, the Kinsey Institute for Research in Sex, Gender, and Reproduction at Indiana University. Founder Alfred Kinsey's books, *Sexual Behavior in the Human Male* (1948) and *Sexual Behavior in the Human Female* (1953), helped demystify sex and make public discussion acceptable. The Kinsey Institute is an independent corporation and, during Reinisch's 11-year tenure as director, federal and private research grant funding increased tenfold; the library, archives, art collections, and research and administrative spaces were expanded, modernized, and renovated; the Institute's research became multidisciplinary in focus; and a public education program was instituted. A series of international multidisciplinary conferences led to the publication of four scholarly volumes on sex differences, adolescence and puberty, sexual orientation, and AIDS and sexuality, and "The Kinsey Report" regular column was published to inform the public of Institute research.

Increased public awareness brought increased criticism, and Reinisch defended the Institute from attacks by conservative political and religious forces as well as from academic critics. In the late 1980s, a university committee issued an unfavorable review of Reinisch's programs and requested her resignation. The board of trustees, however, supported her tenure as director for five more years and, after an investigation, the president of Indiana University apologized to Reinisch and publicly supported the accomplishments of her directorship. She retired in 1993 with the titles of Director Emerita and Senior Research Fellow.

Reinisch initially planned to be an elementary school teacher, but held a variety of jobs before returning to school to obtain a master's degree in psychology to enhance her career as a music business executive. After reading **Eleanor Maccoby**'s *The Development of Sex Differences* (1966), she became fascinated by the discussion of the effects of prenatal hormones on the development of gender and sex differences, and decided to pursue advanced studies at Columbia University. She taught at Rutgers University before moving to Indiana University as professor of psychology and psychiatry, and was chosen as the third director of the Kinsey Institute in 1982. She has published scientific articles in many leading journals as well as a book, *The Kinsey Institute New Report on Sex* (1990), which was translated into several languages. The volume was based on a national survey of American sexual knowledge, and addressed the public's questions with the most current scientific information. After leaving the Institute, Reinisch continued to work as an independent consultant and researcher for a variety of organizations, including the Institute of Preventive Medicine at Copenhagen University Hospital, Denmark and the Museum of Sex in New York City.

Among Reinisch's many awards are the Morton Prince Award from the American Psychopathological Association (1976), the Dr. Richard J. Cross Award for Outstanding Contributions to the Field of Human Sexuality from Robert Wood Johnson Medical School (1991), and an Award for Contributions to Sexology of the Society for the Scientific Study of Sex (1993). She is a fellow of the American Association for the Advancement of Science, American Psychological Association, and American Psychological Society, and a member of the American Association of Sex Educators, Counselors and Therapists, International Academy of Sex Research, International Society of Psychoneuroendocrinology, International Society for Research on Aggression, International Society for Developmental Psychobiology, World Research Network on the Sexuality of Women and Girls, Behavior Genetics Association, and Society for Research in Child Development.

Further Resources

The Kinsey Institute. http://www.kinseyinstitute.org.

Reskin, Barbara F.

Sociologist

Education: B.A., sociology, University of Washington, 1968, M.A., sociology, 1970, Ph.D., sociology, 1973

Professional Experience: acting assistant professor, University of California, Davis, 1971–1972; assistant professor to associate professor, sociology, Indiana University, Bloomington, 1973–1983; professor, sociology and women's studies, University of Michigan, Ann Arbor, 1983–1985; professor, sociology, and Director of Graduate Studies, University of Illinois, Urbana-Champaign, 1985–1991; professor, sociology, Ohio State University, 1991–1997; professor, sociology, Harvard University, 1997–2002; S. Frank Miyamoto Professor of Sociology, University of Washington, Seattle, 2002–

Concurrent Positions: study director, Committee on Women's Employment and Related Social Issues, National Research Council/National Academy of Sciences, Washington, D.C., 1981–1982; visiting scholar, Institute for Research on Women and Gender, Stanford University, summer, 1987; visiting professor, sociology, University of North Carolina, 1988, University of Notre Dame, 1997, Stockholm University, 1999, Manchester University, 1999

Barbara Reskin specializes in the sociology of work, including sexual and racial inequality in the workplace. In addition to numerous articles on topics related to affirmative action, gender and promotion, gender and management, racial segregation among female workers, and the effect of family responsibilities on women's careers (in particular, many of her early publications, including her dissertation, focused on the professional advancement of women scientists), among the books she has authored, co-authored, or edited are *Sex Segregation in the Workplace: Trends, Explanations, Remedies* (1984), *Women's Work, Men's Work: Sex Segregation on the Job* (1986), *The Realities of Affirmative Action* (1998), and *Women and Men at Work* (1994, 2nd ed., 2002). Reskin has been an invited lecturer at universities and organizations in the United States and abroad, and she has consulted with corporations and on legal cases on issues related to employment discrimination. Her research has been supported by grants from academic, government, and professional organizations, including the National Science Foundation, Economic Policy Institute, Institute for Women's Policy Research, Rockefeller Foundation, and National Institute of Mental Health.

Reskin was born in Minnesota, and her parents had ties to radical political and labor groups. Her father died when she was only seven years old, and her mother worked a series of clerical jobs to support the family. Barbara also worked a variety of clerical and manual jobs before attending Reed College. She left Reed and moved to Cleveland, Ohio, where she was introduced to the Congress on Racial Equality (CORE) during the civil rights movement of the mid-1960s. She was involved in actions such as organizing strikes, sit-ins, and a summer Freedom school. She attended a sociology night class at Case Western Reserve in Ohio, and she went on to receive her degree in sociology from the University of Washington, Seattle in 1968. She continued on for graduate study and became active in the feminist movement as well, helping organize a Reproductive Counseling Center and co-authoring a pamphlet about birth control for college women. She received a master's degree and then a Ph.D. in 1973 with her dissertation on "Sex Differences in the Professional Life Chances of Chemists." She began her teaching career at the University of California, Davis while still a graduate student and later served on the faculty of several Midwestern universities. In 1981, she took a year off from teaching to direct a study of sex segregation in the workplace for the National Academy of Sciences (NAS) in Washington, D.C. This experience galvanized her commitment not only to feminist social science research but also to applying research to social justice policy.

Reskin was elected a fellow of the NAS in 2006 and has served on numerous NAS and National Research Council committees, including the Committee on the Education and Employment of Women in Science and Engineering (1978–1982). She is a fellow of the American Academy of Arts and Sciences and the Sociological Research Association, and has served as vice president (1990) and president (2002) of the American Sociological Association (ASA). Her numerous awards and honors include a Distinguished Scholar Award of the ASA Section on Sex and Gender (1995), an SWS Mentorship Award (1998), and a DuBois Distinguished Scholarly Career Award from the ASA (2008).

Further Resources

University of Washington. Faculty website. http://www.soc.washington.edu/people/faculty_detail.asp?UID=reskin.

American Sociological Association. "Barbara F. Reskin. President 2002." http://www2.asanet.org/governance/reskin.html.

Resnik, Judith A.

1949–1986
Electrical Engineer, Astronaut

Education: B.S., Carnegie Mellon University, 1970; Ph.D., electrical engineering, University of Maryland, 1977

Professional Experience: electrical engineer, RCA Corporation, 1970–1974; biomedical engineer, Laboratory of Neurophysiology, National Institutes of Health, 1974–1977; senior systems engineer, Xerox Corporation, 1977–1978; astronaut,

Electrical engineer Judith Resnik was one of the first six women selected as astronauts in 1978. (AP/Wide World Photos)

National Aeronautics and Space Administration (NASA), 1978–1986, missions STS 41-D (1984) and STS 51-L (1986)

Judith Resnik was one of the first six women to be selected as astronauts by NASA in 1978. She was the second woman in the United States to fly in space, and she was among the crew who died when the space orbiter *Challenger* exploded on January 28, 1986, just after the launch from Cape Canaveral, Florida. Earlier, in 1984, she was a member of the crew of the Earth orbiter *Discovery* and was responsible for operating the Remote Manipulator System (RMS) on that mission. The RMS is the huge robotic arm that can lift satellites out of the orbiter and bring them back again. The crew was nicknamed "Icebusters" because they were able to use the arm to remove ice particles from the orbiter. When the RMS was first tested in space in 1981, it was another female astronaut, **Sally Ride**, who assisted from Mission Control. Although Resnik was an expert on using the shuttle arm, her initial flight assignment did not call for that specialty but instead required a great deal of photographic work. Later, when the flight was changed to include the shuttle arm, she had the opportunity to use her expertise as an electrical engineer.

Many of the male astronauts and employees of NASA were vehemently opposed to adding women to the program, but most reluctantly admitted that the women were qualified for their jobs. The first men selected had all been military test pilots because they had experience flying at high altitudes. However, in the late 1970s, the space program was shifting toward developing an orbiting space station, which required crewmembers with more scientific backgrounds, and most of the women astronauts held degrees in engineering or physics. All of the astronauts trained in multiple assignments in order to expand their capabilities to the maximum. After her death, Resnik became a hero and role model and was profiled in various news magazine and books about the women astronauts.

Further Resources

Kevles, Bettyann H. 2003. *Almost Heaven: The Story of Women in Space*. New York: Basic Books.

National Aeronautics and Space Administration. "Judith A. Resnik (Ph.D.)." http://www.jsc.nasa.gov/Bios/htmlbios/resnik.html.

Richardson, Jane S.

b. 1941
Biochemist

Education: B.A., philosophy, Swarthmore College, 1962; M.A., philosophy, Harvard University, 1966

Professional Experience: technical assistant, chemistry, Massachusetts Institute of Technology, 1964–1969; general physical scientist, Laboratory of Molecular Biology, National Institute of Arthritis and Metabolic Diseases, 1969; associate, anatomy, Duke University, 1970–1984, medical research associate professor, biochemistry and anatomy, 1984–1988, biochemistry, 1988–1991, James B. Duke Professor of Biochemistry, 1991–

Concurrent Positions: co-director, Molecular Graphics and Modeling Shared Resource, Duke Comprehensive Cancer Center, Duke University, 1988–

Jane Richardson is a biochemist and crystallographer who studies the three-dimensional structures of proteins, emphasizing the underlying principles of their architecture, aesthetics, interrelationships, and folding mechanism. She and her husband and collaborator, David Richardson, developed a way of mapping protein folding known as the *Richardson diagram*. Her 3-D ribbon diagrams were first published in the journal *Protein Chemistry* in 1981 and have since become the standard images for visualizing protein strands and structures. In 1985, she received a five-year MacArthur "genius" grant, and *Science Digest* chose the Richardsons' work on the first chemical synthesis of the protein betabellin as one of the year's 100 best inventions. Betabellin is a bell-shaped, beta-pleated-sheet protein whose structural properties were accurately predicted. Creating proteins that do not occur in nature can provide scientists with a better understanding of the structure of natural proteins, and the synthesis of proteins may open the way to designing hormones and drugs, and improving myriad industrial products.

Richardson had an early interest in science, and as a teenager in 1958, she won third place in the national Westinghouse Science Talent Search with her project on calculations of the satellite *Sputnik*'s orbit made from her own observations. She studied philosophy, mathematics, and physics at Swarthmore College, and went on to receive a master's degree from Harvard. Although she never earned a Ph.D. or M.D., she worked at the National Institutes of Health in the Laboratory of Molecular Biology for several years before moving to Duke University, where she advanced in rank to become the James B. Duke Professor of Biochemistry, and became co-director of a lab and of the Molecular Graphics and Modeling Shared Resource at the Duke Comprehensive Cancer Center. She has received honorary doctorates from Swarthmore and from the University of North Carolina.

Richardson was elected to membership in the National Academy of Sciences in 1991 and the Institutes of Medicine in 2006. She has been a member of the National Center for Research Resources and the National Institutes of Health, and an industrial consultant for Upjohn Company, Hoffman-LaRoche Company, Allied Chemical Corporation, Becton Dickinson, and NutraSweet. She is a member of the American Academy of Arts and Sciences, Biophysical Society, American Crystallographic Association, and Protein Society Office.

Further Resources

Duke University. "Richardson Laboratory." http://kinemage.biochem.duke.edu/.

Ride, Sally Kristen

b. 1951
Physicist, Astronaut

Education: B.A., English and B.S., physics, Stanford University, 1973, M.S., 1975, Ph.D., physics, 1978

Professional Experience: researcher, Department of Physics, Stanford University, 1978; trainee, National Aeronautics and Space Administration (NASA), 1978–1979, astronaut, 1979–1987, missions STS-2 (1981), STS-3 (1982), STS-7 (1983), STS 41-G (1984), special assistant, long-range and strategic planning, 1987; science fellow, Stanford University Center for International Security and Arms Control, 1987–1989; professor, physics and director, California Space Institute, University of California, San Diego, 1989–

Sally Ride was the first American woman to be sent into outer space in 1983 and the first American woman to make two space flights. Ride's first flight was in the

Sally Ride, America's first woman astronaut, communicates with ground controllers from the flight deck during the six day mission of the Challenger in June, 1983. (National Archives)

space shuttle *Challenger* in June 1983. Among the team's missions were deployment of international satellites and numerous research experiments supplied by a number of groups—ranging from a naval research lab to high school students. While operating the shuttle's robot arm, she handled the first satellite deployment and retrieval, the first time such an arm had been used in space during flight. Her second flight was also in the *Challenger* in October 1984. This time, the robot arm was used to readjust a radar antenna on the shuttle as well as to deploy and capture a satellite. Objectives on this mission covered scientific observations of the Earth and demonstrations of potential satellite-refueling techniques. Ride was chosen for a third scheduled flight, but it was canceled after the *Challenger* exploded in January 1986. She was the only astronaut chosen for the commission investigating the mid-launch explosion of the *Challenger*, which killed all crewmembers aboard.

Ride created NASA's Office of Exploration, and she was also the first woman astronaut to leave the space program when she quietly resigned in 1987 to join the Stanford Center for International Security and Arms Control. She went on to become director of the California Space Institute and physics professor at the University of California, San Diego. Ride has always been committed to science education, and during her tenure at NASA, she regularly addressed students at high schools and colleges about careers in science and engineering. In 2001, she created Sally Ride Science, an organization that encourages girls to study science, and she established an interactive educational Internet site, Space.com. She has published three children's books about space: *To Space and Back* (1986), *Voyager: An Adventure to the Edge of the Solar System* (1992), and *The Third Planet: Exploring the Earth from Space* (1994).

Ride was appointed a member of the Presidential Commission of Advisors on Science and Technology in 1994, and she has received the Jefferson Award for Public Service from the American Institute for Public Service (1984) and two National Spaceflight Medals (recognizing her shuttle missions of 1983 and 1984). At the National Air and Space Museum in Washington, D.C., there is a model of Sally Ride in her space uniform honoring her as the first American woman in space.

Further Resources

Kevles, Bettyann H. 2003. *Almost Heaven: The Story of Women in Space*. New York: Basic Books.

National Aeronautics and Space Administration. "Sally K. Ride (Ph.D.)." http://www .jsc.nasa.gov/Bios/htmlbios/ride-sk.html.

University of California, San Diego. Faculty website. http://cass.ucsd.edu/personal/ sride.html.

Space.com. http://www.space.com.

Riley, Matilda (White)

1911–2004
Sociologist

Education: B.A., Radcliffe College, 1931, M.A., sociology, 1937

Professional Experience: research assistant, Harvard University, 1932; vice president, Market Research Company of America, 1938–1949; research specialist, Rutgers University, 1950, professor, 1951–1973, director, sociology laboratory, and chair, Departments of Sociology and Anthropology, 1959–1973; professor, political economics and sociology, Bowdoin College, 1973–1981; associate director,

National Institute on Aging, Washington, D.C., 1979–1998, scientist emeritus, 1999–2004

Concurrent Positions: chief consulting economist, War Production Board, 1941–1943; summer faculty, Harvard University, 1955; visiting professor, New York University, 1956–1961; associate and director, Aging and Society, Russell Sage Foundation, 1964–1973, staff sociologist, 1974–1977; research fellow, Center for Advanced Study in the Behavioral Sciences, Palo Alto, CA, 1978–1979; senior research associate, Center for Social Sciences, Columbia University, 1978–1980

Matilda Riley was one of the foremost authorities on aging and gerontology. While *aging* refers to the psychological of mental, emotional, or physical development of a person of any chronological age, *gerontology* deals with the aging process and with issues specifically related to later life. In the February 1987 issue of *American Sociological Review*, Riley stated, "I believe that an understanding of age can clarify and specify time-honored sociological propositions, raise new research questions, demand new (as well as the old) methodological approaches, and even enhance the integrative power of our discipline (a power eroded in recent years through pluralism and disputes)." A few years later, she developed a theory about the influence exerted by the lives and experiences of sociologists on social and intellectual structure and change, both in sociology and in society as a whole. She identified examples of this influence in four areas of concern: sociological practice, gender, age, and dynamic social systems. In addition to papers published in journals, she published eight books and edited five more. She co-edited *Age and Structural Lag: Society's Failure to Provide Meaningful Opportunities in Work, Family, and Leisure* (1994), in which the authors argued that the lack of employment opportunities for older people was both unnecessary and modifiable. As the American baby-boomer generation reaches retirement age in the early twenty-first century, Riley's work remains especially relevant.

For much of her career, she had a professional partnership with her husband, Jack Riley, who died in 2002. Even as she approached her nineties, she remained active in her career. Instead of retiring after nearly 25 years at Rutgers University, she moved to Bowdoin College and then spent 20 years as associate director of the National Institute on Aging. Her election to membership in the National Academy of Sciences in 1994 was a long-delayed recognition of her years of research. In 1998, she was named Scientist Emeritus at the National Institutes of Health (NIH), the only social scientist ever given that distinction, and the NIH organized a 2001 lecture series in her honor entitled "Soaring: An Exploration of Science and the Life Course."

Riley received numerous honorary degrees and awards, such as the Commonwealth Award in Sociology (1984), the Distinguished Creative Contribution to

Gerontology Award (1990) and the Kent Award (1992), both of the Gerontological Society of America, and the Radcliffe Alumnae Award (1982). She was a fellow of the Center for Advanced Study in the Behavioral Sciences and a member of the Gerontological Society of America and the American Association for the Advancement of Science. She served as executive secretary of the American Sociological Association (ASA) (1949–1960), at a time when women rarely held office in any professional associations, and eventually served a term as president of the ASA (1986). Both Rutgers University and Bowdoin College have established academic prizes in her name.

Further Resources

Abeles, Ronald P. "Soaring: Celebrating Matilda White Riley (1911–2004)." http://www .asanet.org/footnotes/jan05/indexthree.html.

Rissler, Jane Francina

b. 1946
Botanist

Education: B.A., Shepherd College, 1966; M.A., West Virginia University, 1968; Ph.D., plant pathology, Cornell University, 1977

Professional Experience: fellow, fungal physiology, Boyce Thompson Institute, 1977–1978; assistant professor, plant pathology and botany, University of Maryland; staff scientist, Environmental Protection Agency; National Wildlife Federation; senior staff scientist and deputy director, Agriculture and Biotechnology, Food & Environment Program, Union of Concerned Scientists, 1993–

Jane Rissler is a plant pathologist and activist who has raised public awareness about genetically modified and engineered plants. Although scientists and farmers have practiced plant breeding for centuries, in recent decades it has become possible to transfer genetic material among plants. As a plant pathologist, she is concerned about the possibility of transferring diseases from plant to plant and about the possibility of introducing diseased plants into food crops.

Rissler is a senior scientist and director of the Food & Environment Program for the Union of Concerned Scientists, which conducts and compiles scientific research for the purpose of presenting policy suggestions on issues related to agriculture, biotechnology, pesticides, and the environment. In addition to numerous papers and interviews, she has co-authored two of the books the group has published. *Perils Amidst the Promise: The Ecological Risk of Transgenic Plants in a*

Global Market (1993) was written as a scientific and ecological response to policy issues and the politics of biotechnology. That research was enlarged and revised as *The Ecological Risks of Engineered Crops* (1996), in which the authors acknowledge that applications of biotechnology in crops are already a commercial reality. Rissler and others do not oppose genetic engineering as a component of agriculture as a whole, but only wish to encourage public debate about the potential harmful consequences of transgenic plants and to suggest a risk-assessment methodology.

Rissler grew up in rural West Virginia and received her doctorate in plant pathology from Cornell University in 1977. She taught plant pathology at the University of Maryland and was a policy consultant for the Environmental Protection Agency and the National Wildlife Federation before joining the Union of Concerned Scientists. She has been an important liaison to the public, co-editing a newsletter on genetic engineering and appearing on television and radio shows such as NPR, CNN, and various news outlets. She is a member of the American Phytopathological Society.

Further Resources

"Harvest of Fear. Interviews: Jane Rissler." PBS. (October 2000). http://www.pbs.org/wgbh/harvest/interviews/rissler.html.

Union of Concerned Scientists. "Experts." http://www.ucsusa.org/news/experts/jane-rissler.html.

Rivlin, Alice (Mitchell)

b. 1931
Economist

Education: B.A., Bryn Mawr, 1952; M.A., Radcliffe College, 1955, Ph.D., economics, 1958

Professional Experience: teaching fellow and tutor, economics, Harvard University, 1954–1957; research fellow, Economic Studies, Brookings Institution, 1957–1958, senior staff economist, 1958–1966; Deputy Assistant Secretary for Program Coordination, U.S. Department of Health, Education, and Welfare, 1966–1968, Assistant Secretary for Planning and Evaluation, 1968–1969; senior fellow, Economic Studies, Brookings Institution, 1969–1975; director, U.S. Congressional Budget Office, 1975–1983; director, Economic Studies Program, Brookings Institution, 1983–1987, senior fellow, 1987–1992; professor, Public Policy, George

Mason University, 1992–1993; deputy director, U.S. Office of Management and Budget, 1993–1994, director, 1994–1996; vice chair, Federal Reserve Board, 1996–1999; senior fellow, Economic Studies, Brookings Institution, 1999–; co-director, Brookings-Greater Washington Research Program, 2001–2002; professor, Milano Graduate School of Management and Urban Policy, New School University, 2001–2003; director, Greater Washington Research Program, Brookings Institution, 2002–

Concurrent Positions: staff member, Advisory Commission on Intergovernmental Relations, 1961–1962; visiting professor, J. F. Kennedy School of Government, Harvard University, 1988; chair, District of Columbia Financial Assistance and Management Authority, 1998–2001; visiting professor, Public Policy Institute, Georgetown University, 2003–

Alice Rivlin is an economist who has spent at least half of her career in the federal government, beginning as a member of the staff of the U.S. Department of Health, Education, and Welfare, where she implemented a system of budgeting and programming, and brought economic analyses to bear on the agency's policy decisions. She was the first head of the Congressional Budget Office (CBO) when it

Economist Alice Rivlin has held numerous appointments with the U.S. government, including as first head of the Congressional Budget Office and vice chair of the Federal Reserve Board. (AP/Wide World Photos)

was established in 1975. In 1996, she became vice chair of the Federal Reserve Board. Rivlin would have been in a position to replace the chair of the Federal Reserve, Alan Greenspan, but she resigned in 1999, one year before his retirement. After completing her doctorate, Rivlin obtained a position as an economist with the Brookings Institution in Washington, D.C., a well-known non-profit think tank devoted to independent research, education, and publications on social issues. She has been affiliated with Brookings off and on during her career and in 2002 became the director of their Greater Washington Research Program. Under the auspices of the Brookings Institution, she has published regularly on pressing political and economic issues since the 1960s, providing an economic analysis of education policy, medical care, welfare, elder care, and balancing the budget. Her most recent publications include *Beyond the Dot Coms: The Economic Promise of the Internet* (2001) and *Restoring Fiscal Sanity: How to Balance the Budget* (2004). She has also served as a consultant to a variety of government agencies.

In 1974, Congress passed the Congressional Budget and Impoundment Control Act, which provided for a House budget committee, a Senate budget committee, and the CBO. The last was to be an independent, nonpartisan office that would work with the two congressional committees to assist the members of Congress in analyzing and forming policy on federal spending and income. The CBO was also responsible for monitoring the national economy and its impact on the federal budget, for providing budgetary statistics to Congress, and for proposing alternative budgeting policies. As director of the CBO, Rivlin found herself embroiled in controversy because some of the recommendations of the CBO stepped on the toes of powerful people in Congress who had pet projects they were promoting. Also, her analyses and recommendations often were more negative than those provided by the executive branch, a factor that annoyed a series of U.S. presidents. For example, she found herself in conflict with Ronald Reagan's supply-side economics, and she forecast a deficit for 1984 while his office insisted he would balance the budget. Although her forecast proved to be correct, she was able to keep her job because there was a sharp drop in inflation that defused the argument.

After completing her second term at the CBO, Rivlin resigned her office there and returned to teaching, research, and writing. In 1993, she returned to federal employment when President Bill Clinton appointed her the deputy director of the U.S. Office of Management and Budget, the budgeting agency for the executive branch and the agency whose data she had disagreed with while head of the CBO. She was promoted to director in 1994 but resigned in 1996 to serve as vice chair of the Federal Reserve Board. Rivlin has received numerous honors and distinctions, including an honorary law degree from Hood College (1970), the Radcliffe College Founders Award (1970), a prestigious MacArthur Foundation Fellowship

(1983–1988), and a Lifetime Achievement Award from the D.C. Chamber of Commerce (2004).

Further Resources

Brookings Institution. "Alice M. Rivlin." http://www.brookings.edu/scholars/arivlin.htm.

Olson, Paulette I. and Zohren Emami, eds. 2002. *Engendering Economics: Conversations with Women Economists in the United States*. New York: Routledge.

Roberts, Edith Adelaide

1881–1977
Botanist

Education: A.B., Smith College, 1905; M.S., University of Chicago, 1911, Ph.D., plant physiology, 1915

Professional Experience: instructor to associate professor, botany, Mount Holyoke College, 1915–1917; extension worker with women, U.S. Department of Agriculture (USDA), 1917–1919; associate professor, botany, Vassar College, 1919–1921, professor, 1921–1950; consultant and guest scientist, Massachusetts Institute of Technology (MIT), 1950–retirement

Edith Roberts was recognized by her contemporaries for her research in plant physiology, ecology, germination of seeds, and propagation of native plants. She was a faculty member at Vassar College for more than 30 years and established the first outdoor ecological laboratory in the United States in Dutchess County, New York. The laboratory eventually contained more than 2,000 local native plant species, and Roberts co-authored a botanical history of Dutchess County in 1938. After completing her doctorate from the University of Chicago in 1915, she joined the faculty of Mount Holyoke for three years, then accepted a position with the USDA as an extension worker with women for three years. Sources do not mention the specific work she did for the USDA. Since this was during World War I, it is possible she was involved in gardening projects for women during wartime, when women managed farms for the men who were in service. She was appointed an associate professor of botany at Vassar College in 1919 and promoted to professor in 1921.

With gardener and landscape architect Elsa Rehmann, Roberts wrote a series of articles on plant ecology for *House Beautiful* magazine which were collected into a popular 1929 book, *American Plants for American Gardens: Plant Ecology, the Study of Plants in Relation to Their Environment*; the book was reprinted in a new edition in 1996. Roberts was also the author of *American Ferns:*

How to Know, Grow and Use Them (1935). She retired from Vassar in 1950, but went on to consult for the department of food technology at MIT, researching plant sources for vitamins. Roberts was a member of the Botanical Society of America, the American Forestry Association, and the American Association for the Advancement of Science.

Roberts, Lydia Jane

1879–1965
Nutritionist and Home Economics Educator

Education: teaching credential, Mount Pleasant Normal School (later Central Michigan University), 1909; B.S., home economics, University of Chicago, 1917, M.S., 1919, Ph.D., home economics, 1928

Professional Experience: school teacher, 1899–1915; assistant to associate professor, home economics, University of Chicago, 1919–1930, professor and department chair, 1930–1944; chair, home economics, University of Puerto Rico, 1946–1952

Lydia Roberts was a pioneer in the field of nutrition of children and had a key role in the development of government nutrition standards, such as determining the Recommended Dietary Allowances (RDA) of vitamins and minerals. She entered the University of Chicago at age 36 to begin her formal training in nutrition. She was already teaching at Chicago when received her Ph.D. in home economics in 1928; she was promoted to full professor and became department chair in 1930. At Chicago, she offered a curriculum with a strong basis in scientific research and was able to work on children's nutrition issues in a clinical setting. She conducted surveys of children's feeding and nutrition status for the U.S. Children's Bureau. Her book, *Nutrition Work with Children* (1927), was based on her dissertation research and became a classic in its field, going through several editions. It was as chair of the U.S. government's Food and Nutrition Board (FNB) that she developed the RDA guidelines based on the latest scientific research on human nutrient and vitamin needs. The first RDA report was the result of a committee of more than 40 nutrition scientists, an amazing 25 of whom were women. The findings of the FNB committee were first published by the American Dietetic Association in 1943.

After retiring from the University of Chicago, Roberts accepted a position as chair of the home economics department at the University of Puerto Rico, a position she retained until 1952. During this time she reported on nutrition on the island for the U.S. Department of Agriculture and co-authored a report, *Patterns of Living in Puerto Rican Families* (1949). Even after formally retiring from the University, she remained in Puerto Rico working on issues of nutrition and economic development for a rural community and developing an experimental program that became an island-wide model.

Roberts received the Borden Award of the Home Economics Association (1938). She was a member of the Council of Foods and Nutrition of the American Medical Association and the American Association for the Advancement of Science.

Further Resources

Harper, Alfred E. 2003. "Contributions of Women Scientists in the U.S. to the Development of Recommended Dietary Allowances." *The Journal of Nutrition*. 133: 3698–3702. (November 2003). http://jn.nutrition.org/cgi/content/full/133/11/3698.

Robinson, Julia Bowman

1919–1985
Mathematician

Education: A.B., University of California, Berkeley, 1940, M.A., 1941, Ph.D., mathematics, 1948

Professional Experience: mathematician, Berkeley Statistical Laboratory, 1939–1945; junior mathematician, Rand Corporation, 1949–1950; lecturer, mathematics, University of California, Berkeley, 1960–1964, 1966–1967, 1969–1970, and 1975, professor, mathematics, 1976–1985

Julia Robinson was a mathematician whose research focused on number theoretical decision problems and on recursive functions. She was one of the first American women mathematicians. She showed an early interest in mathematics and was often the only girl in high school taking advanced courses in mathematics and physics. Upon her high school graduation, she received a special medal for excellence in science and math, and entered San Diego State University at the age of 16; she later transferred to and received multiple degrees from the University of California, Berkeley. She was married to a mathematics professor, Raphael M. Robinson, in December 1941, just after the Japanese attack on Pearl Harbor, and during World War II, she worked for the Berkeley Statistical Laboratory on military projects. She was discouraged at being unable to secure a faculty position at Berkeley (due to anti-nepotism rules) and being unable to have children (due to serious health problems since her own childhood), but she continued on to receive her doctorate in 1948.

She worked for the Rand Corporation for two years and then took several years off for her health, including heart surgery. Beginning in 1960, she began teaching one graduate course per quarter at Berkeley. She became interested in a list of unsolved mathematical problems posed in 1900 by German number theorist David Hilbert. Robinson set to work on solving the equation known as "Hilbert's Tenth Problem" and published several papers on the topic. In 1970, she learned that a Russian mathematician had solved the equation based on her hypothesis. She became internationally known for this work and, in 1975, was promoted to full professor at Berkeley. Her sister, Constance Reid, collected Robinson's autobiography as well as several articles about her work for the 1996 volume, *Julia, A Life in Mathematics.*

Robinson was the first woman mathematician elected to the National Academy of Sciences (1975), and she became the first woman elected president of the American Mathematical Society (1983). She received an honorary degree from Smith College (1979) and in 1983 was awarded a prestigious five-year MacArthur Fellowship. She also was a member of the American Academy of Arts and Sciences and the Association for Symbolic Logic.

Further Resources

Reid, Constance, ed. 1996. *Julia, A Life in Mathematics.* Washington, D.C.: Mathematical Association of America.

Agnes Scott College. "Julia Bowman Robinson." Biographies of Women Mathematicians. http://www.agnesscott.edu/lriddle/women/robinson.htm.

Roemer, Elizabeth

b. 1929
Astronomer

Education: B.A., University of California, Berkeley, 1950, Ph.D., astronomy, 1955

Professional Experience: assistant astronomer, University of California, 1950–1952, laboratory technician, Lick Observatory, 1954–1955, research astronomer, 1955–1956; research associate, Yerkes Observatory, University of Chicago, 1956; astronomer, Flagstaff Station, U.S. Naval Observatory, 1957–1966, acting director, 1965; associate professor, astronomy, and member, Lunar and Planetary Laboratory, University of Arizona, Tucson, 1966–1969, professor, 1969–1998, astronomer, Steward Observatory, 1980–1998, emerita

Elizabeth Roemer is renowned as the premier recoverer of "lost" comets, that is, comets whose planned rediscovery is based on predictions from previous returns. She calls her profession "astrometry," which is the branch of astronomy that deals with the measurement of the positions and motions of the celestial bodies. In her lifetime study of comets, she has rediscovered at least 79 returning periodic comets and visual and spectroscopic binary stars, plus computing the orbits of comets and minor planets. Her publications have covered many topics, such as comets and minor planets, astronomy and practical astronomy, computation of orbits, astrometric and astrophysical investigations of comets, minor planets and satellites, and dynamical astronomy. She is regarded by her peers as a contributor to many scientific and astronomical discoveries, and her precise photographic observations of comets have led to a great many cometary orbits of importance.

In 1965, a colleague named Asteroid 1657 "Roemera" in her honor. Although each comet and asteroid is assigned a number in an international database, not all have names; after the sightings have been verified, it is the privilege of the discoverer to name the item or to have it named in their honor. Roemer made her first major rediscoveries while she was working at the U.S. Naval Observatory at Flagstaff, Arizona, and it was at that same time that her photographic records of comets and her notes on their physical characteristics began to earn her national recognition.

She taught adult classes in the local public school system while attending school at the University of California, Berkeley. She also served as an assistant astronomer and later as a laboratory technician at the Lick Observatory. She worked briefly for the university after graduation and was also a research associate at the Yerkes Observatory of the University of Chicago. She then joined the staff of the Flagstaff Station of the U.S. Naval Observatory and later moved to the University of Arizona as an associate professor and a member of the Lunar and Planetary Laboratory before becoming a full professor of astronomy. She retired in 1998 but continues her research on comets and asteroids.

She has received numerous prizes, such as the B. A. Gould Prize of the National Academy of Sciences (1971), the Donohoe lectureship of the Astronomical Society of the Pacific (1962), the National Aeronautics and Space Administration (NASA) Special Award (1986), and the Dorothea Klumpke Roberts Prize of the Astronomical Society of the Pacific (1950) named for another American astronomer who was recognized for her work in charting and cataloging stars in the late nineteenth century. Roemer is a fellow of the American Association for the Advancement of Science and a member of the American Astronomical Society, American Geophysical Union, Astronomical Society of the Pacific, International Astronomical Union, British Astronomical Association, and Royal Astronomical Society of London.

Rolf, Ida P.

1896–1979
Biochemist, Physical Therapist

Education: B.S., Barnard, 1916; Ph.D., biological chemistry, College of Physicians and Surgeons, Columbia University, 1920

Professional Experience: associate, chemotherapy and organic chemistry, Rockefeller Institute, 1920–1928; independent practitioner and physical therapist, 1930–1979; founder, Rolf Institute of Structural Integration, Boulder, Colorado, 1971–1979

Ida Rolf was a biochemist and physical therapist who created a unique and controversial treatment of "structural integration," also termed "Rolfing." Her method is a vigorous program of physical manipulations to release anger and tensions, and to restore the free flow of fluids, nerve impulses, and energy through the body. It was based on the belief that the body is plastic and not a fixed unit, as medical science would hold. A key feature of the treatment is that the structure, particularly the alignment, of the body would often be changed, and the patient might look much different because he or she was standing, moving, and walking in a new way. Some sources attribute Rolf's interest in physical therapy and natural medicine to a condition suffered by one her children; others report that Rolf herself was seeking alternative treatments after suffering from illnesses resulting from being kicked by a horse as a young woman.

After receiving her doctorate in biological chemistry, Rolf worked for the Rockefeller Institute. She eventually inherited some money and left the Institute to independently study various methods of physical therapy. She traveled to Switzerland, where she studied physics and homeopathic medicine and returned to the United States to study chiropractic medicine and yoga therapy. Around 1940, she began to develop her own theories of the mind–body connection based on the idea that both psychological and physical histories shape, and sometimes deform, people's bodies, thickening connective tissue and tightening muscles in response to psychological as well as physical injury, and revealing past tensions and unexpressed angers. These abnormal tightenings and thickenings interfere with the flow of fluids and can sometimes block the free passage of nerves and nerve impulses through the body. She began traveling throughout the United States, Canada, and Europe, lecturing and demonstrating her method of "structural integration." In the mid-1960s, she was invited to give demonstrations at the Esalen Institute in California, a community that tried to integrate elements of Eastern cultures, such as Zen Buddhism, and radical therapy systems, such as Gestalt psychotherapy. Although Rolf did not approve of the Esalen lifestyle, which included nudism and drugs, the Institute provided a base of operations for her for a few years. In 1971, based on the success of her workshops at Esalen, she organized the Guild for Structural Engineering, later renamed the Rolf Institute of Structural Integration, in Boulder, Colorado.

The Institute continues to train Rolf practitioners to carry on her work, based on the principles laid out in her 1977 book, *Rolfing: The Integration of Human Structure*. There are presently more than 1,500 certified Rolf practitioners worldwide, and the method has received regular attention in the popular press. Most recently, the Rolfing method was featured on *The Oprah Winfrey Show* in April 2007. In some circles, it is still seen as controversial because Rolf did not have a medical degree, but also because the emotional release common at the sessions can be overwhelming for a patient without concurrent psychiatric treatment.

Further Resources

The Rolf Institute of Structural Integration. http://www.rolf.org/.

Roman, Nancy Grace

b. 1925
Astronomer

Education: B.A., Swarthmore College, 1946; Ph.D., astronomy, University of Chicago, 1949

Professional Experience: assistant, Swarthmore College Observatory, 1943–1946; graduate assistant, astronomy and astrophysics, University of Chicago, 1946–1948, research associate, stellar astronomy, Yerkes Observatory, University of Chicago, 1949–1951, instructor, 1951–1954, assistant professor, 1954–1955; astronomer, Radio Astronomy Branch, U.S. Naval Research Laboratory, 1955–1956, head, Microwave Spectroscopy Section, 1956–1957, consultant, 1958–1959; head, observational astronomy program, National Aeronautics and Space Administration (NASA), 1959–1960, chief astronomer, 1960–1979, program scientist, Hubble Space Telescope, 1979–1980, principal scientist, Astronomical Data Center, Goddard Space Flight Center, 1981–1997, astronomer/programmer, Sigma Data and MA/Com, 1981–1986, principal scientist, Hughes STX, 1986–1996, head, Astronomical Data Center, 1995–1997, chief scientist, 1997; teacher training, Montgomery College, Maryland, 1997–1999

Concurrent Positions: consultant, ORI, Inc. 1980–1989; senior professional, Space Systems Division, McDonnell Douglas, 1988–1994

Nancy Roman is renowned for developing satellite observatories to explore the universe from a vantage point that is free from atmospheric interference. She pioneered the use of satellites for gamma ray, x-ray, and radio observations, and she has also used traditional Earth-based telescopes to study topics such as stellar motions, photoelectric photometry, and spectroscopy. She is especially noted for the research she conducted at NASA, where for many years she was the highest-ranking woman scientist. The opening of the astronaut program to women in 1978, and the launch of NASA's moon program in 1988, greatly expanded opportunities for women scientists, but Roman and a few others achieved recognition for their work prior to that program. In a 1964 NASA-approved book, *Scientists Who Work with Astronauts,* by Lynn and Gray Poole, astronomers Nancy Roman and **Jocelyn Gill** were the only two women who were profiled.

Roman's association with NASA began in 1959, when she was appointed head of the observational astronomy program. She developed an ambitious plan to observe objects in space by using rocket and satellite observatories, and in the 1960s, she designed instrumentation and made substantial measurements from gamma ray, radio, and visible light satellites, such as the orbiting solar observatories. Her programs provided astronomers with the planetary surface knowledge that led to the successful 1976 *Viking* probes to collect data from Mars. In the 1970s, her papers dealt with new satellite data, but she still did Earth-based observation, such as at Kitt Peak Observatory. Asteroid number 2516 Roman is named after her. In the 1970s and 1980s, she measured x-ray and ultraviolet readings from the successful OAO-3, or *Copernicus*, satellite, and recorded stellar spectra from the U.S. space station *Skylab*, which circled the Earth between 1973 and 1979. She was also the NASA program scientist for a planned space telescope, and the Hubble was eventually launched in 1990. She has also worked as a consulting astronomer for the Astronomical Data Center, editing and documenting astronomical catalogs for electronic archiving.

Roman has also been committed to science education, and during the late 1990s, she team-taught courses for advanced students and K–12 science teachers. Roman has received numerous honorary degrees. She is a fellow of the American Astronautical Society and American Association for the Advancement of Science, and a member of the American Astronomical Society.

Further Resources

Montgomery College. Faculty website. http://www.montgomerycollege.edu/Departments/planet/Nancy/Nancy.htm.

Romanowicz, Barbara

b. 1950
Geophysicist, Seismologist

Education: Ecole Normale Supérieure, "Sèvres", Paris, France, 1970–1974; Maîtrise de Mathématiques Pures, Université Paris, 1972; Agrégation de Mathématiques, Paris, 1973; M.S., applied physics, Harvard University, 1975; Doctorat, astronomy, Université Paris, 1975; Doctorat d'Etat, Université Paris, Spécialité Géophysique, 1979

Professional Experience: Attachée de Recherches, C.N.R.S., Institut de Physique du Globe, Paris, 1978–1979; postdoctoral associate, Massachusetts Institute of Technology (MIT), 1979–1981; Chargée de Recherches, C.N.R.S., I. P. G., Paris, Director, Geoscope Program, 1981–1986, Directeur de Recherches, 1986–1990; professor, geophysics, University of California, Berkeley, and director, Berkeley Seismological Laboratory, 1991–

Barbara Romanowicz is a geophysicist and seismologist who studies earthquakes, plate tectonics, and deep-earth (from the crust to the inner core) structures and movement. She has been involved in the development of special tools and observatories for measuring global seismic activity on land and in the oceans, including as co-founder in 1985 of ORFEUS, a European data center for broadband seismology, and co-founder in 1986 of the Federation of Digital Seismic Networks (FDSN). In 1997, she collaborated on the Monterey Bay Ocean-Bottom International Seismic Experiment (MOISE), and since 2002 has been involved with the Monterey Bay Ocean-Bottom Broadband Seismometer experiment (MOBB) in collaboration with the Monterey Bay Aquarium Research Institute (MBARI). Trained in France, she has been a professor of geophysics and director of the Berkeley Seismological Laboratory in California since 1991. She also served as chair of the Department of Earth and Planetary Science at the University of California, Berkeley between 2002 and 2006.

Romanowicz was elected to the National Academy of Sciences in 2005. She is a fellow of the American Geophysical Union (president, Seismology Section, 1994–1998) and American Academy of Arts and Sciences, and has served on numerous government and scientific research committees, including the National Earthquake Prediction Evaluation Council, National Research Council Committee on the Science of Earthquakes, International Ocean Network Committee, and Advisory Council to the Southern California Earthquake Center. She is the recipient of the French Academy of Sciences Prize (Fonds Doistau-Blutet) (1989), Silver Medal of the Centre National de la Recherche Scientifique (French NSF) (1992), A. Wegener Medal of the European Union of Geosciences (1999), and Gutenberg Medal of the European Geophysical Society (2003).

Further Resources

University of California, Berkeley. Faculty website. http://eps.berkeley.edu/development/view_person.php?uid=8698.

Monterey Bay Aquarium Research Institute. "New Seafloor Observatory Provides Round-the-Clock Monitoring of Ocean and Earth." (18 March 2009). Monterey Ocean-Bottom Broadband Seismometer. http://www.mbari.org/news/homepage/2009/mars-mobb -deimos.html.

Rose, Flora

1874–1959
Home Economist

Education: B.S., Kansas State Agricultural College, 1904; M.A., food and nutrition, Columbia University, 1909; Ped.D., New York State College for Teachers, 1931; Sc.D., Kansas State Agricultural College, 1937

Professional Experience: instructor, food and nutrition, Kansas State College, 1903–1906; lecturer, home economics, Cornell University, 1907–1911, professor, home economics, 1911–1940, co-director, School of Home Economics, 1919–1925, co-director, New York State College of Home Economics, 1925–1932, director, 1932–1940, emeritus professor

Flora Rose was recognized for her research in nutrition, weight control, and the science of homemaking. She received her undergraduate degree from Kansas State College and taught there for four years before attending Columbia University in New York. She earned a master's degree in food and nutrition from Columbia and then spent 30 years as co-director (with **Martha Van Rensselaer**) and then director of the School of Home Economics at Cornell University, later established as the separate New York State College of Home Economics. Rose and Van Rensselaer were reformers who led a campaign to start

Home economist Flora Rose. (Courtesy of Cornell University)

programs in home economics at major universities and were instrumental in persuading the New York legislature to create the program at Cornell. Rose became a lecturer and then professor in home economics through the agriculture department, and then co-director and, after Van Rensselaer's death, director of the New York State College of Home Economics. Rose and Van Rensselaer were the first women faculty members at Cornell to be promoted to full professors and were an inseparable administrative team (one colleague addressed them together as "Miss Van Rose"). They also shared their personal lives as well, living together until Van Rensselaer's death in 1932.

In the first decades of the twentieth century, the U.S. Department of Agriculture instituted programs for vocational education in public schools and for agricultural extension services for adults, including housewives. This opened up an unprecedented amount of funding for jobs for women as teachers and researchers in home economics and nutrition. As a nutrition researcher, Rose focused on dietary needs to fit household budgets, in particular the development of low-cost fortified cereals and the effect of nutrition on health and infant mortality. During World War I, she helped organize food relief program for children in Belgium, activities that earned her recognition with the Order of the Crown. After retiring in 1940, she moved to California, where she continued with her research and teaching through the California State Health Department. Rose authored or co-authored several books, including, *A Manual of Home-Making* (1919), *The New Butterick Cook-Book* (1924), and *Pioneers in Home Economics* (1948). She was a member of the American Association for the Advancement of Science.

Further Resources

Cornell University Library. Faculty biography. http://rmc.library.cornell.edu/homeEc/bios/florarose.html.

Rose, Mary Davies Swartz

1874–1941
Chemist and Nutritionist

Education: Litt.B., Denison University, 1901; diploma, home economics, Mechanics Institute, 1902; B.S., Columbia University, 1906; Ph.D., physiological chemistry, Yale University, 1909

Professional Experience: high school teacher, 1899–1905; assistant, nutrition, Teachers College, Columbia University, 1906–1907, instructor, nutrition and dietetics, 1909–1910, assistant to associate professor, 1911–1923, professor, 1923–1940

Mary Swartz Rose was a pioneer in research on nutrition and dietetics, including the vitamin content of food, protein comparison, effects of nutrients on anemia, metabolism, and trace elements in the diet. She was appointed an assistant professor of nutrition at Teachers College a year after the department was established, and the department became a national university center for training teachers of nutrition. She published more than 40 scientific papers and two widely used textbooks, *A Laboratory Hand-Book for Dietetics* (1912) and *The Foundations of Nutrition* (1927). She also wrote popular books for mothers, *Feeding the Family* (1916) and *Teaching Nutrition to Boys and Girls* (1932). After receiving a certificate from Denison University in 1901 and a diploma in home economics from the Mechanics Institute in Rochester, New York, in 1902, she taught high school home economics for five years. She then enrolled in Teachers College, Columbia University, where she received her undergraduate degree in 1906, remaining another year as an assistant in the household arts department. Since there were no graduate programs in nutrition at the time, she enrolled in Yale, where she received her doctorate in physiological chemistry in 1909. She met Anton Rose when both were graduate students at Yale; they were married in 1910 and had one son, Richard. She returned to Teachers College to become the first full-time instructor in nutrition and dietetics. She organized a program in which students could secure a solid grounding in the scientific aspects of nutrition as well as in the best methods for teaching the subject. She was promoted to assistant professor in 1911, associate professor in 1918, and professor in 1923.

A charter member of the American Institute of Nutrition, she was its president in 1937 and 1938 and associate editor of its publication, *Journal of Nutrition*, from 1928 to 1936. The American Dietetic Association elected her an honorary member in 1919. Teachers College established a scholarship and the Greater New York Dietetic Association established a lectureship in her name. She also was a member of the American Association for the Advancement of Science, the American Society of Biological Chemists, the Society of Experimental Biology and Medicine, the American Home Economics Association, and the American Public Health Association. Her biography is *Mary Swartz Rose: Pioneer in Nutrition* (1979) by Juanita A. Eagles, et.al. Beginning in 2008, the American Society for Nutrition and the Council for Responsible Nutrition co-sponsor two awards in her name, the Mary Swartz Rose Young Investigator Award and the Mary Swartz Rose Senior Investigator Award.

Rosenblatt, Joan (Raup)

b. 1926
Mathematical Statistician

Education: B.A., mathematics, Barnard College, 1946; Ph.D., statistics, University of North Carolina, 1956

Professional Experience: statistical analyst, National Institute of Public Affairs and Bureau of the Budget, 1948; assistant statistician, University of North Carolina, 1953–1954; mathematician, National Bureau of Standards, 1955–1969, chief statistician, Engineering Laboratory, 1969–1978, deputy director, Computer and Applied Mathematics Laboratory, National Institute of Standards and Technology, 1979–1993, director, 1993–1996

Joan Rosenblatt was renowned for her research as a mathematical statistician at the National Institute of Standards and Technology, formerly the National Bureau of Standards. The mission of the institute is to maintain and disseminate the basic units of measurement such as mass, length, temperature, frequency, and electrical units for application in industry and government regulations. Her research includes nonparametric statistical theory, applications of statistical techniques in the physical and engineering sciences, and the reliability of complex systems. At the National Bureau, Rosenblatt provided statistical consulting services to all parts of the institute and researched improved statistical methods for applications in the physical and engineering sciences as well. The research problems with which Rosenblatt was concerned in the early 1990s arose from the proliferation of new federal regulations based on physical measurements of such things as water, air, pesticides, noise, radiation, occupational health and safety, and transportation safety. One of the most difficult problems was how to measure chemical additives in food that the Food and Drug Administration handles to satisfy regulations that bar the use of known cancer-causing additives in food processing.

Rosenblatt grew up in a family that stressed education, and both her parents held Ph.D.s. Her mother, Clara Eliot Raup, a professor of economics at Barnard College, was among the first to promote the study of consumer economics and was also a role model for combining career and motherhood, being the first woman at Barnard to receive an unpaid maternity leave. After receiving her undergraduate degree, Rosenblatt worked for several government agencies before returning to graduate school and completing her doctorate in 1956. She spent 40 years in government employment, retiring in 1996.

Among the awards Rosenblatt has received are the Federal Woman's Award (1971), the Gold Medal of the Department of Commerce (1976), and the Founders Award of the American Statistical Association in 1991. She was a member of the

Committee on Applications and Theoretical Statistics of the National Research Council from 1985 to 1988. She is a fellow of the American Association for the Advancement of Science, American Statistical Association, and Institute of Mathematical Statistics, and a member of the American Mathematical Society and International Statistical Institute.

Further Resources

Murray, Margaret Anne Marie. 2000. *Women Becoming Mathematicians: Creating an Identity in Post–World War II America.* Cambridge, MA: MIT Press.

Rowley, Janet Davison

b. 1925
Cytogeneticist, Geneticist

Education: B.S., anatomy, University of Chicago, 1946, M.D., 1948

Professional Experience: research assistant, University of Chicago, 1949–1950; resident, Marine Hospital, U.S. Public Health Service, Chicago, 1950–1951; physician, Infant Welfare and Prenatal Clinics, Department of Health, Montgomery County, Maryland, 1953–1954; research fellow, Cook County Hospital, Chicago, 1955–1960; instructor, neurology, University of Illinois Medical School, 1961; trainee, radiobiology, Churchill Hospital, England, 1961–1962; research associate, Department of Medicine and Argonne Cancer Research Hospital, University of Chicago, 1962–1969, associate professor, Department of Medicine and Franklin McLean Memorial Research Institute, 1969–1977, professor, 1977–1984, Blum-Riese Distinguished Service Professor, Medicine, Molecular Genetics and Cell Biology, and Human Genetics, 1984–

Janet Rowley is a cytogeneticist internationally renowned for her research on chromosome abnormalities in a form of leukemia and lymphoma. Cytogeneticists investigate the role of cells in evolution and heredity, and Rowley's research has introduced new diagnostic tools for oncologists and opened new avenues to possible gene therapies for cancer. She has helped to pinpoint cancer gene locations and correlate them to chromosome aberrations. During her long career at the University of Chicago, she developed the use of quinacrine and Giemsa staining to identify chromosomes in cloned cells, and thus was able to identify abnormalities that occur in some chromosomes in certain cancers. In 1972, she was the first person to discover the recurring translocation, or shifting, of genetic material, and since that time, more than 70 such translocations have been detected in human malignant cells. Her research indicated that both translocations and deletions of

Cytogeneticist Janet Rowley is presented with the Presidential Medal of Freedom by President Barack Obama, 2009. (AP/Wide World Photos)

genetic material occur in malignancy and that cancer is caused by a complex series of events within a single cell, making some genes overactive and eliminating other genes that would normally suppress growth. Her research revealed that any cell is potentially cancerous.

Starting with her undergraduate studies in the 1940s, Rowley has had a long association with the University of Chicago for the majority of her career. In addition to her numerous scientific papers, she is the author of *Chromosome Changes in Leukemia* (1978) and the editor or co-editor of *Chromosomes and Cancer: From Molecules to Man* (1983), *Genes and Cancer* (1984), *Consistent and Chromosomal Aberrations and Oncogenes in Human Tumors* (1984), and *Advances in Understanding Genetic Changes in Cancer* (1992). She is co-founder and co-editor of the journal, *Genes, Chromosomes and Cancer*.

Rowley was elected to membership in the National Academy of Sciences in 1984 and the Institute of Medicine 1985. She received the National Medal of Science in 1999 and in 2009 received the Presidential Medal of Freedom, the nation's highest civilian honor. She has received almost every major cancer-research award, including the Esther Langer Award (1983), the Kuwait Cancer Prize (1984), the A. Cressy

Morrison Award from the New York Academy of Sciences (1985), the Judd Memorial Award from the Sloan-Kettering Cancer Center (1989), the Charles S. Mott Prize from General Motors Research Foundation (1989), the G. H. A. Clowes Memorial Award from the American Association for Cancer Research (1989), the Robert de Villiers Award from the Leukemia Society of America (1993), the Gairdner International Prize (1996), the Albert Lasker Clinical Medicine Research Prize (1998), the Franklin Medal of the American Philosophical Society (2003), and the Genetics Prize of the Peter and Patricia Foundation (2009). She is a member of the American Academy of Arts and Sciences, American Philosophical Society, American Society of Human Genetics (president, 1993), American Society of Hematology, and American Association for Cancer Research. She has received honorary doctorates from the University of Arizona, the University of Pennsylvania, Knox College, the University of Southern California, and Harvard University.

Further Resources

University of Chicago. Faculty website. http://experts.uchicago.edu/experts.php?id=212.

Wasserman, Elga. 2002. *The Door in the Dream: Conversations with Eminent Women in Science*. Washington, D.C.: Joseph Henry Press.

Roy, Della Martin

b. 1926
Geochemist, Materials Scientist

Education: B.S., chemistry, University of Oregon, 1947; M.S., mineralogy, Pennsylvania State University, 1949, Ph.D., mineralogy, 1952

Professional Experience: assistant, mineralogy, Pennsylvania State University, 1949–1952, research associate, geochemistry, 1952–1959, senior research associate, 1959–1969, associate professor, materials science, 1969–1975, professor, Materials Research Laboratory, 1975–1992, emerita

Della Roy is known for her research in materials science, which is concerned with the uses of new materials and their applications to existing processes and products. Her research includes phase equilibria, materials synthesis, crystal chemistry and phase transitions, crystal growth, cement chemistry, hydration and microstructure, concrete durability, biomaterials, special types of glass, radioactive waste management, geological isolation, chemically bonded ceramics, and waste management science. Much of the new materials-science research arose from the aeronautics and nuclear energy programs starting in the 1950s, around the time Roy received her Ph.D.

Although Roy's background is in mineralogy, she has worked with many types of materials, including ceramics, biomaterials, and concrete. She founded a journal for *Cement and Concrete Research* in 1971. Although her husband, Rustum Roy, is internationally known for his research in materials science, science policy, and alternative medicine, and the two collaborated on scientific papers, Della Roy maintained her own research programs and, in addition to receiving four patents, authored or co-authored hundreds of scientific papers. Both she and her husband had minerals named after them: *dellaite* and *rustumite*. After receiving her doctorate from Pennsylvania State University, she was appointed as an assistant in mineralogy and progressed through the ranks to senior research associate and then full professor in the Materials Research Laboratory. She formally retired in 1992, but continues her affiliation and research at Pennsylvania State University.

Roy was elected to the National Academy of Engineering in 1987. She has been a member of the Highway Research Board of the National Academy of Sciences, was chair of a National Academy of Sciences Research Committee on Concrete (1980–1983), and was a member of the Committee on Concrete Durability (1986–1987). She has received numerous awards, including the Jepson Medal (1982) and the Copeland Award (1987) of the American Ceramic Society, and the Slag Award of the American Concrete Institute (1989). She was made an honorary fellow of the Institute for Concrete Technology in 1987 and is a fellow of the Mineralogical Society of America, American Concrete Society, American Ceramic Society, and American Association for the Advancement of Science, and a member of the Materials Research Society, Geochemical Society, Clay Minerals Society, Concrete Society (UK), American Nuclear Society, American Society for Testing and Materials (ASTM), and Society of Women Engineers.

Further Resources

Pennsylvania State University. Faculty website. http://www.mri.psu.edu/faculty/dmr.asp.

University of Oregon College of Arts & Sciences. "Alumni. Della Roy '47: Cement Paves the Way to Illustrious Career." Alumni & Development. (19 June 2007). http://uoregon.edu/~wits/wits/files/pdf/della-roy.pdf.

Rubin, Vera (Cooper)

b. 1928
Astronomer, Cosmologist

Education: B.A., Vassar College, 1948; M.A., physics, Cornell University, 1951; Ph.D., astronomy, Georgetown University, 1954

Professional Experience: instructor, mathematics and physics, Montgomery County Junior College, 1954–1955; research associate astronomer, Georgetown University, 1955–1965, lecturer, 1959–1962, assistant professor, astronomy, 1962–1965; staff member, Department of Terrestrial Magnetism, Carnegie Institution of Washington, 1965–

Concurrent Positions: associate editor, *Astronomical Journal*, 1972–1977; associate editor, *Astrophysical Journal Letters*, 1977–1982; editorial board, *Science Magazine*, 1979–1987; Beatrice Tinsley visiting professor, astronomy, University of Texas, Austin, 1988

Vera Rubin is a specialist in the branch of astronomy called *cosmology*, which deals with the general structure and origin of the universe. She is one of America's foremost astronomers, and has spent her career observing galactic structure, rotation, and dynamics. Her pioneering research in the 1970s demonstrated the possible existence of a large percentage of matter in the universe that is invisible to the naked eye, and astronomers now estimate that up to 90% of the universe may be composed of this "dark matter." She studied physics at Cornell, where, for her master's thesis, she analyzed the motion of 108 galaxies and discovered that they shared a large-scale, systematic motion in addition to motion resulting from the expansion of the universe. When she presented her findings at a meeting of the American Astronomical Society in 1950, the scientific community was not prepared to believe in large-scale motions, and her work generated great controversy. Several years later, she was vindicated when a noted cosmologist agreed with her theory. Her doctoral advisor was applying nuclear physics to Big Bang cosmology, and her dissertation, again ahead of her time, showed that instead of being randomly distributed, galaxies tend to clump together.

Rubin did not start doing observational astronomy until the 1960s and, with colleague Kent Ford, found evidence that a large group of galaxies,

Astronomer and cosmologist, Vera Rubin. (Courtesy of the Carnegie Institution)

Support for Iraqi Women Scientists

Through the U.S. National Academy of Sciences' Committee on Human Rights, several prominent American women scientists helped launch a "twinning project" in 2007 to provide professional and social support to Iraqi women scientists and engineers whose careers and research have been disrupted by war. The program is chaired by **Maxine Singer**, **Vera Rubin**, and **Myriam Sarachik**, and invites women members of the NAS, the National Academy of Engineering (NAE), and the Institute of Medicine (IOM) to be paired with Iraqi colleagues and provide them with information, news and scientific papers in their field, and moral support. The network reaches out to women living in Iraq as well as those who have fled to other countries and are attempting to continue their work. The program works in consultation with human-rights groups, nongovernmental organizations, and private groups, and does not rely upon funding from the U.S. government.

---

including the Earth's Milky Way, are moving rapidly with respect to the rest of the universe. Although the theory was immediately controversial, this time the astronomy community took "the Rubin-Ford effect" seriously. Rubin and Ford found that stars at the outer margins of galaxies travel as rapidly as stars closer to the galaxy center. This indicates that there must be a large amount of invisible matter, even at the fringe of a galaxy, where the number of visible stars dwindles, because matter is necessary to accelerate the outer stars in their rapid orbits. Rubin theorized that a huge reservoir of extra material that is invisible to the telescope must be part of each galaxy, and her team has analyzed 200 galaxies in pursuit of this research. Her work on spiral galaxies was discussed on the public television show, *Stephen Hawking's Universe: On the Dark Side*, in 1997.

Rubin was elected to membership in the National Academy of Sciences (NAS) (1981) and received the National Medal of Science (1993). She was also awarded the NAS James Craig Watson Medal (2004) "for her seminal observations of dark matter in galaxies, large-scale relative motions of galaxies, and for generous mentoring of young astronomers, men and women." She has received numerous honorary degrees and other awards, including the Russell Lecturer Prize of the American Astronomical Society (1994), Gold Medal of the Royal Astronomical Society (1996), Women and Science Award of the Weizmann Institute (1996), Cosmology Prize of the Peter Gruber Foundation (2002), Bruce Medal of the Astronomical Society of the Pacific (2003), and Distinguished Achievement Award of Vassar College (2007). In 2008, she became co-chair (with geneticist **Maxine Singer** and physicist **Myriam Sarachik**) of an NAS project to pair women scientists in the United States with Iraqi women scientists for mentoring and career support.

Further Resources

Carnegie Institution. Faculty website. http://www.dtm.ciw.edu/rubin/.

National Academy of Sciences. 2008. "International Twinning Project for Iraqi Women Scientists, Engineers, and Health Professionals." Committee on Human Rights. (March 2008). http://sites.nationalacademies.org/PGA/humanrights/PGA_044086.

Wasserman, Elga. 2002. *The Door in the Dream: Conversations with Eminent Women in Science*. Washington, D.C.: Joseph Henry Press.

Byers, Nina and Gary A. Williams. 2006. *Out of the Shadows: Contributions of Twentieth-Century Women to Physics*. New York: Cambridge University Press.

Rudin, Mary Ellen (Estill)

b. 1924
Mathematician

Education: B.A., University of Texas, Austin, 1944, Ph.D., mathematics, 1949

Professional Experience: instructor, mathematics, Duke University, 1950–1953; assistant professor, University of Rochester, 1953–1957; lecturer, University of Wisconsin, Madison, 1959–1970, professor, mathematics, 1971–1991, emerita

Mary Ellen Rudin is renowned for her contributions to set-theory topology in mathematics, particularly the construction of counterexamples. Topology is an abstract geometry that looks at the properties of mathematical spaces. She entered the University of Texas with no specific plans for an area of study, but was mentored by an unorthodox mathematics research professor by the name of R. L. Moore. At the time she completed her Ph.D., in 1949, many universities were under pressure to hire women mathematicians, and Rudin found a job as an instructor at Duke University. She met and married mathematician Walter Rudin at Duke, and they moved to the University of Rochester and then on to the University of Wisconsin, Madison. At Madison, she was only able to work as a lecturer due to an anti-nepotism rule, but after the rules were changed in 1971, she was promoted to full professor. Each summer at the University of Wisconsin, there were mathematics conferences and collaborations, and in 1974, Rudin gave a series of 10 lectures that were subsequently published as *Lectures on Set Theoretic Topology* (1975). Rudin was considered one of the best-known female mathematicians of her generation. In 1991, a conference, "The Work of Mary Ellen Rudin," was held in her honor at Madison, and the proceedings were published in the *Annals of the New York Academy of Sciences* (1993).

Rudin's family was committed to the value of education. Both of her parents and even both of her grandmothers had college degrees, and her family insisted that girls as well as boys should have the opportunity for further education. She has exhibited outstanding dedication and service to her profession and has written more than 90 research papers or book chapters. She has been recognized nationally as well as internationally, serving on advisory boards for the National Academy of Sciences and National Science Foundation, and as a visiting professor at institutions in New Zealand, Mexico, and China. Rudin is a member of the American Mathematical Society, Mathematical Association of America, the Association for Women in Mathematics, and the Association for Symbolic Logic, and a fellow of the American Academy of Arts and Sciences. In 1995, she was elected to the Hungarian Academy of Sciences and has been honored by the Mathematical Society of The Netherlands.

Further Resources

Henrion, Claudia. 1997. *Women in Mathematics: The Addition of Difference*. Bloomington: Indiana University Press.

Murray, Margaret Anne Marie. 2000. *Women Becoming Mathematicians: Creating an Identity in Post–World War II America*. Cambridge, MA: MIT Press.

Rudnick, Dorothea

1907–1990
Embryologist

Education: B.A., languages, University of Chicago, 1928, Ph.D., zoology, 1931

Professional Experience: fellow, Yale University, 1931–1934; research fellow, University of Rochester, 1934–1937; assistant instructor, genetics, Storrs experiment station, University of Connecticut, 1937–1939; instructor, zoology, Wellesley College, 1939–1940; assistant professor, biology, Albertus Magnus College, New Haven, Connecticut, 1940–1948, professor, 1948–1977; research associate, Yale University, 1940–1971, associate fellow, 1969–1977

Concurrent Positions: Guggenheim fellow, 1952–1953; U.S. Public Health Service special fellow, 1965–1966

Dorothea Rudnick was recognized for her research in embryology, which focused on experimental embryology of the chick and rat; developmental genetics of the chick; and enzymatic development in the liver, brain, and retina of the chick. She discovered an interest in the sciences as an undergraduate taking courses in zoology and embryology. While still in graduate school, her research on chick embryos was published in a scientific journal. After receiving her doctorate from the University of Chicago in 1931, she was a fellow of the Osborne Zoological Laboratory at Yale and then a National Research Fellow at the University of Rochester. In 1937, she accepted a position as assistant instructor at the University of Connecticut and as instructor in zoology at Wellesley in 1939. She was appointed assistant professor of biology at Albertus Magnus in 1940, and advanced to full professor in 1948. Albertus Magnus was a small liberal-arts college with very limited laboratory facilities, so she maintained a lab at nearby Yale University, where she conducted studies on the embryology of the chick and rat. In 1952, she won a Guggenheim fellowship to conduct research and lecture in Europe. She retired from Yale in 1977.

Rudnick came from a family of scientists; her father was a chemist and her two brothers became physicists. She originally studied languages in college, however, but later combined her fluency in several European languages and her interest in science by writing book reviews of English, French, and German works and translating a biography of a German scientist, *Theodor Boveri: Life and Work of a Great Biologist*. For several years, she served as editor of the symposia of the Society for the Study of Growth and Development, and secretary and editor at the Connecticut Academy of Arts and Sciences. She was a member of the American Association for the Advancement of Science, American Society of Zoologists, American Association of Anatomists, Society for Developmental Biology, Tissue Culture Association, and International Institute of Embryology.

Russell, Elizabeth Shull

1913–2001
Geneticist

Education: A.B., zoology, University of Michigan, 1933; A.M., Columbia University, 1934; Ph.D., genetics, University of Chicago, 1937

Professional Experience: assistant, zoology, University of Chicago, 1935–1937; independent investigator, Roscoe B. Jackson Laboratory, Bar Harbor, Maine, 1937–1940, research associate, 1946–1957, senior staff scientist, 1957–1978, emeritus senior scientist, 1978–1988

Concurrent Positions: Guggenheim fellow, 1958–1959

Elizabeth Russell was a pioneer in genetics who bred and distributed millions of mice for scientific research around the world. Her early research was on fruit flies, but she became interested in mammalian genetics and began working with mice, the genetic makeup of which is 95 percent identical to that of humans. Her research focused on pigmentation, hereditary anemia, muscular dystrophy, cancer, and the genetic effects on aging. Her work on marrow transplants in mice had implications for later human treatments. She spent much of her career at the Roscoe B. Jackson Laboratory in Bar Harbor, Maine, which is known internationally for its research in breeding mice (now known by the brand name JAX Mice) to represent specific genetic conditions. When a fire broke out at the laboratory in 1947, it destroyed thousands of mice, and Russell was in charge of rebuilding the stock, a task that took another 10 years.

Russell came from a strong family background in science. Her father, Aaron Franklin Shull, taught zoology and genetics at the University of Michigan, and her mother, Margaret Jeffrey Buckley, had a master's degree in zoology and had taught at Grinnell College in Iowa. After receiving her master's degree from Columbia in 1934, Elizabeth Shull joined the University of Chicago as an assistant in zoology while she completed her doctorate. She received her Ph.D. in 1937 and, that same year, married fellow graduate student William L. Russell. The couple began working at the Jackson Laboratory, although Elizabeth was unable to secure a full-time permanent position due to anti-nepotism rules. She began her career as an independent investigator at the lab, and took several years off in the 1940s, probably due to family responsibilities, as the couple had four children. The couple divorced in 1947, however, and Elizabeth returned to work at Jackson as a research associate, then senior staff scientist. She organized a conference at Jackson of scientists from around the globe and subsequently received a Guggenheim fellowship to pursue her research on mammalian genetics. She formally retired in 1988 after several years as an emeritus senior scientist and traveled twice to Liberia, West Africa as a visiting instructor at Cuttington College.

Russell was elected to the National Academy of Sciences in 1972. She was a trustee for several colleges in Maine and in 1991 was inducted into the Maine Women's Hall of Fame. She was a member of the American Academy of Arts and Sciences, the Genetics Society of America (vice president, 1974; president, 1975–1976), the American Philosophical Society, the American Society of Naturalists, the Council of the National Institute on Aging, the Society for Developmental Biology, and the Union of Concerned Scientists.

Further Resources

Barker, Jane E. and Willys K. Silvers. 2002. "Elizabeth S. Russell. 1913–2001." *Biographical Memoirs*. 81. Washington, D.C.: National Academy Press. http://www.nap.edu/html/biomems/erussell.pdf.

S

Sabin, Florence Rena

1871–1953
Anatomist

Education: B.S., Smith College, 1893; M.D., Johns Hopkins University, 1900

Professional Experience: school teacher, 1894–1895; assistant, zoology, Smith College, 1896; intern, Johns Hopkins Hospital, 1900–1901, assistant to associate, anatomy, School of Medicine, Johns Hopkins University, 1902–1905, associate professor, 1905–1917, professor, histology, 1917–1925; head, Department of Cellular Studies, Rockefeller Institute for Medical Research, New York, 1925–1938; head, Committee on Health, Colorado, 1945–1947; manager, Department of Health and Charities, Denver, Colorado, 1947

Florence Sabin is regarded as one of the outstanding woman scientists in the medical field in the first half of the twentieth century. As a medical student, she had shown great interest in research, and she published her first paper, on the nuclei of cochlear and vestibular nerves, during her second year. She received her medical degree in 1900 and chose to continue in research and teaching. She was the first female faculty member at Johns Hopkins Medical School when she was appointed in 1902 and was the first woman to advance to full professor in 1917. Her major areas of research were the origin of the lymphatic vessels, the study of red and white corpuscles, and the pathogenesis of tuberculosis. Her first research efforts were in a controversial field, the origin of lymphatic vessels. By using the approach of injecting lymphatic channels with India ink, she demonstrated that the vessels derived from the venous system. This work caused considerable controversy but ultimately was acclaimed as a highly significant contribution. Other important contributions included the development of supravital staining techniques for living cells and the identification of the monocyte as a definitive type of white blood cell.

Sabin left Johns Hopkins and accepted a position as a member of the Rockefeller Institute in 1925, where she conducted significant research on tuberculosis before retiring as emeritus member in 1938. After she retired, she returned to her native Colorado and continued her involvement with public health issues. She began what was, effectively, another career when she was asked to head the Colorado State

Committee on Health, established after the end of World War II. Her first project was to conduct health surveys of Colorado residents, and she drafted several pieces of public-health legislation to address the high mortality rate and poor healthcare systems in the state. In 1947, she was appointed head of Health and Charities for the city of Denver, and set out to improve public health and hygiene standards in hospitals and restaurants, and to promote preventative healthcare in identifying tuberculosis and other contagious diseases.

Sabin was the first woman elected to the National Academy of Sciences (1925). She was also the first woman president of the American Association of Anatomists (1924–1926). She received honorary degrees from a dozen universities. Among her other honors and awards were the National Achievement Award (1932), M. Carey Thomas Prize (1935), Trudeau Medal of the National Tuberculosis Association (1945), and Albert Lasker Public Service Award (1951). A bronze statue was placed in her honor in Statuary Hall in Washington, D.C. In addition to her numerous scientific papers, she was the author of *An Atlas of the Medulla and Mid-Brain* (1901) and *Biography of Franklin Paine Mall* (1934). She was a member of the American Association for the Advancement of Science, the American Physiological Society, the Society for Experimental Biology and Medicine, the Harvey Society, and the National Tuberculosis Association, and an honorary member of the New York Academy of Sciences.

Further Resources

National Institutes of Health. "The Florence R. Sabin Papers." Profiles in Science, National Library of Medicine. http://profiles.nlm.nih.gov/RR/.

Sager, Ruth

1918–1997
Geneticist

Education: B.S., University of Chicago, 1938; M.S., plant physiology, Rutgers University, 1944; Ph.D., genetics, Columbia University, 1948

Professional Experience: Merck postdoctoral fellow, Rockefeller Institute for Medical Research, 1949–1951, assistant biochemist, 1951–1955; research associate, zoology, Columbia University, 1955–1960, senior research associate, 1960–1966; professor, biology, Hunter College, 1966–1975; professor, cellular genetics and chief, Division of Cancer Genetics, Dana-Farber Cancer Institute, Harvard Medical School, 1975–1988

Ruth Sager was a geneticist who pioneered the development of experimental material for the analysis of nonchromosomal heredity, called *non-Mendelian inheritance* or *cytoplasmic inheritance*. Her research interests included organelle genetics and biogenesis, mammalian cell genetics, genetic mechanisms of carcinogenesis, tumor suppressor genes, and breast cancer. Sager began her scientific career as a graduate student at Columbia University, where she studied plant genetics and was heavily influenced by the work of contemporary renowned geneticist **Barbara McClintock**. Sager held a postdoctoral fellowship to work with a microbiologist at the Rockefeller Institute for Medical Research, and she concentrated her own research on chloroplast DNA. She returned to Columbia as a research associate and collaborated with Professor Francis Ryan on their book on *Cell Heredity*, published in 1961.

Geneticist Ruth Sager, 1964. (AP/Wide World Photos)

Sager's work changed the way biologists think about cell heredity. Still, there was a long delay in recognizing her achievements in academia, and she was not appointed a full professor until she moved to Hunter College in 1966, 18 years after receiving her doctorate. She had moved toward cancer research when she spent a year in London at the Imperial Cancer Research Fund Laboratory as a Guggenheim fellow. In 1975, she was invited to head the Dana-Farber Cancer Institute at Harvard, where she and her colleagues researched the growth of cancer cells and the search for tumor-suppressor genes, and had some success working with breast cancer cells. She retired from Harvard in 1988 but continued to be a voice for cancer research. She died of bladder cancer in 1997.

Sager was elected to the National Academy of Sciences in 1977 and the Institute of Medicine in 1992. Among her honors and awards, she received the Gilbert Morgan Smith Medal of the National Academy of Sciences (1988). She was a member of the American Society for Cell Biology, International Society for Cell Biology, Genetics Society of America, American Academy of Arts and Sciences, American Society of Biological Chemists and Molecular Biologists, American

Society of Naturalists, American Association of Cancer Research, and American Society of Human Genetics.

Further Resources

"Ruth Sager, Faculty of Medicine—Memorial Minute." *Harvard University Gazette.* (4 November 2004). http://news.harvard.edu/gazette/2004/11.04/16-mm.html.

Saif, Linda

b. 1947
Microbiologist, Animal Scientist

Education: B.A., College of Wooster, Ohio, 1969; M.S., microbiology, Ohio State University, 1971, Ph.D., microbiology and immunology, 1976

Professional Experience: postdoctoral research associate, Ohio Agricultural Research and Development Center (OARDC), Department of Veterinary Science, Ohio State University, 1976–1979, assistant to associate professor, Food Animal Health Research Program, 1979–1990, Distinguished University Professor, 1990–

Linda Saif is a microbiologist whose work on animal viruses gained international attention during the global SARS (Severe Acute Respiratory Syndrome) outbreak of 2002 and 2003. Her research focuses on animal digestive and respiratory viral infections, bases for immunities, the development of vaccines, and foodborne illnesses. Saif, a professor of food animal sciences at the OARDC, was called upon by the World Health Organization (WHO) and the U.S. Centers for Disease Control as head of one of the only laboratories in the world that had conducted research on a deadly coronavirus that had caused a global lethal infection of pigs; the virus could be transmitted from animals to humans, and was believed to be the pathogen responsible for SARS. Few biologists studying human viruses had encountered coronaviruses, and Saif and her husband, Mo Saif (also affiliated with the OARDC), consulted on ways to detect, respond to, and stop the spread of the SARS virus that killed nearly 1,000 people in less than one year. Her lab was invited to join the WHO's elite network of International Reference Laboratories, an affiliation that has attracted graduate students and grant money to support her work.

Saif was raised in Ohio and spent much of her childhood in and around her grandparents' farm. She earned her bachelor's degree from the College of Wooster in 1969. She briefly attended Case Western Reserve University before receiving her master's and doctorate in microbiology and immunology from Ohio State University in 1971 and 1976, respectively. She stayed on at the university, rising through the ranks from postdoctoral researcher to assistant professor, and in 2002 became a Distinguished University Professor, the highest faculty honor.

Saif was elected to the National Academy of Sciences in 2003. That same year, she received an honorary doctorate from Ghent University in Belgium. She has also received the Beecham Laboratories Award for Research Excellence (1989), a Distinguished Veterinary Immunologist Award from the American Association of Veterinary Immunologists (1995), and University Distinguished Scholar Awards from the Ohio State University (1995 and 2002). She is an honorary diplomate of the American College of Veterinary Microbiologists, an elected fellow of the American Association for the Advancement of Science, and a member of the Conference of Research Workers in Animal Diseases, American Society of Virology, and American Association of Veterinary Immunologists.

Further Resources

Ohio State University. "Dr. Linda Saif Laboratory." http://www.oardc.ohio-state.edu/lsaiflab/.

Sammet, Jean Elaine

b. 1928
Computer Scientist

Education: B.A., mathematics, Mount Holyoke College, 1948; M.A., mathematics, University of Illinois, 1949

Professional Experience: teaching assistant, mathematics, University of Illinois, 1948–1951; dividend technician, Metropolitan Life Insurance Company, 1951; teaching assistant, mathematics, Barnard College, 1952–1953; engineer, Sperry Gyroscope Company, 1953–1958; section head, MOBIDIC Programming, Sylvania Electric Products Company, 1958–1959, staff consultant, program research, 1959–1961; Boston advanced program manager, International Business Machines (IBM) Corporation, 1961–1965, program language technical manager, 1965–1968, program technology planning manager, 1968–1979, division software technical manager, 1979–1983, program language technology manager, 1983–1986, senior technical staff member, 1986–1988; consultant, 1989–

Concurrent Positions: lecturer, Adelphi College, 1956–1958, Northeastern University, 1967, University of California, Los Angeles, 1967–1972, Mount Holyoke College, 1974

Jean Sammet is renowned for her professional contributions to the use of computers for nonnumerical mathematics and for developments in the theory of high-level programming languages. She is most famous for her work on the design and development of COBOL and FORMAC, the most widely used programming language in the world from the late 1960s through the 1970s, primarily for commercial applications. She studied mathematics in college and graduate school, and began work on computers at Sperry Gyroscope in 1955 as supervisor of their first scientific programming group. At the same time, she was a lecturer on digital computer programming at Adelphi College, where she also taught one of the earliest courses on FORTRAN in the United States. She moved to Sylvania Electric Products in 1958 and oversaw the development of software for MOBIDIC, the U.S. Army computer system. During her Sylvania years, she was involved in the initial creation of COBOL.

In 1961, Sammet began her long association with IBM to organize and manage the Boston Programming Center. She initiated the concept, and directed the development of, FORMAC (FORmula MAnipulation Compiler), the first widely used general language and system for manipulating nonnumeric algebraic expression. In 1965, she became programming language technology manager and then moved to the IBM Federal Systems Division in 1968, where she held various positions involving planning, internal consulting, and lecturing on programming languages. In 1969, she published a book, *Programming Languages: History and Fundamentals*, recognized as the "standard work on programming languages" and "an instant computer classic." In 1979, she began work on "Ada," the first programming language developed for the U.S. Department of Defense. In 1986, she was named a senior technical staff member; she formally retired from IBM in 1988, but continued to consult for the company.

Sammet was elected to membership in the National Academy of Engineering in 1977 and in 1978 received an honorary doctorate from her alma mater, Mount Holyoke College. Among her numerous other awards are IBM's Outstanding Contribution Award (1965), Mount Holyoke College Alumnae Association Centennial Award (1972), Association for Computing Machinery (ACM) Distinguished Service Award (1985), Augusta Ada Lovelace Award of the Association for Women in Computing (1989), Distinguished Service Award of the ACM Special Interest Group on Programming Languages (SIGPLAN) (1997), and a Fellow Award of the Computer History Museum (2001). She is a member of the Mathematical Association of America and the Association for Computing Machinery, serving as president from 1974 to 1976. She was also a member of the board of directors of the Computer Museum in Boston (1983–1993).

Sarachik, Myriam Paula (Morgenstein)

b. 1933
Physicist

Education: B.A., physics, Barnard College, 1954; M.S., Columbia University, 1957, Ph.D., physics, 1960

Professional Experience: research assistant, solid-state physics, International Business Machines (IBM) Watson Laboratory, Columbia University, 1955–1960, research associate, 1960–1961; member of technical staff, AT&T Bell Laboratories, 1962–1964; assistant to associate professor, physics, City College of New York, 1964–1971, professor, 1971–1995, distinguished professor, 1995–

Concurrent Positions: principal investigator, U.S. Air Force research grant, 1965–1972, National Science Foundation grant, 1972–1974; executive officer, graduate program in physics, City College of New York, 1975–1978

Myriam Sarachik is an experimental condensed-matter physicist who is renowned for her research on superconductivity, disordered metallic alloys, metal-insulator transitions in doped semiconductors, hopping transport in solids, properties of strongly interacting electrons in two dimensions, and spin dynamics in molecular magnets. She was born Myriam Morgenstein in Antwerp, Belgium, and when she was just seven years old, her family began their escape from the Nazis that would take them to France, then Cuba, and on to New York. She attended the prestigious Bronx High School of Science and then majored in physics at Barnard College. While at Columbia working on a master's and then doctorate in physics, she worked as a research assistant and then a research associate in the IBM Watson Laboratory. After receiving her Ph.D. she worked on the technical staff of AT&T Bell Laboratories for two years, and then became an assistant professor of physics at the City College of New York. She rose through the ranks to become full professor in 1971 and then distinguished professor (the highest faculty rank) in 1995, a position she still holds.

Sarachik was elected to membership in the National Academy of Sciences (NAS) in 1994 and in 2008 was elected to the NAS 17-member Governing Council. Among her other most recent awards are the New York City Mayor's Award for Excellence in Science and Technology (1995), the Sloan Public Service Award from the Fund for the City of New York (2004), the Oliver E. Buckley Prize in Condensed Matter Physics (2005), and the L'Oréal-UNESCO Award for Women in Science (2005). She has served on numerous national and international boards and advisory panels, including for the National Science Foundation,

U.S. Department of Energy, American Institute of Physics, National Research Council, Zernike Institute for Advanced Materials of the University of Groningen, Netherlands, and Science Advisory Committee of the Hong Kong University of Science and Technology. She is a fellow of the American Physical Society (vice president, 2001; president, 2003), New York Academy of Sciences, American Academy of Arts and Sciences, and American Association for the Advancement of Science. In 2008, she became co-chair (with astronomer **Vera Rubin** and geneticist **Maxine Singer**) of a National Academy of Sciences project to pair women scientists in the United States with Iraqi women scientists for mentoring and career support.

Further Resources

National Academy of Sciences. 2008. "International Twinning Project for Iraqi Women Scientists, Engineers, and Health Professionals." Committee on Human Rights. (March 2008). http://sites.nationalacademies.org/PGA/humanrights/PGA_044086.

Wasserman, Elga. 2002. *The Door in the Dream: Conversations with Eminent Women in Science*. Washington, D.C.: Joseph Henry Press.

Byers, Nina and Gary A. Williams. 2006. *Out of the Shadows: Contributions of Twentieth-Century Women to Physics*. New York: Cambridge University Press.

Savitz, Maxine (Lazarus)

b. 1937
Organic Chemist, Electrochemist

Education: B.A., chemistry, Bryn Mawr College, 1958; Ph.D., organic chemistry, Massachusetts Institute of Technology, 1961

Professional Experience: National Science Foundation fellow, University of California, Berkeley, 1961–1962; instructor, chemistry, Hunter College, 1962–1963; research chemist, Electric Power Division, U.S. Army Engineering Research and Development Laboratory, Fort Belvoir, 1963–1968; associate professor, chemistry, Federal City College, 1968–1971, professor, 1971–1972; professional manager, Research Applied to National Needs, National Science Foundation, 1972–1973; chief, buildings conservation policy research, Federal Energy Administration, 1973–1975; division director, buildings and industrial conservation, Energy Research and Development Administration, 1975–1976, division director, buildings and community systems, 1976–1979, Deputy Assistant Secretary of Conservation, Department of Energy, 1979–1983; president, Lighting Research Institute, 1983–1985;

assistant to vice president of engineering, Ceramic Components Division, Garret Corporation, 1985–1987; general manager, Ceramic Components Division, AlliedSignal Aerospace Company (now Honeywell, Inc.), 1987–2000, general manager, Technology Partnerships, 2000–2006; vice president, National Academy of Engineering, 2006–

Maxine Savitz is an organic chemist who is recognized for her expertise in research management in both government and industry. Her research includes free radical mechanisms, anodic hydrocarbon oxidation, fuel cells, more efficient use of energy in buildings, community systems, appliances, agriculture and industrial processes, transportation, batteries and other storage systems, new materials, and advanced structural ceramic materials. She has spent recent years serving as general manager of different divisions at Honeywell, Inc. (formerly AlliedSignal) working on ceramics for aerospace applications. Earlier, she was an executive with the U.S. Department of Energy, establishing energy-saving guidelines for buildings during the oil crises of the 1970s. Recommendations of her team in the areas of longer-burning lighting, new batteries, and new technologies, and development of alternative fuels for vehicles and improved public transportation, were among the measures mandated by the Energy Conservation and Production Act of 1976.

Savitz was elected to membership in the National Academy of Engineering (NAE) in 1992, and in 2006 was elected to a four-year term as vice president of the NAE. Among her committee appointments are the Energy and Engineering Board of the NAE, the Office of Technical Assessment of the U.S. Congress Energy Demand Panel, the natural materials advisory board, National Research Council, the advisory committee of the division of ceramics/materials, Oak Ridge National Laboratory, the advisory board for the secretary of Energy, and the Defense Science Board. She is one of the directors of the Washington Advisory Group, and in 2009, she was appointed by President Obama to the Council of Advisors on Science and Technology (PCAST). Savitz is a member of the American Association for the Advancement of Science and American Ceramic Society, and a fellow of the California Council on Science and Technology (CCST).

Scarr, Sandra (Wood)

b. 1936
Psychologist

Education: B.A., sociology, Vassar College, 1958; M.A., Harvard University, 1963, Ph.D., psychology, 1965

Professional Experience: instructor, University of Maryland, 1964–1965, assistant professor, psychology, 1965–1966; lecturer, University of Pennsylvania, 1967–1968, assistant to associate professor, 1968–1971; associate professor, University of Minnesota, 1971–1973, professor, 1973–1977; professor, Yale University, 1977–1983; Commonwealth Professor of Psychology, University of Virginia, 1983–1995; chief executive officer and chair, KinderCare Learning Centers, Inc., 1995–1997, director, 1997–1999

Concurrent Positions: visiting associate professor, Bryn Mawr College, 1969; fellow, Center for Advanced Studies in the Behavioral Sciences, Stanford University, 1976–1977; visiting professor, Gothenburg, Stockholm, and Uppsala Universities, Sweden, 1993–1994

Sandra Scarr is renowned for her research on how genetics, psychology, and environment can inform public policy debates. Her research interests include genetic variability in human behavior, particularly intelligence and personality, and the effects of variation in the quality of home and childcare environments on children's development. She has investigated how the family influences personality development, intelligence, and school achievement, and what effects interventions such as preschool programs have on children. She is considered an expert on daycare systems and adoption with her studies on the correlation between environment and intelligence. Scarr attracted national attention with her research on controversial topics such as daycare, racial differences in IQ and school performance, and the effects of lead exposure on the IQs of children.

Scarr became interested in child development as an undergraduate studying sociology at Vassar in the 1950s, at a time when the subject was only beginning to garner serious intellectual consideration. After graduation, she worked for social service agencies, where she began to differentiate between the psychological and economic needs of different client groups. She went on to study psychology in graduate school at Harvard and, in 1967, began researching why black children perform so poorly in school and on intelligence tests. After 10 years of research, she concluded that such performance was owing to sociocultural disadvantage; her work was published as *Race, Social Class, and Individual Differences in IQ* (1981). As a career woman with four children, she combined her background in child development with research on daycare to develop expertise on the controversial subject. Recognizing that childcare is an important social and economic necessity that allows women to participate in the labor force, she argued against the concern that childcare may have an adverse effect on the emotional development of children, looking at the role of parental attachment and anxiety over separation. Her book, *Mother Care/Other Care* (1984), received the National Book Award of the American Psychological Association, and she has published hundreds of articles and reviews on the topic. In the mid-1990s, she was chief executive officer and then director of a national childcare chain, KinderCare.

Scarr's awards and recognitions include the Distinguished Contribution to Research on Public Policy of the American Psychological Association (1988), James M. Cattell Award of the American Psychological Society (1993), and Dobzhansky Award for Lifetime Achievement from the Behavior Genetics Association. She is a fellow of the American Association for the Advancement of Science and American Academy of Arts and Sciences, and a member of the American Psychological Association, Behavior Genetics Association (president, 1985–1986), and Society for Research in Child Development (president, 1989–1991), and a founding member of the American Psychological Society (president, 1996). She is identified in some sources as Scarr-Salapatek.

Further Resources

O'Connell, Agnes N. and Nancy Felipe Russo, eds. 2001. *Models of Achievement: Reflections of Eminent Women in Psychology.* Vol. 3. Mahwah, NJ: Lawrence Erlbaum Associates.

Scharrer, Berta Vogel

1906–1995
Neuroendocrinologist

Education: Ph.D., zoology, University of Munich, 1930

Professional Experience: assistant, Research Institute of Psychiatry, University of Munich, 1932–1934; guest investigator, Neurological Institute, Frankfurt, 1934–1937; guest investigator, Department of Anatomy, University of Chicago, 1937–1938; guest investigator, Rockefeller Institute, 1938–1940; senior instructor, Western Reserve University, 1940–1946; instructor to assistant professor, University of Colorado, 1946–1954; professor, anatomy, Albert Einstein Medical College, 1955–1995

Concurrent Positions: Guggenheim fellow, 1947–1948

Berta Scharrer and her husband, Ernst Scharrer, pioneered the research on neurosecretion that helped to create a new discipline in physiology, that of neuroendocrinology. Neurosecretion is the theory that nerves secrete hormones into the blood. Among the most important of the couple's findings was the discovery that, in both mammals and insects, there were two completely analogous neuroendocrine organ systems, each of which controlled a variety of non-nervous processes. Her other research interests included comparative endocrinology, ultra-structure, and neuroimmunology. Berta Scharrer concentrated on invertebrates while her husband studied vertebrates and, therefore, even though they worked together, they produced few joint publications. Although she held several presti-gious research positions at institutions in the United States and Europe, due to anti-nepotism rules, she was unable to obtain a full-time faculty appointment until the couple joined the Albert Einstein Medical College. In several interviews, Scharrer said the situation was to her advantage because she could concentrate on research without the burden of administrative responsibilities and pressure to publish. Scharrer remained at Albert Einstein for 40 years, formally retiring just months before her death in 1995.

Scharrer was elected to the National Academy of Sciences in 1967. She received honorary degrees from eight universities and numerous awards, including the Kraepelin Gold Medal Award (1978), the Koch Award (1980), the Henry Gray Award (1982), the Schleiden Medal (1983), and the National Medal of Science from the National Science Foundation (1983). She was elected president of the American Association of Anatomists (1978–1979), and was an honorary member of the American Society of Zoologists and the International Society of Neuro-endocrinology. She was a member of the American Academy of Arts and Sciences and several German and other European scientific academies.

Further Resources

Purpura, Dominick P. 1998. "Berta V. Scharrer. December 1, 1906–July 23, 1995." *Bio-graphical Memoirs*. 74: 288–307. Washington, D.C.: National Academy Press. http://books.nap.edu/openbook.php?record_id=6201&page=288.

Schwan, Judith A.

1925–1996
Chemical Engineer

Education: B.S., chemical engineering, University of Cincinnati, 1948; M.S., physical chemistry, Cornell University, 1950

Professional Experience: research chemist to senior chemist, Emulsion Research Division, Eastman Kodak Laboratories, 1950–1965, laboratory head, 1965–1968, assistant director, 1968–1971, director, 1971–1975, assistant director, Kodak Research and Development Laboratories, 1975–1987

Judith A. Schwan was a chemical engineer who helped develop new types of film during her more than 35-year career at Eastman Kodak Laboratories in Rochester, New York. She eventually received more than 20 patents for her new research processes and development of new products related to Kodachrome and Kodacolor brand color negative films, print films, and Ektachrome motion picture films. Schwan was part of a generation of women who entered into engineering during the post–World War II boom in technological and scientific careers. She completed her undergraduate work at the University of Cincinnati, and went on to graduate school at Cornell University in New York, where she majored in physical chemistry and took numerous courses in chemical engineering. She began working on new product development at the Kodak Emulsion Research laboratory immediately after receiving her master's degree in 1950. She rose to senior research chemist and laboratory head, and held a variety of management positions, including assistant director and director of the Research and Development Laboratories.

Schwan was elected to the National Academy of Engineering in 1982. She received a Distinguished Alumnus Award from the University of Cincinnati, the Athena Award of Rochester, New York Chamber of Commerce, and the Technicolor-Herbert T. Kalmus Gold Medal Award of the Society of Motion Picture Engineers (1979). She was on the council of the Industrial Research Institute (1979–1981) and was a member of the Society of Motion Picture and Television Engineers, American Chemical Society, and Society of Photographic Scientists and Engineers.

Further Resources

Thomas, Leo J. 2002. "Judith A. Schwan. 1925–1996." *Memorial Tributes: National Academy of Engineering.* 10: 206–209. Washington, D.C.: National Academy Press. http://books.nap.edu/openbook.php?record_id=10403&page=206.

Schwarzer, Theresa Flynn

b. 1940
Geologist, Petroleum Geologist

Education: B.S., Rensselaer Polytechnic Institute, 1963, M.S., 1966, Ph.D., geology, 1969

Professional Experience: instructor, geology, State University of New York at Albany, 1969; research fellow, Rice University, 1969–1972; senior research geologist, Exxon Company, 1972–1974, research specialist, 1974–1976, senior research specialist, 1976–1978, senior explorer geologist, Gulf Coast Division, Exxon USA, 1978–1980, project leader, Texas Offshore Division, 1980–1981, district production geologist, East Texas Division, 1981–1983, senior supervisor, Exxon Production Research Company, 1983–1987, geological advisor, Exxon USA, 1987–

Theresa Schwarzer is a geologist who has been recognized for her expertise in petroleum exploration. Her research interests include inorganic and organic geochemistry; remote sensing; multivariate statistical techniques; and interpretation and integration of geophysical, geological, and geochemical data for hydrocarbon exploration. For more than 35 years, she has worked for Exxon Corporation in increasing levels of responsibility for research in hydrocarbon exploration. Among her achievements are the discovery of commercial oil and gas deposits, and research on and development of unconventional exploration methods. As a geologist, she relies upon detailed maps, soil and rock analyses, soundings, and other details to conduct remote sensing of a potential site for exploration. Diminishing energy, mineral, and water resources, and increasing environmental and political concerns over drilling for oil, have placed a premium on the unique qualifications of geoscientists.

Schwarzer served as chair of the women geoscientists committee of the American Geological Institute from 1973 to 1977. She is a member of the Geological Society of America, American Association of Petroleum Geologists, Society for Exploration Geo-Physicists, and Geochemical Society.

Scott, Juanita (Simons)

1936–2001
Developmental Biologist

Education: A.A., Clinton Junior College, 1956; B.S., biology, Livingstone College, North Carolina, 1958; M.S., biology, Atlanta University, 1962; Ed.D., science education, University of South Carolina, 1979

Professional Experience: high school teacher, 1958–1960; instructor, biological science, Benedict College, 1963–1964; instructor, Morris College, 1965–1967; assistant to associate professor, Benedict College, 1968–1981, professor, 1981–1987, head, Division of Mathematics and Natural Sciences, 1987–1994, head, Department of Biological and Physical Science, 1992–1994, dean, Division of Arts and Sciences, 1994–2001

Juanita Scott was a developmental biologist known for her research on problems of water pollution in the rivers and streams of South Carolina. She started her research in the late 1960s on the pollutants being dumped into the waters by industrial firms, and by the mid-1980s, she had become interested in the microscopic characteristics of individual cells. She studied how pollutants, such as lead, cadmium, and mercury, act on different structures within a cell. Her research indicated that parts of a frog's skin cells are more likely to react to metal contamination than other parts of the skin cells. She and her team of student researchers found that a frog's skin not only repels some toxic compounds but also has some antibiotic properties. After receiving a doctorate in science education from the University of South Carolina in 1979, Scott did additional postdoctoral studies in biology, microbiology, and human sexuality at North Carolina State University, Columbia College, Clark College, and New York University, continuing her own research and publishing papers on environmental and cellular biology. She began teaching biology at Benedict College in Columbia, South Carolina, in 1963 and spent her entire career there in various teaching and administrative positions, including overseeing 10 academic departments as dean of the Division of Arts and Sciences.

Scott grew up on a farm near Columbia, South Carolina, that had no running water or electricity. Although there were 15 children in the family, her parents placed great emphasis on education. She graduated from high school at 16 and, although she did not have any particular ambition to be a scientist, had a good capacity for learning and did particularly well in her science courses. Influenced by her biology teacher, she decided to major in biology at Livingstone College, but also completed the courses for a teaching certificate. She then taught in a high school that, although it was relatively new, was segregated and not well-funded; for instance, there was no scientific equipment available in the laboratory.

After becoming a college instructor and administrator, she remained concerned about the quality of science teaching in middle and junior high schools, and found that many students arrived at college or university with little knowledge of the sciences and frequently had the attitude that all science courses were too hard. She developed summer science project workshops for middle school students (now operating as the Juanita S. Scott Middle School Summer Enrichment Program [MSSEP]), and worked with elementary and high school teachers under a National Science Foundation grant to develop math, science, and technology curricula, and improve the quality of instruction at each level by assuring that teachers understand the basic scientific concepts. For several years, she was involved with directing research, teaching biology, and conducting in-service training classes for teachers.

Further Resources

Benedict College. "Juanita Simons Scott, Ed.D." http://www.benedict.edu/news/accomplishments/bc-news-faculty_n_staff_accomplishments-juanita_simons_scott-20070515.html.

Seddon, Margaret Rhea

b. 1947
Physician, Astronaut

Education: B.A., physiology, University of California, Berkeley, 1970; M.D., University of Tennessee, 1973

Professional Experience: general surgery resident; medical doctor with a specialty in medical nutrition; astronaut program, National Aeronautics and Space Administration (NASA), 1978–1997; Assistant Chief Medical Officer, Vanderbilt University Medical Center, Nashville, Tennessee, 1997–2007; patient safety expert, Lifewing Partners LLC, 2007–

Margaret Rhea Seddon is a physician and retired astronaut who flew on three space shuttle flights. She was one of the six women who were first selected for the NASA astronaut program in 1978. She was the first woman to complete her training in 1979, but when plans for the first shuttle program were near completion, she learned she was pregnant and was unable to begin the training program; it was **Sally Ride** who became the first woman astronaut in space in 1983. Seddon was assigned to later missions as a payload specialist, launch-and-rescue helicopter physician, technical assistant to the director of flight-crew operations, and member of the Aerospace Medical Advisory Committee. While at NASA, she also worked part-time, when possible, as an emergency-room physician. All together, she logged more than 700 hours in space on three different missions (STS 51-D *Discovery* [1985], STS-40 *Columbia* [1991], and STS-58 *Columbia* [1993]); on these missions, she conducted experiments on the effects of gravity and on the effects of space flight on the cardiovascular, metabolic, musculoskeletal, and other systems. She retired from NASA in 1997.

After leaving NASA, Seddon became Assistant Chief Medical Officer at Vanderbilt University Medical Center (VUMC) in Tennessee. She was terminated from Vanderbilt in 2007 and subsequently filed a (still-pending) gender-discrimination lawsuit, claiming that VUMC "has not made a concerted effort . . . to recruit, encourage and attract high-level female physicians to key clinical leadership positions." Seddon also claims she did not receive supplemental pay as a faculty member, as did male colleagues in the same position. Seddon currently works with Lifewing Partners LLC, which provides patient-safety training to hospitals. Her work has been profiled in mainstream newspapers and magazines, and her research published in medical journals such as the *Journal of the American College of Surgeons* and the *American Journal of Clinical Nutrition*.

A recipient of many NASA and scientific awards, Seddon was named a Laurel Legend for her lifetime contributions to aviation by *Aviation Week and Space Technology* magazine in 2004, and in 2005, she was inducted into the Tennessee Aviation Hall of Fame.

Further Resources

Kevles, Bettyann H. 2003. *Almost Heaven: The Story of Women in Space*. New York: Basic Books.

National Aeronautics and Space Administration. "Margaret Rhea Seddon (M.D.)." http://www.jsc.nasa.gov/Bios/htmlbios/seddon.html.

Edgemon, Erin. 2008. "Ex-Astronaut Files Suit against Vanderbilt Medical Center." *Murfreesboro Post*. (19 August 2008). http://www.murfreesboropost.com/news.php?viewStory=12539.

Sedlak, Bonnie Joy

b. 1943
Cell Biologist, Developmental Biologist

Education: B.A., Northwestern University, 1965; M.A., Case Western Reserve University, 1968; Ph.D., biology, Northwestern University, 1974

Professional Experience: instructor, biology, Northwestern University, 1971–1972; research associate, biochemistry, Rush Medical College, 1974–1975; assistant professor, biology, Smith College, 1975–1977; assistant professor, biology, State University of New York, Purchase, 1977–1981; associate research scientist, University of California, Irvine, 1981–1985; sales representative, North American Science Associates, Irvine, 1986–1987; program manager, Microbics Corporation, 1987–1988; sensor analyst, Fritzsche, Pambianchi Associates, 1988–1990; biotechnology consultant, 1990–1991; business development and licensing manager, Becton Dickinson Advanced Cellular Biology, 1991–1992; licensing officer for technical transfer, University of California, Alameda, 1992–1994; independent consultant, 1994–

Bonnie Sedlak is a cell biologist whose early research focused on using the electron microscope to study cellular aspects of development and endocrine control in insects. She left research and teaching to work in industry as a business development and licensing manager and biotechnology consultant. Her clients have included healthcare research companies as well as universities and industry involved in medical and technical engineering. She received a doctorate in biology in 1974 and, after teaching and conducting research at several universities, accepted a position as a sales representative for North American Science Associates. She eventually worked in several locations as a licensing manager, overseeing the patent process for research and negotiating license agreements with companies, government agencies, and other universities in order to move research findings to marketable products.

Sedlak is a member of the American Society for Cell Biology, American Association for the Advancement of Science, Society for Developmental Biology, and Electron Microscopy Society of America.

Seibert, Florence Barbara

1897–1991
Biochemist

Education: A.B., Goucher College, 1918; Ph.D., physiological chemistry, Yale University, 1923

Professional Experience: chemist, Hammersley Paper Mill, 1918–1920; instructor, pathology, University of Chicago, and assistant, Sprague Memorial Institute, 1924–1928, assistant professor, biochemistry, 1928–1932; assistant to associate professor, Henry Phipps Institute, University of Pennsylvania, 1932–1955, professor, 1955–1959; director, cancer research laboratory, Mound Park Hospital Foundation, 1964–1966

Concurrent Positions: Guggenheim fellow, Sweden, 1937–1938; visiting lecturer, various schools, 1946–1948

Florence Seibert was a biochemist who purified the tuberculin PPD that is used worldwide in skin tests to detect tuberculosis, or TB. Her research interests also included intravenous therapy and blood transfusions, and the isolation of specific bacteria to give some immunity in cancer. After receiving her undergraduate degree from Goucher College, she originally considered medical school, but a professor helped her get a job as a chemist for a paper mill, possibly due to the shortage of male chemists during World War I. She returned to school to continue her graduate studies at Yale, where she received her doctorate in 1923. At Yale, her main breakthrough was development of a distillation method for removal of bacteria that could contaminate protein solutions used in blood transfusions. Previously, persons receiving such medical interventions were at high risk of infections and fevers due to bacteria. She held a postdoctoral fellowship at the University of Chicago that led to an instructorship in pathology, then a faculty position in biochemistry. At Chicago, she developed her method for purifying the proteins used in the TB test, which not only protected patients, but provided better diagnostic results. She was also affiliated with the Sprague Memorial Institute, and her

Biochemist Florence Seibert, ca. 1948. She developed the protein substance used for the tuberculosis skin test. (National Library of Medicine)

work was supported with funds from the National Tuberculosis Association (now the American Lung Association). Her method was later adopted as the standard by the World Health Organization.

In 1932, Seibert followed her mentor and collaborator and accepted a faculty position in biochemistry at the University of Pennsylvania. In 1937, she spent a year conducting research at the University of Upsala in Sweden, under a prestigious Guggenheim fellowship. Despite her continued achievements in improving the existing tuberculin skin test, she did not advance to full professor until 1955. Even after her formal retirement in 1959, she continued her research and volunteer activities on behalf of cancer research. In 1968, she published an autobiography, *Pebbles on the Hill of a Scientist*. She lived most of her life with her sister, Mabel, who served as her longtime research assistant.

Seibert was the author or co-author of dozens of scientific papers and articles. She was the recipient of five honorary degrees, as well as the Trudeau Medal from the National Tuberculosis Association (1938), the Garvan Medal of the American Chemical Society (1942), the Gimbal Award (1945), the Scott Award (1947), and the John Eliot Memorial Award of the American Association of Blood Banks

(1962). She was also a member of the American Association for the Advancement of Science. In 1990, she was inducted in the National Women's Hall of Fame.

Semple, Ellen Churchill

1863–1932
Geographer

Education: A.B., Vassar College, 1882, A.M., 1891; University of Leipzig, 1891–1892, 1895

Professional Experience: founder and teacher, Semple Collegiate School, 1893–1895; lecturer, geography, University of Chicago, 1906–1920 (intermittently); lecturer, anthropogeography Clark University, 1921–1923, professor, 1923–1932

Concurrent Positions: lecturer, Oxford University, 1912, 1922; Wellesley College, 1914–1915; University of Colorado, 1916; Columbia University, 1918

Ellen Semple was recognized by her contemporaries as one of the outstanding geographers of her time. After attending the University of Leipzig, she and her sister opened a private school in which she taught history. She combined this experience with her interest in geography in her book, *American History and Its Geographic Conditions* (1903). The publication of this work resulted in invitations to teach in the new department of geography at the University of Chicago. Her second book, *Influences of Geographic Environment, on the Basis of Ratzel's System of Anthropo-Geography* (1911), was viewed as one of the most scholarly books on geography at that time. A third book, published shortly before her death, was *The Geography of the Mediterranean Region: Its Relation to Ancient History* (1931).

After Semple received her undergraduate degree from Vassar in 1882, she returned home to teach in a private school. She opened her own school in 1893 after returning from Europe. Throughout her career, she rode into the backcountry of Kentucky to study the influence of geographic isolation on the life of the people there. Her research papers received favorable reviews. She taught off and on at the University of Chicago and also lectured at Oxford University, Wellesley College, the University of Colorado, and Columbia University. In 1921, she obtained a tenure-track faculty appointment at the new graduate department of geography at Clark University and was quickly promoted to professor. Semple received the Cullum Medal of the American Geographical Society (1914) and the Gold Medal of the Geographic Society of Chicago (1932). She received an honorary degree from the University of Kentucky in 1923, and in 1921, she was the first woman to be elected president of the Association of American Geographers. She also was a member of the American Geographical Society.

Shalala, Donna Edna

b. 1941
Political Scientist

Education: B.A., Western College, 1962; M.S., Syracuse University, 1968, Ph.D., political science, 1970

Professional Experience: volunteer (Iran), Peace Corps, 1962–1964; graduate research fellow, Maxwell School of Citizenship and Public Affairs, Syracuse University, 1966–1968; lecturer, social science and assistant to dean, 1968–1970; assistant professor, political science, Bernard M. Baruch College, City University of New York, 1970–1972; associate professor and chair, Program in Politics and Education, Teachers College, Columbia University, 1972–1979; Assistant Secretary for Policy Development and Research, Housing and Urban Development (HUD), Washington, D.C., 1977–1980; professor, political science and president, Hunter College, 1980–1987; professor, political science and chancellor, University of Wisconsin, Madison, 1987–1993; Secretary, U.S. Department of Health and Human Services, 1993–2001; professor, political science and president, University of Miami, 2001–

Political scientist Donna Shalala was appointed president of the University of Miami in 2001, after serving as Secretary of the Department of Health and Human Services under President Bill Clinton. (U.S. Department of Health and Human Services)

Concurrent Positions: visiting professor, Yale Law School, 1976; co-chair, Advisory Commission on Consumer Protection and Quality in the Health Care Industry, 1996–

Donna Shalala occupied one of the most influential offices in Washington, D.C., as Secretary of the U.S. Department of Health and Human Services under President Bill Clinton between 1993 and 2001. The agency is one of the largest in government and has one of the largest budgets, including funds for scientific research. In that capacity, she oversaw some of the most important government departments related to public health and policy, such as the National Institutes of Health, Centers for Disease Control, Food and Drug Administration, Social Security Administration, and the Indian Health Service, among others. One of her early actions was to escalate the budgets for cancer prevention at the National Cancer Institute and the Centers for Disease Control, with a special emphasis on breast cancer. She was concerned that women's issues were underfunded, underdiagnosed, and undertreated. Another of her goals was to shield scientific research from political pressure and excessive bureaucratic burdens. She also questioned the social values portrayed in many television programs and their effect on our society.

Shalala is a prominent political scientist who has held a variety of successful positions in both government and academic settings. She held professorships at several universities and is past president of Hunter College and chancellor of the University of Wisconsin, Madison, and is currently president of the University of Miami. Her first political position was as assistant secretary for policy development and research for HUD. During the early 1970s, she wrote four books: *Neighborhood Governance* (1971), *City and the Constitution* (1972), *Property Tax and the Voters* (1973), and *Decentralization Approach* (1974). She has been a member of the Committee on Economic Development (1991–1993), a member of the board of directors of the Institute of International Economics (1981–1993), a member of the Children's Defense Fund (1980–1993), and a trustee of the Brookings Institution (1989–1993). In 2006, she chaired the Committee on Maximizing the Potential of Women in Academic Science and Engineering, which investigated the absence and obstacles to women in high-level research positions in the sciences. In 2007, she was called upon by President George W. Bush to head a commission investigating allegations about conditions at Walter Reed Army Medical Center. Shalala received the Distinguished Service Medal of Teachers College, Columbia University, in 1989. She is a member of the American Political Science Association, American Society for Public Administration, and National Academy of Public Administration.

Further Resources

University of Miami. "President Donna E. Shalala's Biography." http://www.miami.edu/index.php/about_us/leadership/office_of_the_president/president_donna_e_shalalas_biography/.

Shapiro, Lucille (Cohen)

b. 1940
Molecular Biologist

Education: B.A., Brooklyn College, 1962; Ph.D., molecular biology, Albert Einstein College of Medicine, 1966

Professional Experience: assistant to associate professor, molecular biology, Albert Einstein College of Medicine, 1967–1977, professor, 1977–1986; Eugene Higgins Professor and chair, microbiology, College of Physicians and Surgeons, Columbia University, 1986–1989; Joseph D. Grant Professor and chair, developmental biology, School of Medicine, Stanford University, 1989–; co-founder and director, Anacor Pharmaceuticals, 2002–

Lucille "Lucy" Shapiro has had a distinguished career as a molecular biologist working on the genetics and biochemistry of the bacterial cell cycle and unicellular differentiation. After receiving her doctorate at Albert Einstein College of Medicine, she continued on as an assistant professor and rose through the ranks

Molecular biologist Lucille Shapiro. (Courtesy of the Stanford University News Service Library)

to full professor. She remained at Albert Einstein for 20 years before becoming professor and chair of microbiology at Columbia University. In 1986, she moved to Stanford in California, where she has served as chair and now director of the Beckman Center for Molecular and Genetic Medicine. In 2002, she co-founded Anacor Pharmaceuticals, a biopharmaceutical company developing new antimicrobial treatments for bacterial and fungal diseases and infections.

Shapiro's research has been published in numerous medical journals, including *Journal of Bacteriology, Journal of Molecular Biology, Cell, Molecular Biology of the Cell, Trends in Genetics*, and *Science*, and she has been a distinguished lecturer at a number of universities. Her expertise has been sought as a board member and scientific advisor in academia, government, and corporate settings, including for G. D. Searle Company, Massachusetts General Hospital, SmithKline Beecham, the Helen Hay Whitney Foundation, Whitehead Institute of the Massachusetts Institute of Technology, Harvard University, Howard Hughes Medical Institute, the president's council of the University of California, Silicon Graphics, Inc., and, most recently, Gen-Probe, a medical research company in San Diego. She has twice served as an American Cancer Society Established Investigator and was a nonexecutive director of GlaxoSmithKline (2001–2006).

Shapiro was elected a member of the Institute of Medicine of the National Academy of Sciences in 1991. She has received numerous awards, including the Alumna Award of Honor of Brooklyn College (1983), an Excellence in Science Award of the Federation of American Societies for Experimental Biology (FASEB) (1994), and the Selman Waksman Award of the National Academy of Sciences (2005). She is a fellow of the American Association for the Advancement of Science, the American Philosophical Society, and the California Council on Science and Technology (CCST), and a member of the American Society of Biochemistry and Molecular Biology, American Society for Microbiology, American Society for Cell Biology, Genetics Society of America, and New York Academy of Sciences.

Further Resources

Stanford University. Faculty website. http://med.stanford.edu/profiles/devbio/faculty/Lucille_Shapiro/.

Anacor Pharmaceuticals. http://www.anacor.com/.

Shaw, Jane E.

b. 1939
Physiologist, Clinical Pharmacologist

Education: B.S., University of Birmingham, England, 1961, Ph.D., physiology, 1964

Professional Experience: staff scientist, Worcester Foundation for Experimental Biology, 1964–1970; senior scientist, Alza Research, 1970–1972, principal scientist, 1972–; president, Alza Research Division, and chair of the board, Alza Ltd., 1985, executive vice president, Alza Corporation, 1985–1987, president and chief operating officer, 1987–1994; founder and consultant, Stable Network, 1994–; chair and chief executive officer, Aerogen, Inc. (now Nektar Therapeutics), 1998–2005; chairman of the board, Intel, 2009–

Concurrent Positions: director and committee chair, McKesson Corporation, 1992–; nonexecutive chairman and committee chair, Intel Corporation, 1993–; director, OfficeMax, 1994–2006; director, Talima Therapeutics, Inc.

Jane Shaw is renowned for research that led to the development of transdermal drug patches, such as those used for motion sickness. Her research includes elucidation of the physiological role of the prostaglandins, mechanism of action of analeptics, mechanism of gastric secretion, and physiology and pharmacology of the skin. As a graduate student at the University of Birmingham, England, she worked with Peter Ramwell identifying prostaglandins. After graduation, she and several other members of the research team followed Ramwell to the Worcester Foundation for Experimental Biology in Massachusetts, part of the much-publicized brain drain in England in the 1960s.

In 1970, she was invited to join Alza Corporation, a private company that manufactures pharmaceutical products and conducts commercial research and development on drug-delivery systems for human and veterinary use. Shaw holds several patents for technology that allows a patient to absorb a prescription drug through the skin from a bandage-like patch. Transdermal therapeutic systems for drug delivery are advantageous in chronic conditions such as hypertension because patients may forget to take medication when they have no symptoms. They are also advantageous when medications have to be given very frequently. Beginning as senior scientist, she moved quickly through the ranks to become president of the research division, executive vice president of Alza Corporation and board chair of the parent company, Alza Ltd., and then president and chief operating officer until 1994. She next founded her own biopharmaceutical firm, Stable Network, and served as a consultant. Between 1998 and her retirement in 2005, she served as chief executive officer at Aerogen, Inc. (now Nektar Therapeutics), a firm that develops drug-delivery devices for respiratory ailments. Shaw personally holds several patents in this area of research.

Shaw has consulted for numerous pharmaceutical research companies and has been a savvy businesswoman as well, serving on the boards of corporations such as OfficeMax and, most recently, chair at computer semiconductor manufacturer, Intel. She has published more than 100 professional articles and received an honorary doctorate from Worcester Polytechnic Institute in 1992. She is a member of the American Association for the Advancement of Science, New York Academy of Sciences, American Physical Society, American Society of Clinical Pharmacology and Therapeutics, American Association of Pharmaceutical Scientists, and American Pharmaceutical Association.

Shaw, Mary M.

b. 1943
Computer Scientist

Education: B.A., mathematics, Rice University, 1965; Ph.D., computer science, Carnegie-Mellon University, 1971

Professional Experience: systems programmer and researcher (part-time), Rice University Computer Project, 1962–1968; assistant professor, computer science, Carnegie-Mellon University, 1972–1977, senior research computer scientist, 1977–1982, associate professor, 1982–1986, chief scientist, Software Engineering Institute, 1984–1987, professor, 1986–, Alan J. Perlis Professor of Computer Science, 1995–

Computer scientist Mary Shaw. (Courtesy of the Carnegie Mellon University)

Concurrent Positions: member, Human Computer Interaction Institute, Carnegie-Mellon University, 1994–, fellow, Center for Innovation and Learning, 1997–1998, member scientist, Institute for Software Research, 1999–, co-director, Sloan Software Industry Center, 2001–2006

Mary Shaw is a renowned expert in computer software and a leading proponent of developing software engineering as a discipline. Her research includes software architecture, programming language design, abstraction techniques for advanced programming, software engineering, and computer-science education.

She has made major contributions to the analysis of computer algorithms as well as to abstraction techniques for advanced programming methodologies, programming language architecture, evaluation methods for software, performance and reliability of software, and software engineering. She developed computer programs called "abstract data types" as a method for organizing the data and computations used by a program so that related information is grouped together, and she created a programming language called "Aphard" that implemented those abstract data types. She thus made programs more user-friendly for the scientists who are using them to manipulate their research data.

Shaw grew up during the Cold War era of scientific and technological advances, and her father, a civil engineer and government economist, encouraged her interests in science and math. As a high school student, she participated in an after-school program that included a visit to an International Business Machines (IBM) facility and introduction to an early IBM computer program. For several summers during high school, Shaw worked at the Research Analysis Corporation of the Johns Hopkins University Operation Research Office, which gave her the opportunity to explore fields outside the normal school curriculum. Although there were no courses in computer science when she attended Rice University, she found a small group called the Rice Computer Project that had built a computer, the Rice I, under the direction of an electrical engineering faculty. Shaw joined the group and worked on a programming language, writing subroutines and studying how to make an operating system run more rapidly. She received her undergraduate degree in mathematics at Rice and went on to study computer science at Carnegie Mellon in Pennsylvania. After receiving her doctorate in 1971, she joined the faculty as the first female member of the Computer Science Department.

Shaw has been instrumental in developing innovative undergraduate and graduate computer-science curricula and degree programs. She was one of the early scientists to see the need for software engineering as a separate discipline. She even helped develop a curriculum for IBM to offer its own employees and founded the Software Engineering Institute at Carnegie Mellon. She has contributed to several books and published hundreds of scientific papers and reports.

For her contributions to software and systems development and education, Shaw has received the Warnier Prize (1993), the Stevens Award (2005), the Software Engineering Institute Award of Excellence (2006), and the Nancy Mead Award for Excellence in Software Engineering Education (2010). She is a fellow of the Association for Computing Machinery (ACM), the Institute for Electrical and Electronics Engineers (IEEE), and the American Association for the Advancement of Science (AAAS), and a member of the New York Academy of Sciences and the International Federation of Information Processing Societies (IFIPS).

Further Resources

Carnegie Mellon University. Faculty website. http://spoke.compose.cs.cmu.edu/shaweb/.

Sherman, Patsy O'Connell

1930–2008
Chemist

Education: B.A., Gustavus Adolphus College, 1952

Professional Experience: chemical researcher, Minnesota Mining and Manufacturing (3M), 1952–1992

Patsy Sherman was a chemist who, along with colleague Sam Smith, invented Scotchgard Fabric Protector, a moisture and stain repellent, while employed at 3M in the 1950s. The discovery of the substance was largely by accident, when someone in the lab spilled a new latex material onto a shoe and Sherman discovered it could not be washed off. She began to think of new possible applications for such a waterproof material, along with her supervisor, Smith, and in 1955, Sherman and Smith introduced Scotchgard, a protective coating for fabrics and other materials. Sherman began working for 3M immediately after graduating from college and remained there for 40 years, until her retirement in 1992. She rose through the ranks from research specialist to manager of the chemical resources division to head of technical development, and held several patents for fluorochemical polymers and processes.

The Scotchgard product made 3M a household name and earned the company millions of dollars, but in 2002 it was announced that 3M would remove Scotchgard from the market over environmental concerns. The property that made its chemical makeup attractive as a fabric protector, its insolvency or inability to be broken down, also made it potentially dangerous. Although tests of potential toxicity to humans and to the water supply remain inconclusive, elevated levels of perfluorochemicals have been found in the blood of company employees as well as in studies of certain animal species. In light of this research, 3M chose to exercise what they called "responsible environmental management" in phasing out the current chemical process used to create Scotchgard products. The product is still available as the company experiments with alternative formulas, and government organizations will continue to monitor the potential environmental and health effects of perfluorochemicals.

Sherman was committed to science education and was often an invited speaker to serve as a role model for young students. She received the Joseph M. Biedenbach

Distinguished Service Award of the American Society for Engineering Education in 1991. In 2001, she was inducted into the National Inventors Hall of Fame, one of only a handful of women to be acknowledged, and in 2002, she was one of 37 inventors who appeared at a celebration of the 200th anniversary of the U.S. Patent and Trademark Office. She was a longtime member of the American Chemical Society.

Shields, Lora Mangum

1912–1996
Biologist

Education: B.S., biology, University of New Mexico, 1940, M.S., 1942; Ph.D., botany, University of Iowa, 1947

Professional Experience: associate professor, biology, New Mexico Highlands University, 1947–1954, professor and department head, 1954–1978, director, Environmental Health Division, 1971–1978; researcher and visiting professor, Navajo Community College, Shiprock, New Mexico, 1978–retirement

Lora Shields was been recognized for her research on the effects of nuclear bomb testing on Southwestern plants and vegetation, and the human health hazards from mining uranium. She was the first Native American (Navajo) to receive a doctorate in botany. She studied at the University of New Mexico, and after receiving her Ph.D. from the University of Iowa, she returned to New Mexico as associate professor of biology at New Mexico Highlands University. She was promoted to full professor and department head in 1954, and named director of the Environmental Health Division in 1971. A few years later, she took a position as appointed researcher and visiting professor at Navajo Community College. By the 1970s, the U.S. government was mining uranium almost exclusively from Southwestern Native American lands, and Shields was committed to examining the health and environmental impact of this development. Her research focused on nuclear effects on vegetation, birth anomalies in the Navajo uranium district among miners and other inhabitants, effects of radiation exposure on plants, and streptococcal disease among the Navajo Indian population. Her work was supported by grants from the National Institutes of Health, National Science Foundation, March of Dimes Birth Defects Foundation, and Minority Biomedical Research Support, among others. She also received research grants from some of the agencies that recently declassified data regarding the effects of nuclear testing on humans, such as the Atomic Energy Commission, and from pharmaceutical companies.

Dedicated to science education at all levels, Shields was involved throughout her career with the New Mexico Academy of Science (NMAS) as secretary-treasurer (1951–1953), president (1954), and recipient of the NMAS Distinguished Scientist Award (1965); she also served as state representative to the National Association of Academies of Science (1960–1984) and became president of the NAAS (1976). For many years, she was editor of the *New Mexico Journal of Science*. She was a member of the American Association for the Advancement of Science and the Ecological Society of America.

Shipman, Pat

b. 1949
Paleoanthropologist

Education: B.A., Smith College, 1970; M.A., New York University, 1974, Ph.D., anthropology, 1977

Professional Experience: visiting lecturer, anthropology, Jersey City State College, 1974; adjunct instructor, Fordham University, 1975; editor and research associate, American Institutes for Research, 1976–1978; associate research scientist, Department of Earth and Planetary Sciences (joint appointment, Department of Cell Biology and Anatomy), Johns Hopkins University, 1978–1981, assistant professor, cell biology and anatomy, 1981–1986, assistant dean, Academic Affairs, School of Medicine, 1985–1990, associate professor, 1986–1995; independent author, 1990–

Concurrent Positions: editor, *Anthroquest*, 1990–1992; adjunct professor, biological anthropology, Pennsylvania State University, 1995–

Pat Shipman is a paleoanthropologist who spent many years in Kenya as a research scientist, excavating paleontological and archaeological sites, and examining fossils stored there. Her research focused on trying to deduce the environmental context in which our earliest ancestors evolved and what their lifestyles and adaptations were like. She is particularly interested in the history of science and how scientific information is used. She is the co-author of *The Neandertals: Changing the Image of Mankind* (1993), which focuses on how the interpretations of these finds have fluctuated through the gradual accumulation of information on both the anatomical characteristics and the geographical distribution of the remains. The central theme is how scientific opinion on the Neandertals has tended to shift between two extreme positions: the people who see them as being in the main course of human evolution, and those who see them as representing a sideline of human population.

Shipman's next book, *The Evolution of Racism: Human Differences and the Use and Abuse of Science* (1994), traces the attempts of scientists from the mid-nineteenth century to the present to grapple with the issues of race, from evolution, to eugenics, to intelligence testing and debates about immigration. In *Taking Wing: Archaeopteryx and the Evolution of Bird Flight* (1998), she draws on diverse scientific fields to give a comprehensive analysis of the ideas that explain how the adaptations needed for animal flight came about. Since leaving a full-time academic position in 1995, Shipman has been committed to bringing scientific information and debates to the general public. She has published on numerous scientific topics in popular science magazines and appeared on several television documentaries, such as "In Search of Human Origins" in 1997. In addition to her numerous articles, she has authored or co-authored more than 10 books on scientists and the history of the science, the most recent including *The Man Who Found the Missing Link: Eugene Dubois' Lifelong Quest to Prove Darwin Right* (2001), *To the Heart of the Nile: Lady Florence Baker and the Exploration of Central Africa* (2004), and *The Ape in the Tree: An Intellectual and Natural History of Proconsul* (with Alan Walker, 2005).

In 2005, the Center for Research into the Anthropological Foundations of Technology (CRAFT) and the Stone Age Institute at Indiana University acknowledged Shipman for her "lifetime contributions to paleoanthropology and taphonomy." She is a member of the American Association of Physical Anthropologists, Society for American Archaeology, American Society of Mammalogists, Society of Vertebrate Paleontology, and American Association for the Advancement of Science.

Further Resources

Pennsylvania State University. Faculty website. http://www.anthro.psu.edu/faculty_staff/shipman.shtml.

Shockley, Dolores Cooper

b. 1930
Pharmacologist

Education: B.S., pharmacy, Xavier University, Louisiana, 1951; M.S., pharmacology, Purdue University, 1953, Ph.D., pharmacology, 1955

Professional Experience: assistant, pharmacology, Purdue University, 1951–1953; assistant professor, pharmacology, Meharry Medical College, Nashville, Tennessee, 1957–1967, associate professor, 1967–

Concurrent Positions: Fulbright fellowship, University of Copenhagen, 1955–1956; visiting assistant professor, Albert Einstein Medical College, 1959–1962

Dolores Shockley is known for her research in pharmacology, which is the science dealing with research on the preparation, uses, and especially the effects of drugs. Her research interests are the consequences of drug action on stress, the effects of hormones on connective tissue, the relationship between drugs and nutrition, and the measurement of nonnarcotic drugs. When she entered undergraduate school, she planned to become a pharmacist and operate her own drugstore, but during college, her interest shifted to research. She was the first African American woman to earn a doctorate in pharmacology in the United States and the first black woman to earn any doctorate from Purdue. After completing postdoctoral research at the University of Copenhagen, Shockley returned to the United States as an assistant professor at Meharry Medical College, a historically black medical school in Nashville, Tennessee. At first, she was uncertain that she had made a wise choice because some of the men thought she was just working there temporarily, but she soon proved she was there to stay and became a respected member of the faculty. She was promoted to associate professor in 1967, and later served as chair of the departments of microbiology and of the graduate program in pharmacology, the first African American woman to chair a department of pharmacology in the United States.

Shockley's awards and honors include the Lederle faculty award (1963–1966), and she was named Distinguished Alumni at the Purdue University School of Pharmacy and Pharmaceutical Sciences (2009). She is a member of the American Pharmaceutical Association and the American Association for the Advancement of Science. Vanderbilt University School of Medicine designated the Dolores C. Shockley Lectureship and Mentoring Award in her honor.

Further Resources

Jordan, Diann. 2006. *Sisters in Science: Conversations with Black Women Scientists on Race, Gender and Their Passion for Science.* West Lafayette, IN: Purdue University Press.

Shoemaker, Carolyn (Spellmann)

b. 1929
Planetary Astronomer

Education: B.A., Chico State College, 1949, M.A., history and political science, 1950

Professional Experience: visiting scientist, astrogeology, U.S. Geological Survey (USGS), Flagstaff, Arizona, 1980–; research professor, astronomy, Northern Arizona University, 1989–; staff member, Lowell Observatory, Flagstaff, 1993–

Concurrent Positions: research assistant, California Institute of Technology (CalTech), 1981–1985; guest observer, Mt. Palomar Observatory, 1982–1994

Carolyn Shoemaker has discovered more than 30 comets and 800 asteroids, more than any living astronomer. She first became known to the general public when the periodic comet Shoemaker-Levy 9 (named for Carolyn and husband, Gene Shoemaker, and their colleague David

Astronomer Carolyn Shoemaker has discovered more comets and asteroids than any living astronomer. (AP/Wide World Photos)

Levy) impacted on Jupiter in July 1994, and she was interviewed on television programs. However, she was already renowned in the scientific community because of the number of comets she had identified. Shoemaker uses the 18-inch Schmidt telescope at Mt. Palomar, ultra-fine-grain film, and a stereomicroscope. She worked with her husband, founder of the USGS Center for Astrogeology in Flagstaff, Arizona, in all of the discoveries except one, but he created the search program for comets and Earth-crossing asteroids that they used. Another area in which she has worked is in identifying Earth-approaching asteroids. For two weeks each month, during the dark of the moon, search teams gather at Mt. Palomar in California to track asteroids and meteorites that are close enough to impact the Earth. Such objects regularly fall to Earth throughout the world, and a large one could cause severe damage. Shoemaker has identified a record 500 asteroids, including 41 Earth-approachers.

Carolyn Shoemaker came to her scientific research later in life. Her husband was a world expert on impact craters, both on Earth and on other planets, and he trained the astronauts who landed on the moon in the basics of geology. Carolyn taught school, but after their own children were grown, she started accompanying her husband as an unpaid field assistant on his studies of craters on the Earth and then helped with his work surveying the moon. She got a position reviewing films of the night sky at

CalTech and soon became expert in identifying the tiny dark smudges on the films. She discovered her first comet in 1983, at the age of 54, without a degree in astronomy.

Shoemaker has received numerous honors, including a National Aeronautics and Space Administration (NASA) Exceptional Achievement Medal (1996), Woman of Distinction Award of the National Association for Women in Education (1996), and Distinguished Alumna of California State University, Chico (1996). With her husband Gene (who died in 1997 while on a research trip to Australia) she has been the co-recipient of the Rittenhouse Medal (1988) and the James Craig Watson Medal (1998); in 1995, the two were also named Scientists of the Year. She is the author of the report on Shoemaker-Levy 9 in the USGS *Yearbook* (1994), and her work has been featured in the media, such as on public television programs. She is a fellow of the American Academy of Arts and Sciences and a member of the Astronomical Society of the Pacific. She received an honorary doctorate from Northern Arizona University in 1990.

Further Resources

U.S. Geological Survey. "Carolyn Shoemaker." http://astrogeology.usgs.gov/About/People/CarolynShoemaker/.

Shotwell, Odette Louise

1922–1998
Organic Chemist

Education: B.S., chemistry, Montana State University, 1944; M.S., University of Illinois, 1946, Ph.D., organic chemistry, 1948

Professional Experience: teaching assistant, inorganic chemistry, University of Illinois, 1944–1948; research chemist, Northern Regional Research Laboratory, U.S. Department of Agriculture (USDA), 1948–1977, research leader, mycotoxin analysis and chemical research, 1975–1989

Concurrent Positions: consultant, Bureau of Veterinary Medicine, Food and Drug Administration, 1981–1986; consultant, Canadian Health and Welfare Department, 1983–1989; consultant and collaborator, USDA

Odette Shotwell was a chemist who made significant contributions to environmental science, and was recognized for her work in developing a cancer-producing toxin from molds. She held three patents, and her work led to or contributed to the development of several new antibiotics. Her research included synthetic organic chemistry; the chemistry of natural products, including isolation, purification, and characterization; microbial insecticides; and mycotoxins. Her own father was a research entomologist and, in one instance, she conducted research on the chemistry of Japanese beetles as part of a government effort to stop the spread of the pests. Shotwell suffered from polio as a child and was confined to a wheelchair for most of her life. Still, she left home in Colorado to study chemistry at Montana State College (now University of Montana). She went on to pursue graduate studies at the University of Illinois and, after receiving her doctorate, joined the Northern Regional Research Laboratory of the USDA in Peoria, Illinois, in 1948. She was promoted to research leader in mycotoxin analysis and chemical research in 1975 and research leader in mycotoxin research in 1985, retiring from the agency (now known as the Northern Center for Agricultural Utilization Research) in 1989. Before and even after retirement, she consulted for the USDA and other government agencies in both the United States and Canada.

Among the awards Shotwell received were the Outstanding Woman Alumna of the Year from the city of Bozeman, Montana (1961), Outstanding Handicapped Federal Employee Award (1969), and Harvey W. Wiley Award of the American Oil Chemical Society (1982). She was elected a fellow of the Association of Official Analytical Chemists, and was a member of the American Association for the Advancement of Science, American Chemical Society, and American Association of Cereal Chemists.

Shreeve, Jean'ne Marie

b. 1933
Inorganic Chemist

Education: B.A., University of Montana, 1953; M.S., analytical chemistry, University of Minnesota, 1956; Ph.D., inorganic chemistry, University of Washington, 1961

Professional Experience: teaching assistant, chemistry, University of Minnesota, 1953–1955; assistant, University of Washington, 1957–1961; assistant professor, chemistry, University of Idaho, Moscow, 1961–1965, associate professor, 1965–1967, professor, 1967–1973, acting chair, Department of Chemistry, 1969–1970 and 1973, head of department and professor, 1973–1987, vice provost of research and graduate studies and professor, chemistry, 1987–

Jean'ne Shreeve is internationally known and nationally recognized for her contributions to the understanding of synthetic fluorine chemistry. Her research includes

Inorganic chemist Jean'ne Marie Shreeve.
(Courtesy of the University of Idaho)

synthesis of inorganic and organic fluorine-containing compounds. The major emphasis of her research has been the synthesis, characterization, and reactions of fluorine compounds that contain nitrogen, sulfur, and phosphorus. She and her students made a significant find when they discovered the compound perfluorourea, which is an oxidizer ingredient. She has also developed new synthetic routes to several important compounds, including chlorodifluoroamine and difluoradiazine. These compounds are used in synthesizing rocket oxidizers, but preparation by previously known techniques was hard to accomplish.

At the time she started her appointment at the University of Idaho, the chemistry department was poorly equipped to support research. However, the state had just designated the campus at Moscow as Idaho's research university and had given it permission to grant doctoral degrees; because of her prominence in research, she was able to contribute to the growth of the chemistry department and its curriculum. She advanced rapidly through the ranks to full professor, head of the department, and then vice provost for research and graduate studies. She has devoted her life to educating other chemists, and she has drawn many exceptional students into graduate studies. Her own interest in chemistry developed when she was an undergraduate at the University of Montana because of an exceptional teacher.

Shreeve's work as a fluorine chemist earned her the 1972 Garvan Medal of the American Chemical Society (ACS) for outstanding achievements by American women chemists. The honor cited her contributions to the fundamental understanding of the behavior of inorganic fluorine compounds and to the synthesis of important new fluorochemicals. She has served on numerous committees in the ACS and the American Association for the Advancement of Science, and she has received numerous awards, including the Distinguished Alumni Award, University of Montana (1970); Outstanding Achievement Award, University of Minnesota (1975); Senior U.S. Scientist Award, Alexander Von Humboldt Foundation (1978); Fluorine Award of the ACS (1978); Excellence in Teaching

Award, Chemical Manufacturers Association (1980); and an honorary doctorate from the University of Montana (1982). She began serving on the board of Governors of Argonne National Laboratory in 1992. She is a fellow of the American Association for the Advancement of Science and a member of the American Chemical Society and American Institute of Chemists.

Further Resources

University of Idaho. Faculty website. http://www.webpages.uidaho.edu/~jshreeve/.

Simmonds, Sofia

1917–2007
Biochemist

Education: B.A., Barnard College, 1938; Ph.D., biochemistry, Cornell University, 1942

Professional Experience: assistant biochemist, medical college, Cornell University, 1941–1942, research associate, 1942–1945; instructor, physiological chemistry, School of Medicine, Yale University, 1945–1946, microbiologist, 1946–1949, assistant to associate professor, biochemistry and microbiology, 1949–1962, biochemist, 1962–1969, molecular biophysicist and biochemist, 1969–1975, professor; 1976–1988, lecturer and dean of undergraduate studies, 1990–1991

Sofia Simmonds has been recognized for her research on bacteria amino acid metabolism, in particular of the *E. coli* bacteria. She spent years in administrative posts at Yale's medical school, some of which continued after her retirement. After receiving her undergraduate degree from Barnard (the women's college of Columbia University) in 1938, she attended Cornell University, where she received her doctorate in biochemistry in 1942. She continued working there as a research associate until 1945 when she accepted an appointment as instructor of physiological chemistry in the school of medicine at Yale; she rose through the ranks at Yale, becoming a full professor in 1976. During her tenure there, she also served as associate dean and then dean of undergraduate studies, a position she continued even after formal retirement in 1988.

Simmonds's husband, Joseph S. Fruton, was also a biochemistry professor at Yale, and together they published *General Biochemistry* (1953), the first comprehensive textbook in the field. Their work has been reissued in several editions and has been translated into Japanese and several European languages. In 2005, the couple established the Joseph S. and Sofia S. Fruton Teaching and Research

Fund for the History of Science at Yale. After more than 70 years of marriage, the couple died within days of each other in July 2007.

Simmonds received the Garvan Medal of the American Chemical Society in 1969. She was also a member of the American Society of Biological Chemists.

Further Resources

"In Memoriam: Biochemists Joseph Fruton and Sofia Simmonds." *Yale Bulletin & Calendar.* 36(2). (14 September 2007). http://www.yale.edu/opa/arc-ybc/v36.n2/story22.html.

Simon, Dorothy Martin

b. 1919
Physical Chemist

Education: A.B., chemistry, Southwest Missouri State College, 1940; Ph.D., physical chemistry, University of Illinois, 1945

Professional Experience: research chemist, E. I. du Pont de Nemours & Company, 1945–1946; chemist, Clinton Laboratory, 1947; associate chemist, Argonne National Laboratory, 1948–1949; aeronautical research scientist, Lewis Laboratory, National Advisory Committee on Aeronautics, 1949–1953, assistant chief, chemical branch, 1954–1955; Rockefeller fellow, Cambridge University, 1953–1954; group leader in combustion, Magnolia Petroleum Company, 1955–1956; principal scientist and technical assistant to president, research and advanced development, Avco Corporation, 1956–1962, director of corporate research, 1962–1964, vice president, defense and industrial products, 1964–1968, corporate vice president and director of research, 1968–1985; founder, Simon Associates consulting firm

Dorothy Simon is a chemist who spent most of her career as a distinguished researcher in the aerospace industry. Her research interests included combustion, aerothermochemistry, and research management and strategic planning. After receiving her doctorate from the University of Illinois in 1945, where she completed some of the earliest work on radioactive fallout, she went to work as a research chemist for a variety of corporations and government agencies, including E. I. du Pont de Nemours, Clinton Laboratory, Argonne National Laboratory, the National Advisory Committee on Aeronautics (the predecessor of the National Aeronautics and Space Administration [NASA]), and Magnolia Petroleum Company. In 1953, she received a prestigious Rockefeller Foundation fellowship to conduct research at key laboratories in England, France, and The Netherlands; upon returning to the United States, she spent the remainder of her career in research and administrative positions at Avco Corporation; in 1968, she was named vice president of research, the company's first female corporate officer. At Avco, she emerged as an international expert in the field of combustion and high-temperature composite materials for aircraft and missile systems.

Her father was head of the chemistry department at Southwest Missouri State College (now Missouri State University), where she received her undergraduate degree in 1940 and where she later established the Dr. Robert W. Martin Research Fellowship for chemistry majors in her father's honor. The university recognizes her as the first student to graduate with a perfect 4.0 grade point average. She has received two honorary doctorates, from Worcester Polytechnic Institute (1971) and Lehigh University (1978). Over the course of her career, she has served on prestigious national and international committees, including the NASA Space Systems and Technology Advisory Committee and the President's Committee on the National Medal of Science (1978–1981). She has also served on the boards of major corporations and was a trustee for two universities. She received the Rockefeller Public Service Award (1953) and the Society of Women Engineers Achievement Award (1966), and was named by *Business Week* magazine as one of the top 100 women in corporate America (1976). She was elected a fellow of the American Institute of Aeronautics and Astronautics and the American Institute of Chemists, and was a member of the American Association for the Advancement of Science, the American Chemical Society, and the Combustion Institute.

Simpson, Joanne Malkus (Gerould)

1923-2010
Meteorologist

Education: B.S., University of Chicago, 1943, M.S., 1945, Ph.D., meteorology, 1949

Professional Experience: instructor, meteorology, New York University, 1943–1944; instructor, meteorology, University of Chicago, 1944-1945; instructor, physics and meteorology, Illinois Institute of Technology, 1946–1949, assistant professor, 1949–1951; research meteorologist, Woods Hole Oceanographic Institution, 1951–1960; professor, meteorology, University of California, Los Angeles, 1960–1965; head, experimental branch, Atmospheric Physics and Chemistry Laboratory, Environmental Science Service Administration, 1965–1971;

director, experimental meteorology laboratory, National Oceanic and Atmospheric Administration (NOAA), Department of Commerce, 1971–1974; professor, environmental science and member, Center of Advanced Studies, University of Virginia, 1974–1976, W. W. Corcoran Professor, 1976–1981; head, Severe Storms Branch, Goddard Space Flight Center, National Aeronautics and Space Administration (NASA), 1979–1988, chief scientist, meteorology and earth sciences, 1988–1992, science director, 1992–1998

Concurrent Positions: adjunct professor, University of Miami, 1971–1974; project scientist, tropical rainfall measuring mission, Goddard Space Flight Center, 1986–; member, board of directors, Atmospheric Sciences and Climatology of National Research Council (NRC) and National Academy of Sciences (NAS), 1990–; chief scientist, Simpson Weather Associates, 1974–1979

Joanne Simpson was the first woman in the world to receive a doctorate in meteorology, and she had a distinguished career as a meteorologist in academia, government, and private business. She started college just at the beginning of World War II, and seized the opportunity to enter the meteorology training program on the University of Chicago campus. Meteorology is the science that deals with the atmosphere and its phenomena, including weather and climate, and after nine months of training, she and the other women in the program trained weather forecasters for the military services. At the end of the war, the women were expected to return to their families or get married, and some faculty members were openly hostile to women students who planned to continue their educations. Simpson had difficulty finding a faculty supervisor but eventually worked with a professor studying clouds and tropical meteorology, the subject of her later book, *Cloud Structure and Distributions over the Tropical Pacific Ocean* (1965). Without a fellowship, she had to work part-time to support herself and obtained a position teaching physics and meteorology at the Illinois Institute of Technology while completing the coursework for her doctorate.

Between subsequent academic appointments, Simpson held high-level positions with government research institutions, such as director of an experimental meteorology laboratory at Coral Gables, Florida, for NOAA, and, later, head of the severe storms division of NASA. She devised and developed a new concept of cloud-seeding experiments aimed at modifying the dynamics of cumulus clouds. When she was a faculty member in the Environmental Sciences Department of the University of Virginia, she and her husband, Robert Simpson, formed a private meteorology consulting service, Simpson Weather Associates. She was for many years the lead project scientist for NASA's Tropical Rainfall Measuring Mission (TRMM).

Simpson was elected to membership in the National Academy of Engineering in 1988. She received an honorary doctorate from the State University of New York, Albany (1991). Among her numerous honors are the Meisinger Award of the American Meteorological Society (1962) and the highest award of the American Meteorological Society, the Rossby Research Medal (1983). Other awards include the Silver Medal (1967) and Gold Medal (1972) of the Department of Commerce, the V. J. Schaefer Award of the Weather Modification Association (1979), the Exceptional Science Achievement Medal of NASA (1982), and the International Meteorological Organization Price (2002). She was a fellow of the American Meteorological Society (AMS) and served as the first female president of the AMS in 1989. She was a member of the American Geophysical Union and the Ocean Society, and a fellow of the American Academy of Arts and Sciences.

Further Resources

Weier, John. "Joanne Simpson (1923–2010)." Earth Observatory. NASA. (23 April 2004, updated 2010). http://earthobservatory.nasa.gov/Features/Simpson/simpson.php.

Center for Science and Technology Policy Research, Cooperative Institute for Research in Environmental Sciences, University of Colorado, Boulder. 2002. "Women in the Atmospheric Sciences—Astounding Progress since World War II: Personal Viewpoint of Joanne Simpson in 2002." *WeatherZine*. 34. (June 2002). http://sciencepolicy .colorado.edu/zine/archives/34/editorial.html.

Singer, Maxine (Frank)

b. 1931
Biochemist, Geneticist

Education: B.A., Swarthmore College, 1952; Ph.D., biochemistry, Yale University, 1957

Professional Experience: postdoctoral fellow, Public Health Service, National Institutes of Health (NIH), 1956–1958, research biochemist, National Institute of Arthritis, Metabolism, and Digestive Diseases, 1958–1974; chief, Nucleic Acid Enzymology Section, Biochemistry Lab, National Cancer Institute, 1975–1980, chief, Biochemistry Lab, 1980-1987, scientist emeritus; president, Carnegie Institution of Washington, 1988–2002; chair, Board of Directors, Whitehead Institute for Biomedical Research, Massachusetts Institute of Technology (MIT), 2003–

Concurrent Positions: visiting scientist, Weizmann Institute of Science, Israel, 1971; instructor, University of California, Berkeley, 1980

Maxine Singer is renowned as a leading scientist in the field of human genetics. Her research laboratory helped to decipher the genetic code, and she is a strong advocate for responsible use of genetics research. During the controversy in the 1970s over the use of recombinant DNA (deoxyribonucleic acid) techniques to alter genetic characteristics, she advocated a cautious approach, and she helped develop guidelines to balance the desire for unfettered research on genetics with designing research programs that make medically valuable discoveries and still meet goals to protect the public from possible harm. She spent her early career conducting research at the NIH, where scientists were learning how to take DNA fragments from one organism in order to insert them into the living cells of another. This new research potentially could lead to the discovery of cures for serious diseases, aid in the development of new crops, and otherwise benefit humanity. In 1972, Singer's colleague, Paul Berg of Stanford University, was the first to create recombinant DNA molecules. Later, he voluntarily stopped conducting studies involving DNA manipulation in the genes of tumor-causing viruses because some scientists feared that a virus with unknown properties might escape from the laboratory and spread into the general population.

In an unprecedented action in 1973, a group of scientists composed a public letter to the president of the National Academy of Sciences and published it in *Science* magazine. They warned that organisms of an unpredictable nature could result from the new technique and suggested that the academy recommend guidelines. The NIH began formulating guidelines for recombinant DNA research, and Singer was instrumental in preparing these guidelines. She also wrote a series of editorials and articles on the topic in *Science* over a period of about five years. She was a strong supporter of the first genetically engineered foods, such as "the Flavr Savr tomato," which reached American supermarket shelves in the 1990s.

In 1988, she became president of the Carnegie Institution, a research organization that conducts high-level biological, earth science, and astronomical research. She retired from Carnegie in 2002 and now serves on the Board of Directors for the Whitehead Institute for Biomedical Research at MIT. She is still affiliated with and conducts regular research at the National Cancer Institute. Singer and Paul Berg published two books on genetics, both of which have received positive reviews: *Genes and Genomes: A Changing Perspective* (1990), a graduate-level textbook on molecular genetics, and *Dealing with Genes: The Language of Heredity* (1992). Although not a textbook, it is a summary of the mechanisms of heredity and the ways in which biologists study and alter the microscopic structure of organisms.

Singer was elected to the National Academy of Sciences in 1979. She has been awarded more than 15 honorary doctorates and has been an advisor or committee member for many academic, governmental, and private organizations. In 1992, she received the National Medal of Science. Her numerous other awards include a Distinguished Service Medal from the U.S. Department of Health and Human Services (1983) and a Public Service Award from the NIH (1995). Her work in bringing science education to inner-city children through her "First Light" weekend science program and through the Carnegie Academy for Science Education earned her the 2007 Public Welfare Medal from the National Academy of Sciences. She is a fellow of the American Academy of Arts and Sciences, and a member of the American Society of Biological Chemists, American Philosophical Society, and American Association for the Advancement of Science. In 2008, she became co-chair (with astronomer **Vera Rubin** and physicist **Myriam Sarachik**) of a National Academy of Sciences project to pair women scientists in the United States with Iraqi women scientists for mentoring and career support.

Further Resources

National Academy of Sciences. 2008. "International Twinning Project for Iraqi Women Scientists, Engineers, and Health Professionals." Committee on Human Rights. (March 2008). http://sites.nationalacademies.org/PGA/humanrights/PGA_044086.

Wasserman, Elga. 2002. *The Door in the Dream: Conversations with Eminent Women in Science*. Washington, D.C.: Joseph Henry Press.

Carnegie Institution. Faculty website. http://www.carnegieinstitution.org/singer.

Sinkford, Jeanne Frances (Craig)

b. 1933
Physiologist

Education: B.S., Howard University, 1953, D.D.S., 1958; M.S., Northwestern University, 1962, Ph.D., physiology, 1963

Professional Experience: research assistant, psychology, U.S. Department of Health, Education, and Welfare, 1953; instructor, College of Dentistry, Howard University, 1958–1960; clinical instructor, dentistry, Northwestern University, 1963–1964; associate professor and chair, prosthodontics, Howard University, 1964–1974, professor, 1968–1991, associate dean, College of Dentistry, 1967–1974, dean, 1975–1991, professor, physiology, Graduate School of Arts and Science, 1976–1991; director, Center for Equity and Diversity, American Dental Education Association, 1991–

Concurrent Positions: attending staff, Howard University Hospital; Children's Hospital, National Medical Center; District of Columbia General Hospital; trustee advisor, American Fund for Dental Health, 1975–1984

Lucy Hobbs Taylor, First Woman Dentist

Lucy Beaman Hobbs Taylor (1833–1910) was the first female professional dentist in the United States. She began her career as a schoolteacher but dreamed of attending medical school. Denied admission to both medical school and the dental college because of her sex, she studied privately with a professor from the Ohio College of Dental Surgery and began practicing in Cincinnati without a diploma in 1861. She gained membership in professional organizations and attended conferences before returning to the Ohio College to complete her formal education, finally earning her doctorate in dentistry in 1865 and paving the way for more women to enter the field. She later married and moved to Kansas, where she and her husband operated a successful joint dental practice. By the year 2003, women made up 17% of practicing dentists and more than 40% of dental students.

Jeanne Sinkford is a physiologist known for her research on dental issues, including endogenous anti-inflammatory substances, chemical healing agents, gingival retraction agents, hereditary dental defects, oral endocrine defects, and neuromuscular problems. She has the distinction of being the first black woman in the United States to become head of a university department of dentistry. She was born in Washington, D.C., and has spent most of her career at Howard University. She studied chemistry and psychology as an undergraduate and received her D.D.S. from Howard in 1958. She taught prosthodontics at the Howard dental school for two years before moving to Chicago, where she received a master's degree and then doctorate in physiology from Northwestern University. She returned to Howard University in 1964, where she rose through the ranks to full professor and, in 1975, became the first female dean of a dental college in the United States.

For many years, she also continued her dental practice by serving on the staffs of various local hospitals. She left Howard University in 1991 and now serves as Director of the Center for Equity and Diversity (formerly the Office of Women and Minorities) at the American Dental Education Association (ADEA) in Washington, D.C. When Sinkford began her career in the 1960s, only about 2% of dentists were female; the field is still heavily male-dominated, with only about 17% of practicing dentists female, but the numbers of women in dental schools is steadily increasing. At the ADEA and in other areas of her professional life, Sinkford has been committed to increasing the numbers of women and minorities in dentistry and the health professions in general.

Sinkford is an elected member of the Institute of Medicine of the National Academy of Sciences. She has received a number of honorary degrees and awards, including the College of Dentistry Alumni Award for Dental Education and Research (1969) and the Alumni Federation Outstanding Achievement Award (1971), both from Howard University; an Alumni Achievement Award from Northwestern University (1970); a Certificate of Merit from the American Prosthodontic Society (1971); the Candace Award of the National Coalition of 100 Black Women (1982); a Trailblazer Award from the National Dental Association (2007); and the Herbert W. Nickens Award of the Association of American Medical Colleges (AAMC) (2009). She is a member of the board of directors for the NIH and in 1974 was inducted into the International College of Dentists. She is a fellow of the American College of Dentists and a member of the American Dental Association, International Association for Dental Research, American Association for the Advancement of Science, and New York Academy of Sciences.

Sitterly, Charlotte Emma Moore

1898–1990
Astronomer, Astrophysicist

Education: A.B., mathematics, Swarthmore College, 1920; Ph.D., astronomy, University of California, Berkeley, 1931

Professional Experience: computer, Princeton Observatory, 1920–1925; computer, Mount Wilson Observatory, Los Angeles, 1925–1928; computer, Princeton University, 1928–1929, assistant spectroscopist, 1931–1936, research associate, 1936–1945; physicist, atomic physics division, National Bureau of Standards, 1945–1968; assistant, Office of Standard Reference Data, 1968–1970; assistant, Space Science Division, U.S. Naval Research Laboratory, 1971–1978

Charlotte Moore Sitterly was an astrophysicist recognized for her work on major projects concerning atomic spectra, atomic energy levels, and spectroscopic data for more than 50 years. After studying mathematics at Swarthmore College, she worked as a "computer" at Princeton University and at Mount Wilson Observatory in Los Angeles, analyzing solar images, before completing her doctorate in astronomy at the University of California, Berkeley. She returned to Princeton as a research associate for several years, during which time she met and married physicist Bancroft W. Sitterly. She joined the National Bureau of Standards in 1945 and spent more than 20 years there compiling standard wavelengths and atomic spectral tables, which are still useful reference tools. She also authored or co-authored eight books, including *The Infrared Solar Spectrum* (1947), *Atomic Energy Levels* (1949–1958), and *An Ultraviolet Multiples Table* (1950–1962).

She served on numerous scientific committees, including as a member of the National Research Council, member of the International Astronomical Union, and consultant to a variety of organizations.

Sitterly received the Annie J. Cannon Prize (1937), the Silver Medal (1951) and Gold Medal (1960) of the U.S. Department of Commerce, the first Federal Woman's Award of the U.S. government (1961), the William F. Meggers Award of the Optical Society of America (1972), and the Bruce Medal of the Astronomical Society of the Pacific (1990). She received honorary doctorates from her alma mater, Swarthmore College (1962), University of Kiel in Germany (1968), and University of Michigan (1971). The Asteroid 2110 Moore-Sitterly is named in her honor. She was elected a fellow of the American Physical Society and the Optical Society of America, and a foreign associate of the Royal Astronomical Society of London. She also was a member of the American Association for the Advancement of Science and the American Astronomical Society.

Slye, Maud Caroline

1879–1954
Pathologist

Education: A.B., Brown University, 1899; University of Chicago, 1906, 1908–1911

Professional Experience: professor, psychology and pedagogy, Rhode Island State Normal School, 1899–1905; staff member, Sprague Memorial Institute, 1911–1944, instructor, pathology, University of Chicago, 1919–1922, assistant professor, 1922–1926, associate professor and director, Cancer Laboratory, 1926–1944

Maud Slye was a pioneer in the study of the inheritance of cancer in mice and how it relates to human cancers. The popular press called her the "American Curie" for her contributions. Her theories on cancer later were proven to be incorrect. At first, she theorized that susceptibility to cancer was limited to the presence of a single recessive characteristic, but she later modified her ideas to agree that more than one gene was involved. A tireless worker, she raised and kept pedigrees on over 150,000 mice during her career. She held a prestigious directorship, although she did not have a doctorate. After receiving her undergraduate degree from Brown University in 1899, she was appointed a professor of psychology and pedagogy at the Rhode Island State Normal School for seven years. She accepted an appointment as member of the staff at the new Sprague Memorial Institute (later affiliated with the University of Chicago) in 1911, retiring in 1944. During this time, she held a joint appointment as a faculty member in pathology at the University of

Chicago, rising to the rank of associate professor and director of the Cancer Laboratory in 1926.

Slye received many honors for her contributions to cancer research, including a Gold Medal from the American Medical Association (1914), a Gold Medal from the American Radiological Society (1922), and the Ricketts Prize of the University of Chicago (1915). Brown University granted her an honorary degree in 1937. In addition to her scientific papers, she wrote two books of poetry: *Songs and Solaces* (1934) and *I in the Wind* (1936). She was a member of the American Medical Association and the New York Academy of Sciences.

Pathologist Maud Slye was an early cancer researcher. (National Library of Medicine)

Further Resources

McCoy, Joseph J. 1977. *The Cancer Lady: Maud Slye and Her Hereditary Studies*. Nashville, TN: Thomas Nelson Books.

Rader, Karen Ann. 2004. *Making Mice: Standardizing Animals for American Biomedical Research, 1900–1955*. Princeton, NJ: Princeton University Press.

Small, Meredith F.

Anthropologist, Primatologist

Education: A.B., anthropology, San Diego State University, 1973; M.A., physical anthropology, University of Colorado, Boulder, 1975;Ph.D., anthropology, University of California, Davis, 1980

Professional Experience: assistant professor, anthropology, Cornell University, 1988–1991, associate professor, 1991–1997, professor, 1997

Meredith Small is an anthropologist and primatologist who specializes in biological and cultural anthropology, evolutionary biology, and human and primate behavior.

She began her career observing both wild and captive macaques and focused on female sexual behavior and care of offspring. She has been a professor of anthropology at Cornell University since 1988 and published her first book, *Female Choices: Sexual Behavior of Female Primates*, in 1993. *Female Choices* was a groundbreaking and controversial look at the different sexual choices made by female primates, showing that females are active participants in sexual and mating relationships.

Since the 1990s, Small has also been a prominent figure in the media with her articles for popular science magazines and websites on issues related to childrearing, sexuality, DNA analysis, and other issues. Her books include *What's Love Got to Do with It? The Evolution of Human Mating* (1995), the immensely popular *Our Babies, Ourselves; How Biology and Culture Shape the Way We Parent* (1998), *Kids: How Biology and Culture Shape the Way We Raise Our Children* (2001), and *The Culture of Our Discontent; Beyond the Medical Model of Mental Illness* (2006). In each of these works, Small has examined the intersection between biology and culture, and looked for lessons from nonhuman primates to explain human behavior, especially with regard to mating and parenting. Small characterized her work in *Our Babies, Ourselves* as a contribution to the new field of ethnopediatrics, or the cross-cultural study of childhood and childrearing that combines the fields of anthropology, psychology, child development, and pediatrics. In the book, Small argued that there is no right or wrong way to raise children and that our ideas about feeding, sleeping with, bonding with, and disciplining children has as much to do with culture as it does with natural instinct, and may not even always be what is "best" for children. In 2005, Small's efforts in bringing scientific research to the general public were honored with an Anthropology in Media Award from the American Anthropological Association.

Further Resources

Cornell University. Faculty website. http://falcon.arts.cornell.edu/anthro/faculty/small.html.

Smith, Elske (van Panhuys)

b. 1929
Astronomer, Environmental Scientist

Education: B.S., astronomy, Radcliffe College, 1950, M.A., astronomy, 1951, Ph.D., astronomy, 1956

Professional Experience: research fellow, Harvard Observatory Solar Project, Sacramento Peak Observatory, Sunspot, New Mexico, 1955–1962; visiting fellow,

Joint Institute for Laboratory Astrophysics, Boulder, Colorado, 1962–1963; associate professor, astronomy, University of Maryland, College Park, 1963–1975, assistant provost, Division of Mathematics and Physical Science and Engineering, 1973–1978, professor, astronomy, 1975–1980, assistant vice chancellor of academic affairs, 1978–1980; dean, College of Humanities and Science, and professor, physics, Virginia Commonwealth University, 1980–1992, interim director, Center for Environmental Studies, 1992–1995, emerita professor, physics

Concurrent Positions: research associate, Lowell Observatory, Flagstaff, Arizona, 1956–1957; consultant, Goddard Space Flight Center, National Aeronautics and Space Administration (NASA), 1963–1965; lecturer, Osher Lifetime Learning Institute

Elske van Panhuys Smith is a solar physicist whose research included active regions on the sun, especially flares and plages; solar chromosphere; interstellar polarization; and solar physics. She was on the faculty at the University of Maryland for more than 15 years and was dean and director of the Center for Environmental Studies at Virginia Commonwealth University, an academic program she helped establish and for which she taught courses on Earth's atmosphere and on energy. In addition to her numerous scientific papers, she co-authored two books: *Solar Flares* (1963) and *Introductory Astronomy and Astrophysics* (1973; 3rd ed., 1992). She retired in 1995 and moved to Massachusetts, where she has been active in the community, and has lectured and taught continuing-education courses on astronomy, cosmology, archaeology, and environmental issues at the Osher Lifelong Learning Institute at Berkshire Community College.

As a scientist, teacher, and administrator, Smith was concerned with factors preventing women from pursuing careers in the sciences. In 1977, she participated in a symposium at the American Association for the Advancement of Science national meeting, the papers for which were collected and published as *Covert Discrimination and Women in the Sciences* (1978, edited by **Judith A. Ramaley**). As an administrator at the University of Maryland, Smith gained insight into the factors that are involved in hiring and promoting faculty members. She interviewed a number of women scientists in both academia and government positions throughout the country, and uncovered deliberate as well as covert discrimination, including discrimination against married women.

Smith is a fellow of the American Association for the Advancement of Science and a member of the International Astronomical Union and American Astronomical Society (founding member and first treasurer of Solar Physics Division). The Elske Smith Distinguished Lecturer Award at Virginia Commonwealth University is named in her honor.

Solomon, Susan

b. 1956
Atmospheric Chemist

Education: B.S., Illinois Institute of Technology, 1977; M.S., University of California, Berkeley, 1979, Ph.D., chemistry, 1981

Professional Experience: research chemist, National Oceanic and Atmospheric Administration (NOAA), 1981–

Concurrent Positions: adjunct instructor, University of Colorado, Boulder, 1983–; member, committee on solar and space physics, National Aeronautics and Space Administration (NASA), 1983–1986, space and earth science advisory committee, 1985–1988; head project scientist, National Ozone Expedition to McMurdo Sound, Antarctica, 1986–1987

Susan Solomon led expeditions to McMurdo Sound, Antarctica, to examine the "hole" in the ozone layer. Her theory was that chlorofluorocarbons (CFCs) could lead to Antarctic ozone destruction when CFCs encounter large masses of stratospheric clouds. CFCs are human-made gases that were widely used in refrigerators, air conditioners, aerosol spray cans, and the manufacture of semi-conductors. In 1985, British scientists reported an ozone hole in the Southern Hemisphere over the South Pole during the pole's spring month of October. The hole was located between the altitudes of about 32,000 and 74,000 feet (the strato-sphere), which normally shields the Earth from the sun's ultraviolet radiation. Sci-entists suspected the damage had been caused by CFCs but were unable to explain the process, but Solomon hit on the solution while attending a lecture on polar stratospheric clouds. She theorized that CFC derivatives react on the cloud surfa-ces. She volunteered to lead the otherwise all-male expedition to McMurdo Sound in 1986, with a follow-up trip in 1987, and her research supported the theory. Her explanation for the cause of the ozone hole is now generally accepted by scientists, and this research led many countries to pass legislation curtailing or outlawing the production and use of CFCs. Solomon continues to study the atmospheric chemistry of ozone in Antarctica as well as in the Arctic in the Northern Hemisphere.

A project during her senior year of college turned Solomon's attention toward atmospheric chemistry. The project involved measuring the reaction of ethylene and hydroxyl radical, a process that occurs in the atmosphere of Jupiter. The summer before entering graduate school, she worked on a study of ozone in the upper atmosphere at the National Center for Atmospheric Research (NCAR) in Boulder, Colorado. At NOAA, she first worked in the Aeronomy Laboratory developing computer models of ozone in the upper atmosphere (*aeronomy* is the

Susan Solomon is an atmospheric scientist for the National Oceanic and Atmospheric Administration who helped explain the hole in the ozone layer over Antarctica. (NASA)

study of chemical and physical phenomena in the upper atmosphere). Although she was concentrating on theoretical studies, the McMurdo Sound expeditions provided an opportunity to take up experimental work in measuring chlorine dioxide in the atmosphere. In addition to her scientific papers, she is co-author (with Guy Brasseur) of *Aeronomy of the Middle Atmosphere: Chemistry and Physics of the Stratosphere and Mesosphere* (1984, 2nd ed., 1986).

Solomon was elected to membership in the National Academy of Sciences in 1992. She has received several awards, including the J. B. MacElwane Award of the American Geophysical Union (1985) and the Gold Medal for exceptional service from the U.S. Department of Commerce (1989). She was named Scientist of the Year in 1992 by *R&D Magazine*. In 2000, President Clinton honored Solomon with a National Medal of Science, and in 2004, she received the prestigious Blue Planet prize for her contributions to finding "solutions to global environmental problems." In 2007, she took on an even more public role in the debate over global warming as co-leader of the United Nations and World Meteorological Organization's massive new report on global climate change. She is a member of the Royal Meteorological Society, the American Geophysical Union, and the American Meteorological Society.

Further Resources

Morell, Virginia A. 2007. "Ahead in the Clouds." *Smithsonian*. 82–85. (February 2007). http://www.smithsonianmag.com/science-nature/ahead_clouds.html.

Sommer, Anna Louise

1889–1973
Plant Nutritionist

Education: B.S., University of California, Berkeley, 1920, M.S., 1921, Ph.D., plant nutrition and chemistry, 1924

Professional Experience: teaching fellow, botany, University of California, Berkeley, 1922–1924, plant nutritionist, 1924, assistant, 1924–1926; research fellow, University of Minnesota, 1926–1929; associate professor, plant nutrition, and associate soil chemist, Alabama Polytechnic Institute (Auburn University), 1930–1948, professor and soil chemist, 1948–1949

Anna Sommer was one of the earliest women identified as a soil chemist, and was responsible for identifying the essential nature of three different trace or "micronutrient" elements: copper (Cu), zinc (Z), and boron (Bu). Her research on plant nutrition and soil fertility, during what has been termed the "trace nutrient gold rush" of the early twentieth century, contributed to scientists' understanding that certain elements were not only beneficial, but necessary for plant growth. She was able to test the effect of these elements on plant growth and reproduction by isolating them with purified water and salt. She published her findings in journals such as *Science, Plant Physiology*, and the *Soil Science Society of America Proceedings*. Her work led to the development of better fertilizers and other improvements in agricultural efficiency.

Sommer received all of her degrees from the University of California, Berkeley, including her doctorate in 1924. She continued on at Berkeley as a plant nutritionist and an assistant until relocating to the University of Minnesota, where she spent three years as a research fellow. She then accepted a position as an associate professor and soil chemist in the department of agronomy and soils at Alabama Polytechnic Institute in Auburn, Alabama (now Auburn University), where she conducted her experiments on trace elements. She was the only tenured woman in that department and was promoted to full professor in 1948, just one year before her retirement. Sommer was a member of the American Association for the Advancement of Science, the American Society of Plant Physiologists, and the Soil Science Society of America.

Further Resources

Weaver, David. 2002. "Mystery-Solving Woman: Pioneering Female Agronomist Solved Early Riddles of Soil Science." *ASK Magazine*. Alabama Agricultural Experiment Station. http://www.aaes.auburn.edu/comm/pubs/askmagazine/fall02/pioneering woman.html.

McIntosh, Marla S. and Steve R. Simmons. 2008. "A Century of Women in Agronomy: Lessons from Diverse Life Stories." *Agronomy Journal*. 100: S-53–S-69. http:// agron.scijournals.org/cgi/content/full/100/Supplement_3/S-53.

Spaeth, Mary Louise

b. 1938
Physicist

Education: B.S., physics and mathematics, Valparaiso University, 1960; M.S., nuclear physics, Wayne State University, 1962

Professional Experience: technical staff member, later senior scientist and project manager, Hughes Aircraft Company, 1962–1974; physicist, program leader, Atomic Vapor Laser Isotope Separation, Lawrence Livermore National Laboratory, 1975–1990, systems engineering and chief technologist, National Ignition Facility, 1990–

Mary Spaeth is renowned for her work in developing the first tunable dye laser, a laser whose color could be changed in midstream. The term *laser* is an acronym for "light amplification by stimulated emission of radiation," and is the name of a device that produces a nearly parallel, nearly monochromatic, and coherent beam of light by exciting atoms to a higher energy level and causing them to radiate their energy in phase. She stumbled upon the method for the tunable dye laser while working on a government project at Hughes Aircraft Company in the mid-1960s, and the patent was thus owned by the U.S. Army. While the laser was developed for military uses, it also had practical consumer applications, such as the modern supermarket checkout lasers.

Since 1975, Spaeth has been with Lawrence Livermore National Laboratory in Berkeley, California, and is also credited with using the dye laser in isotope separation. The laser is now the primary source for deriving the isotopes used in nuclear reactors, and because different isotopes of the same element absorb light at different frequencies, a properly tuned dye laser can be used to separate and alter the isotopic composition of many elements. Originally, scientists at Livermore worked exclusively on refining plutonium for nuclear weaponry, but now most activity is centered on providing a low-cost means of enriching uranium fuel for light-water nuclear power reactors. One of the most promising applications of the tunable dye laser is as part of a guide star project that will allow ground-based stellar observatories to achieve a resolution comparable to that received through the Hubble Space Telescope, which was launched in 1990.

Spelke, Elizabeth

b. 1949
Psychologist

Education: B.A., social relations, Radcliffe College, 1971; student, Yale University, 1972–1973; Ph.D., psychology, Cornell University, 1978

Professional Experience: professor, psychology, University of Pennsylvania; professor, psychology, Cornell University; professor, Brain and Cognitive Sciences, Massachusetts Institute of Technology (MIT), 1996–2001; professor, psychology, Laboratory for Developmental Studies, Harvard, University, 2001–

Elizabeth Spelke is a cognitive psychologist whose innovative research has focused on the perceptual and cognitive capacities of young infants. Her philosophical interest in the origins of knowledge led to her conclusion that even very young babies have innate understandings of location, physical objects, identity, and even numbers and quantities. Her controversial methods and findings challenge the previously held belief that humans are born with sensory capabilities but no specific knowledge or capabilities for understanding abstract concepts, such as "object permanence." Spelke argues that her experiments have shown babies as young as two and a half months comprehending the physical boundaries of objects, and infants as young as six months distinguishing between different sets of numbers. She sees these capabilities as innate, as part of our evolutionary development, and as the foundation for acquisition of other types of knowledge, including language. Critics charge that she has overestimated infant mental capabilities, or that it is nearly impossible to distinguish between innate knowledge and learned experience, since babies are learning from the moment of birth. Regardless, her research has influenced the course of cognitive development research.

Her research on infants also relates to her interest in the question of gender and cognitive development. She has concluded that there are no innate differences between male babies and female babies, and therefore no biological basis for different aptitudes in, for example, math and science. This subject was one of contentious debate after Harvard University president Lawrence Summers made remarks in 2005 suggesting that there are fewer women faculty members at prestigious universities such as Harvard because there are fewer women interested in or capable of higher-level math and science. Spelke, on the faculty at Harvard since 2001, was one of those scientists who criticized Summers's remarks, backing up the innate similarities between male and female with her own scientific research. She wrote a widely distributed review of the available research, "Sex Differences in Intrinsic Aptitude for Mathematics and Science: A Critical Review." She has collaborated and co-authored other papers with brain and cognitive researcher **Nancy Kanwisher** of MIT, and with her Harvard colleague in the Laboratory for Developmental Studies, **Susan Carey**.

Spelke was elected to the National Academy of Sciences in 1999. She has received honorary doctorates from Umeå University, Sweden (1993), Ecole Pratique des Hautes Etudes, Paris, France (1999), and University of Paris-Descartes (2007). Her numerous other awards and honors include the Boyd McCandless Young Scientist Research Award (1984), a prestigious Guggenheim fellowship (1989), a Cattell Fellowship (1992), the MERIT Award of the National Institutes of Health (1993), the William James Award of the American Psychological Society (2000), a Distinguished Scientific Contribution Award of the American Psychological Association (APA) (2000), the Ipsen Prize in Neuronal Plasticity (2001), and the Jean Nicod Prize (2008). She is a fellow of the Society of Experimental Psychologists, American Academy of Arts and Sciences, and American Association for the Advancement of Science.

Further Resources

Harvard University. Faculty website. http://www.wjh.harvard.edu/~lds/index.html ?spelke.html.

Talbot, Margaret. 2006. "The Baby Lab: How Elizabeth Spelke Peers into the Infant Mind." New America Foundation. *The New Yorker.* (4 September 2006). http://www .newamerica.net/publications/articles/2006/the_baby_lab.

Spurlock, Jeanne

1921–1999
Psychiatrist

Education: student, Spelman College, 1940–1942, Roosevelt University, 1942–1943; M.D., Howard University, 1947

Professional Experience: intern, Provident Hospital, Chicago, 1947–1948; resident, general psychiatry, Cook County Hospital, Chicago, 1948–1950; fellow, child psychiatry, Institute for Juvenile Research, Chicago, 1950–1951, staff sychiatrist, 1951–1953; staff psychiatrist, Women's and Children's Hospital, Chicago, 1951–1953; Adult and Child Psychoanalytic Training, Chicago Institute for Psychoanalysis, 1953–1962; director, Children's Psychosomatic Unit, Neuropsychiatric Institute, Chicago, 1953–1959; assistant professor, psychiatry, University of Illinois College of Medicine, 1953–1959; psychiatrist and chief, Child Psychiatry Clinic, Michael Reese Hospital, Chicago, 1960–1968; chair, Department of Psychiatry, Meharry Medical College, Nashville, 1968–1973; visiting scientist, National Institute for Mental Health, 1973–1974; Deputy Medical Director, American Psychiatric Association, 1974–1991

Concurrent Positions: clinical professor, George Washington University College of Medicine, and Howard University, College of Medicine; private practice in psychiatry, 1951–1968

Jeanne Spurlock was a noted psychiatrist who held many high-level appointments in hospitals and clinics as a specialist in child psychiatry. However, she changed the emphasis of her career in 1974, when she was appointed deputy medical director of the American Psychiatric Association. In that capacity, her work was primarily administrative, although she maintained a small private practice and was also a clinical professor at two local medical schools. She served as a lobbyist to policymakers to ensure funding for medical education and postgraduate education, particularly for minorities. She was involved in the recruitment and training efforts of minorities for research and was in charge of a fellowship program for minority psychiatric residents sponsored by the association.

Spurlock was co-editor of *Black Families in Crisis: The Middle Class* (1988), in which she wrote about stresses in parenting and male–female relationships. She was also co-editor of and wrote a chapter on single mothers for *Women's Progress: Promises and Problems* (1990), which focused on various aspects of mothering, including the changing face of adoption in the United States, the problems of working mothers, the special problems of mothers of disabled children, and homosexuality and parenting. She was co-author (with Ian A. Canino) of *Culturally Diverse Children and Adolescents* (1994), which addresses the mental-health needs of African American, Latino, Asian American, and Native American children and adolescents. In this book, the authors explained how the assessment, diagnostic, and treatment phases of clinical work may need to be modified for cultural relevancy. She was also editor and contributor for a volume on *Black Psychiatrists and American Psychiatry*, published by the American Psychiatric Association in 1999.

Spurlock was a member of the American Academy of Child and Adolescent Psychiatry, which has named two fellowships in her honor: the Jeanne Spurlock Research Fellowship in Drug Abuse and Addiction for Minority Medical Students (in conjunction with the National Institute on Drug Abuse), and the Jeanne Spurlock Minority Medical Student Clinical Fellowship in Child and Adolescent Psychiatry. The American Medical Women's Association recognized her posthumously with their Elizabeth Blackwell Award in 2000.

Further Resources

National Institutes of Health. "Dr. Jeanne Spurlock." Changing the Face of Medicine: Celebrating America's Women Physicians. National Library of Medicine, National Institutes of Health. http://www.nlm.nih.gov/changingthefaceofmedicine/physicians/biography_306.html.

Stadtman, Thressa Campbell

b. 1920
Biochemist

Education: B.S., microbiology, Cornell University, 1940, M.S., microbiology and chemistry, 1942; Ph.D., microbial biochemistry, University of California, Berkeley, 1949

Professional Experience: bacteriologist, Sealright Co., New York, 1941; graduate fellow, bacteriology, Agricultural Experiment Station, Cornell University, 1941–1942, research assistant, 1942–1943; research associate, food technology, University of California, Berkeley, 1943–1946; research assistant, biochemistry, Harvard Medical School, 1949–1950; biochemist, Laboratory of Cellular Physiology and Metabolism, Enzyme Section, National Heart Institute (now National Heart, Lung, and Blood Institute), National Institutes of Health (NIH), Bethesda, Maryland, 1950–1974, chief, Section on Intermediary Metabolism and Bioenergetics, Laboratory of Biochemistry, 1974–1988, senior executive service, chief, 1988–

Concurrent Positions: fellow, Oxford University, England, 1954–1955; Rockefeller Foundation fellow, University of Munich, Germany, 1959–1960; institute of biological and physical chemistry, France, 1961

Thressa Stadtman has been recognized for her work in microbiology at the NIH since 1950. Her research has included amino acid intermediary metabolism, one-carbon metabolism, methane formation, microbial biochemistry, and selenium biochemistry, and her research on vitamin B_{12} led to the discovery of new enzymes. A high school principal helped her get a New York State Regents scholarship to attend Cornell, where she studied bacteriology, receiving her undergraduate and master's degrees in bacteriology. She remained at Cornell as a research assistant at the agricultural experiment station, then moved to the University of California, Berkeley to pursue her doctorate. It was at Berkeley that she met and married colleague Earl Stadtman, as both of them were working in the food-technology department researching food spoilage, a major problem for the military in shipping food rations overseas during World War II.

After receiving their Ph.D.s in biochemistry in 1949, the couple moved to Massachusetts, where Thressa was hired as a researcher at Harvard Medical School and Earl worked at Massachusetts General Hospital. Unable to find joint academic appointments, in 1950, the couple accepted positions as biochemists at the NIH's National Heart, Lung, and Blood Institute in Bethesda, Maryland. Unlike most universities at that time, the NIH did not have strict anti-nepotism rules and so, in the 1940s and 1950s as more women scientists earned doctorates, and as government research programs expanded in the postwar era, many scientist couples were hired and made names for themselves as researchers at the NIH. Earl Stadtman died in 2008, and Thressa remains affiliated with the NIH.

Thressa Stadtman was elected to the National Academy of Sciences in 1981 and has been an invited researcher and fellow at universities in England, Germany, and France. Among the other honors she has received are the Hillebrand Award (1979) and Rose Award (1987), both of the Chemical Society of Washington, the Klaus Schwarz Medal of the International Union of Biorganic Chemists (1988), the L'Oréal/Helena Rubenstein "Tribute to a Life Achievement" Award (France, 2000), the Gabriel Bertrand Prize Medal (Italy, 2001), and the Oxygen Club of Greater Washington's Lifetime Achievement Award (2007). She served as secretary (1978–1981) of the American Society of Biochemistry and president (1998–2001) of the International Society of Vitamins and Related Biofactors, and has been a member of the American Chemical Society, American Society of Microbiology, British Biochemistry Society, Northern Germany Academy of Sciences, and Executive Women in Government.

Further Resources

Park, Buhm Soon. "The Stadtman Way: A Tale of Two Biochemists at NIH." National Institutes of Health. http://history.nih.gov/exhibits/stadtman/.

Stanley, Louise

1883–1954
Chemist and Home Economist

Education: A.B., Peabody College, 1903; B.Ed., University of Chicago, 1906; A.M., Columbia University, 1907; Ph.D., biochemistry, Yale University, 1911

Professional Experience: instructor, home economics, University of Missouri, 1907–1911, professor and department chair, 1911–1923; chief, Bureau of Home Economics, U.S. Department of Agriculture (USDA), 1923–1950; consultant for home economics, Office of Foreign Agricultural Relations, 1950–1953

Louise Stanley was the first woman to direct a bureau in the USDA and was responsible for some of the earliest studies on food nutrition. After receiving her master's degree from Columbia University in 1907, she obtained an appointment as an instructor in the department of home economics at the University of Missouri. Her career coincided with the emergence of home economics as a profession and academic discipline, offering more employment opportunities for women scientists. She earned a doctorate from Yale University and advanced quickly through the ranks at Missouri, to full professor and chair of the home economics department, but left academia for government employment. In 1923, Stanley became the highest-ranking woman scientist in the federal government when she was appointed the first chief of the Bureau of Home Economics, USDA. She retired from the USDA in 1950, but spent three more years as a consultant for the Office of Foreign Agricultural Relations.

At the USDA, Stanley helped development four basic diet plans for families at different economic levels, and she authored a book, *Foods, Their Selection and Preparation* (1935). She directed the first national survey of rural housing and the first survey of consumer purchasing. Under her direction, the bureau also conducted time and motion studies of housekeeping methods and worked toward standardizing clothing sizes. She was the official representative of the USDA to the American Standards Association, and was the first woman to hold such an appointment. She later focused on nutritional needs and public education about nutrition in Latin America, and became involved with the UN Conference for Food and Agriculture.

Stanley received an honorary degree from the University of Missouri (1940), which later dedicated the home economics building in her name. She was a member of the American Chemical Society and the American Home Economics Association, which has named a scholarship fund for her.

Stearns, Genevieve

1892–1997
Biochemist

Education: B.S., Carleton College, 1912; M.S., University of Illinois, 1920; Ph.D., biochemistry, University of Michigan, 1928

Professional Experience: high school teacher, 1912–1918; assistant, chemistry, University of Illinois, 1918–1920; research associate, child welfare research station, University of Iowa, 1920–1925; assistant, biochemistry, University of Michigan, 1926–1927; research associate, pediatrics, University of Iowa College of Medicine, 1927–1930, assistant to associate professor, 1930–1943, research professor, pediatrics, 1943–1954, research professor, orthopedics, 1954–1958

Genevieve Stearns was recognized for her research on the nutritional needs of infants, children, and pregnant and nursing women. Her main areas of research included vitamin and mineral requirements, metabolism, and human growth. In addition to her scientific publications on nutritional requirements, she was a contributing author to the book *Infant Metabolism* (1956). Stearns's career followed the pattern of many women of her generation. After receiving her undergraduate degree from Carleton College in 1912, she was a high school teacher until 1918. She returned to school to receive her master's degree from the University of Illinois in 1920, working at the child welfare research station at the University of Iowa until 1925. She returned to school to receive her doctorate from the University of Michigan in 1928 while continuing to work in pediatrics at Iowa. She spent the remainder of her career at Iowa, where she rose through the ranks as a research professor in pediatrics and orthopedics at the University Hospitals, overseeing all pediatric blood and chemical work. In 1950, Stearns was selected by the UN World Health Organization to attend a series of seminars on metabolism in Europe. She traveled abroad again after her retirement, as the recipient of a prestigious Fulbright fellowship to work at the Women's College of Ein Shams University in Cairo, Egypt (1960–1961).

Stearns was elected a fellow of the American Institute of Nutrition. She was a co-recipient of an Alumni Achievement Award from Carleton College, the Borden Award of the American Home Economics Association (1942) and the Borden Award of the American Institute of Nutrition (1946). She was a member of the American Society of Biological Chemists, American Chemical Society, and American Institute of Nutrition.

Steitz, Joan (Argetsinger)

b. 1941
Biochemist, Molecular Biologist

Education: B.S., chemistry, Antioch College, 1963; Ph.D., biochemistry and molecular biology, Harvard University, 1967

Professional Experience: postdoctoral fellow, molecular biology, Cambridge University, 1967–1970; assistant professor, molecular biophysics and biochemistry, Yale University School of Medicine, 1970–1974, associate professor, 1974–1978, professor, 1978–1992, Henry Ford II Professor of Molecular Biophysics and Biochemistry, 1992–1998, Sterling Professor of Molecular Biophysics and Biochemistry, Yale University 1998–

Concurrent Positions: Josiah Macy Scholar, Max Planck Institut für Biophysikalische Chemie, Göttingen, Germany, and Medical Research Council Laboratory of Molecular Biology, Cambridge, England 1976–1977; Fairchild Distinguished Fellow, California Institute of Technology, 1984–1985; investigator, Howard Hughes Medical Institute, 1986–; scientific director, Jane C. Childs Fund for Medical Research, 1991–2002

Biochemist and molecular biologist Joan Steitz has contributed to research on autoimmune diseases such as lupus. (AP/Wide World Photos)

Joan Steitz is one of the most prominent scientists in the field of molecular genetics, and her research may help in the diagnosis and treatment of autoimmune diseases such as lupus. She discovered small nuclear ribonucleoproteins, or snRNPs, pronounced "snurps." She is working in a field that was only discovered in her lifetime. While in graduate school at Harvard, her thesis advisor was James D. Watson, who with Francis Crick had demonstrated the double-helix structure of DNA in the 1950s, for which he won the Nobel Prize. She pursued postdoctoral studies at Cambridge University, where she worked with Crick on how bacterial ribosomes recognize where to start protein synthesis on messenger RNA (mRNA). The best known of the snRNPs are involved in the processing of mRNA in the cell nucleus of mammals. By a process called *splicing*, the double-stranded DNA is first transcribed into single-stranded RNA; then the sections are eventually rejoined in the same order in which they occurred on the DNA molecule. The team discovered that some patients with rheumatic diseases made antibodies against their own snRNPs, which resulted in the development of the splicing process. When physicians determine which antibodies patients have, they have additional clues to diagnosing certain diseases.

Lacking any female professors or researchers as role models, Steitz originally planned to attend medical school, but a summer job in the laboratory at the University of Minnesota piqued her interest in research and paved the way for her entrance into Harvard's graduate program in biochemistry and molecular biology instead. Among her honors, she considers the Weizmann Woman and Science Award (1994) from the New York Academy of Sciences among the most gratifying because it promotes women scientists, and she strongly believes that the presence of women scientists can be an inspiration to female students. Both she and her husband, Thomas Steitz, are Investigators at the Howard Hughes Medical Institute and hold appointments as Professors of Molecular Biophysics and Biochemistry at Yale University School of Medicine.

Steitz was elected to membership in the National Academy of Sciences in 1983, and in 1986, she was awarded the National Medal of Science. She has received six honorary degrees and numerous other awards, including the Eli Lilly award in biological chemistry (1976), U.S. Steel Foundation award in molecular biology (1982), the triennial Warren Prize of Massachusetts General Hospital (1989), the Discovery Award from the Christopher Columbus Fellowship Foundation for biomedical research (1992), the Weizmann Women in Science Award (1994), and the Gairdner Foundation Prize (2006). She is a fellow of the American Association for the Advancement of Science and a member of the American Society of Biological Chemists, American Academy of Arts and Sciences, American Philosophical Society, and New York Academy of Sciences. In 2005, she was elected to the Institute of Medicine of the National Academy of Sciences.

Further Resources

Wasserman, Elga. 2002. *The Door in the Dream: Conversations with Eminent Women in Science*. Washington, D.C.: Joseph Henry Press.

Howard Hughes Medical Institute. "Joan A. Steitz, Ph.D." http://www.hhmi.org/research/investigators/steitzja_bio.html.

Yale University. Faculty website. http://www.mbb.yale.edu/faculty/pages/steitzj.html.

Stern, Frances

1873–1947
Social Worker and Dietitian

Education: Garland Kindergarten Training School, Boston, 1897; student, Massachusetts Institute of Technology, 1909–1912; student, London School of Economics

Professional Experience: secretary and research assistant for Ellen Richards, Massachusetts Institute of Technology (MIT); industrial health inspector, Massachusetts State Board of Labor and Industries, 1912–1915; Division of Home Conservation, U.S. Food Administration; investigator, U.S. Department of Agriculture (USDA); American Red Cross, France, 1918–1922; founder, Boston Dispensary Food Clinic, 1918

Concurrent Positions: teacher, Simmons College School of Social Work, Tufts College Medical School, MIT, and State Teachers College, Framingham

Frances Stern was recognized as an early teacher of nutrition and dietetics. She had an interest in social reform and became interested in child nutrition due to her early work as a kindergarten teacher. She obtained a position as research assistant and special student of chemist Ellen Richards, founder of the American Home Economics Association, in New York. Stern attended home economics conferences with Richards, which stimulated her desire for further scientific knowledge about the relation of food to sociological problems. She enrolled in courses in food chemistry and sanitation at MIT. She developed a visiting housekeeping program for the Boston Association for the Relief and Control of Tuberculosis and later a similar program for the Boston Provident Association. In 1912, she obtained a position as an industrial health inspector for the State Board of Labor and Industries. During World War I, she worked as a member of the Division of Home Conservation of the U.S. Food Administration and, in the USDA, as an investigator of the adequacy of food for the industrial worker.

After consulting with the USDA and with the Red Cross in France during World War I, Stern studied economics and politics as a special student at the London School of Economics. She returned to Boston to establish the Boston Dispensary Food Clinic, which was based on her USDA research on the dietary needs and habits of the urban poor. At the clinic, she worked with immigrants on adapting their native foods to affordable products that were available in this country. She addressed the needs of her particular clients, including having her dietary charts and nutrition information printed in several different languages.

Stern's clinic established an international reputation, and in 1925, she received funding to establish a nutrition education program to train American and foreign doctors, dentists, social workers, and nurses in dietetics. She taught nutrition and social work at various schools, such as Simmons College, Tufts College Medical School, MIT, and the State Teachers College at Framingham. Stern was awarded an honorary degree from Tufts Medical School, and the Boston Food Clinic was eventually renamed the Frances Stern Nutrition Center at Tufts University.

Stern co-authored the book *Food for the Worker* (1917) to show the need for unifying science, social work, income, and nutrition. Her other co-authored books were

Food and Your Body (1932), *How to Teach Nutrition to Children* (1942), and *Diabetic Care in Pictures* (1946), and she was the sole author of *Applied Dietetics* (1936), which incorporated new information about the role of vitamins in nutrition. Stern was a member of the American Public Health Association, the American Home Economics Association, and the American Dietetic Association.

Further Resources

Tufts University. "Frances Stern Nutrition Center." http://nutrition.tufts.edu/ 1177953850925/Nutrition-Page-nl2w_1177953851896.html.

Jewish Women's Encyclopedia. "Frances Stern: 1873–1947." http://jwa.org/encyclopedia/ article/stern-frances.

Stickel, Lucille Farrier

1915–2007
Zoologist

Education: B.A., Eastern Michigan University, 1936; M.S., University of Michigan, 1938, Ph.D., zoology, 1949

Professional Experience: biologist, Patuxent Wildlife Research Center, U.S. Fish and Wildlife Service, 1943–1947, 1961–1972, director, 1972–1982

Lucille Stickel developed original methods for determining pesticide residue levels in wildlife. Her research included vertebrate population ecology and the ecology and pharmacotoxicology of environmental pollution. In her work in the pioneering field of pesticide research, she studied the significance and levels of chemical residues in animal brain tissue and developed a method still used to determine acceptable levels today. In 1946, she published one of the earliest reports on the pesticide DDT. Wildlife toxicology research has important implications for human health as well, since humans can consume either the polluted water or the contaminated fish and wildlife. She earned a master's degree from the University of Michigan and joined the Patuxent Wildlife Research Center at Laurel, Maryland, as a biologist, in 1943. She returned to Michigan to complete her Ph.D., but then took several years off from her career before returning to Patuxent as a biologist in 1961. She was promoted to director in 1972, a position she held until her retirement in 1982. Her husband, William F. Stickel, was also a researcher at Patuxent, and the two collaborated on studies of the environmental effects of pesticides on birds and eggshell thinning, and on other research related to small mammal populations. In 1989, a chemistry and physiology lab at the Wildlife Research Center was renamed in their honor.

Stickel received the Federal Woman's Award of the Department of the Interior (1968), a Distinguished Service Award of the Department of the Interior (1973), the Aldo Leopold Award of the Wildlife Society (1974), and the Rachel Carson Award of the Society of Environmental Toxicology and Chemistry (1998). In 1974, she received an honorary doctorate from her alma mater, Eastern Michigan University.

Further Resources

Howell, Judd A. "Lucille Farrier Stickel 1915–2007." http://www.pwrc.usgs.gov/what-snew/events/stickel/.

Stiebeling, Hazel Katherine

1896–1989
Food Chemist and Nutritionist

Education: Skidmore College; B.S., Columbia University, 1919, M.A., nutrition, 1924, Ph.D., chemistry, 1928

Professional Experience: school supervisor, home economics, 1915–1918; supervising teacher, home economics, Kansas State Teachers College, 1919–1923; instructor, nutrition, Columbia University, 1924–1926; senior food economist, Bureau of Home Economics, U.S. Department of Agriculture (USDA), 1930–1944, assistant chief, 1943–1944, chief, 1944–1954, director of research, human nutrition and home economics, 1954–1957, director, institute of home economics, 1957–1960, deputy administrator, Agricultural Research Service, 1960–1963

Hazel Stiebeling was a nutritionist noted for her work in developing government dietary guidelines, including the concept of daily allowances of vitamins and minerals, or Recommended Dietary Allowances (RDA). During her long career at the USDA, her research involved the composition and nutritive values of food, energy metabolism, and food consumption habits of different population and income groups. Joining the USDA's Bureau of Home Economics shortly after it was established, she was promoted to assistant bureau chief in 1943 and chief in 1944. Although there were changes in the name of the bureau, she continued as head until 1960, when she was appointed deputy administrator of the Agricultural Research Service, retiring in 1963. Prior to joining the USDA, she had been a school supervisor in home economics, a supervising teacher for home economics at Kansas State, and an instructor in nutrition at Columbia University, where she had received all of her academic degrees, including a Ph.D. in chemistry in 1928. Stiebeling was part of a generation of women who brought scientific rigor to home economics and nutrition studies.

Stiebeling became interested in domestic science and food chemistry while in high school and then in her courses at Skidmore College. She taught school for three years before enrolling at Columbia University Teachers College. She went on for a master's degree and taught food and nutrition courses while completing her doctorate in chemistry. Her early research was on the effect and content of vitamins in the human body, and some of her studies were published in the *Journal of Biological Chemistry*. She was particularly concerned with the ability of low-income families to prepare nutritious food. While working at the USDA, she published the first research on quantitative dietary recommendations for vitamins and minerals, standards that were eventually applied nationally and internationally.

Stiebeling received the Borden Award in 1943, the Distinguished Service Award from the USDA in 1952, and the President's Gold Medal Award for civilian service in 1959. She was a member of the American Statistical Association and the American Home Economics Association, and a fellow of the American Institute of Nutrition. She received several honorary degrees.

Further Resources

Levine, Susan. 2008. *School Lunch Politics: The Surprising History of America's Favorite Welfare Program*. Princeton, NJ: Princeton University Press.

Harper, Alfred E. 2003. "Contributions of Women Scientists in the U.S. to the Development of Recommended Dietary Allowances." *The Journal of Nutrition*. 133: 3698–3702. (November 2003). http://jn.nutrition.org/cgi/content/full/133/11/3698.

Stokey, Nancy

b. 1950
Economist

Education: B.A., economics, University of Pennsylvania, 1972; Ph.D., economics, Harvard University, 1978

Professional Experience: assistant to associate professor, Department of Managerial Economics and Decision Sciences, Kellogg Graduate School of Management, Northwestern University, 1978–1983, professor, 1983–1987, Harold L. Stuart Professor of Managerial Economics, 1988–1990; professor, economics, University of Chicago, 1990–1996, Frederick Henry Prince Professor of Economics, 1997–2004, Distinguished Service Professor, 2004–

Concurrent Positions: visiting lecturer, economics, Harvard University, 1982; visiting professor, economics, University of Minnesota, 1983; visiting professor, economics, University of Chicago, 1983–1984; visiting scholar, Research Department, Federal Reserve Bank of Minneapolis, 2000–2002

Nancy Stokey is an economist who specializes in economic theory and economic development. She is particularly interested in the effect of education and job training on national economic growth. In addition to her numerous articles on global aid, social mobility, free trade, industrialization, development, and taxation, she has authored or co-authored textbooks, including *Recursive Methods in Economic Dynamics* (1989) and *The Economics of Inaction* (2008). After attending the University of Pennsylvania, she went on to receive her Ph.D. in economics from Harvard in 1978. She taught at Northwestern University for 12 years before moving to the University of Chicago in 1990, where she is a Distinguished Service Professor of Economics. In 2004, she was named one of eight economists (and the only woman) on

Economist Nancy Stokey. (Courtesy of the University of Chicago)

the Expert Panel of the Copenhagen Consensus Center, an international think tank that brings together researchers, policymakers, philanthropists, and nongovernmental organizations (NGOs) to address global challenges, such as global warming, terrorism, clean water, and development.

Stokey was elected a member of the National Academy of Sciences in 2004. She is also a fellow of the American Academy of Arts and Sciences and the Econometric Society, and served as vice president of the American Economic Association (1996–1997). She received an honorary doctorate from Northwestern University (2005), and her research has been supported by numerous grants from the National Science Foundation. She has been on the editorial board of the *Journal of Political Economy*, *Journal of Economic Growth*, *Games and Economic Behavior*, and *Journal of Economic Theory*.

Further Resources

Copenhagen Consensus Center. http://www.copenhagenconsensus.com.

University of Chicago. Faculty website. http://home.uchicago.edu/~nstokey/.

Stoll, Alice Mary

b. 1917
Biophysicist

Education: B.A., Hunter College, 1938; M.S., physiology and biophysics, Cornell University, 1948

Professional Experience: assistant, allergy, metabolism, and infrared spectrophotography, New York Hospital and Medical College, Cornell University, 1938–1943, temperature regulation, 1946–1948, physiological research associate, environmental thermal radiation, 1948–1953; physiologist, medical research laboratory, U.S. Naval Air Development Center (NADC), Pennsylvania, 1953–1956, special technical assistant, 1956–1960, head, thermal laboratory, 1960–1964, head, biophysical and bioastronautical division, 1964–1970, head, biophysical laboratory, crew systems department, 1970–1980

Concurrent Positions: U.S. Naval Reserves, 1943–1946; consultant, Arctic Aerospace Medicine Laboratory, Ladd Air Force Base, Alaska, 1952–1953

Alice Stoll was a pioneer in bioengineering with the U.S. Navy, and in particular was responsible for the development of fire-resistant and fire-retardant fibers and fabrics. Her research on the effects of heat and thermal radiation, and the biophysics of and engineering guidelines for thermal safety, led to the development of "Nomex" (manufactured by the Du Pont company), a fabric used in the uniforms worn by firefighters. She also studied the effects of rapid acceleration on the human heart. During the post–World War II era, the armed services were developing supersonic planes that made many physiological demands on crews as well as planes. Her work for the NADC involved assuring that crews can withstand the extremely cold temperature at high altitudes, the physiological stress of breaking the sound barrier at supersonic speeds, and the constant danger of fire in a closed environment. She personally developed and received patents on the specific instrumentation needed for her research.

Stoll received dual master's degrees in physiology and biophysics from Cornell University in 1948 and subsequently worked as a physiological research associate in environmental thermal radiation at the medical school. She was simultaneously in the Naval Reserves and worked as a consultant for other government laboratories, including the Arctic Aerospace Medicine Laboratory in Alaska. She accepted an appointment at the NADC as a physiologist in the medical research laboratory in 1953. She then rose through the ranks as head of the thermal laboratory in 1960, and then head of the biophysical and bioastronautical division in 1964, and head of the biophysical laboratory in the crew systems department in 1970, formally retiring in 1980.

Stoll received the Federal Civil Service Award (1965), an Achievement Award of the Society of Women Engineers (1969), and the Paul Bert Award of the Aerospace Medical Association (1972). She was elected a fellow of the American Association for the Advancement of Science and of the Aerospace Medical Association, and was a charter member of the Biophysical Society. She was also a member of the American Physiological Society and the American Society of Mechanical Engineers.

Stroud-Lee, F. Agnes Naranjo

b. 1922
Radiation Biologist

Education: B.S., University of New Mexico, 1945; Ph.D., University of Chicago, 1966

Professional Experience: research technician, hematology, Los Alamos Scientific Laboratory, 1945–1946; associate cytologist, Argonne National Laboratory, 1946–1969; director, Department of Tissue Culture, Pasadena Foundation for Medical Research, 1969–1970; senior research cytogeneticist, Scientific Data Analysis Section, Jet Propulsion Laboratory, 1970–1975; staff cytogeneticist, health research division, Los Alamos Scientific Laboratory, 1975–1979; independent consultant, radiobiology and cytogenetics

Agnes Stroud-Lee has been recognized for her research in radiobiology and chromosomal abnormalities. Her work has increased scientific understanding of certain birth defects. Her research has included automation of chromosome analysis by computers, effects of radiation on animal tumors, effects of ionizing radiation *in vitro* and *in vivo*, and mammalian radiation biology. During her career, she worked at several of the major research centers in radiobiology, such as Los Alamos and Argonne National laboratories. Stroud-Lee is a member of the Tewa tribe of the Santa Clara Indian Pueblos and was the first Native American woman to hold a research scientist position at a national laboratory.

After receiving her undergraduate degree from the University of New Mexico in 1945, she was employed at Los Alamos for one year before receiving an appointment as associate cytologist at Argonne, where she worked until 1969. During this time, she was also working toward her Ph.D. in biology and zoology from the University of Chicago, which she received in 1966. In 1969, she moved to California as director of the tissue culture program at the Pasadena Foundation for Medical Research. She then accepted an appointment at the Jet Propulsion Laboratory as a senior research cytogeneticist in 1970. She returned to Los Alamos in 1975 as a staff cytogeneticist in the health research division. In 1979, she left to consult in radiobiology and cytogenetics.

Stroud-Lee has received numerous honors and awards, including the Morrison Prize in Natural Sciences of the New York Academy of Sciences (1955), the National Aeronautics and Space Administration (NASA) Certificate of Recognition (1976), and a Diploma of Honor in Cytology at the First Pan-American Cancer Cytology Congress. She has been a member of the Radiation Research Society, the American Society for Cell Biology, the Biophysical Society, and the Tissue Culture Association. She is the subject of a children's book, *Scientist from the Santa Clara Pueblo, Agnes Naranjo Stroud-Lee*, published by the Equity Institute (1985). She was twice married and is listed variously in the sources as Stroud, Stroud-Schmink, or Stroud-Lee.

Stubbe, JoAnne

b. 1946
Chemist

Education: B.S., chemistry, University of Pennsylvania, 1968; Ph.D., chemistry, University of California, Berkeley, 1971

Professional Experience: postdoctoral fellow, chemistry, University of California, Los Angeles (UCLA), 1971–1972; assistant professor, chemistry, Williams College, 1972–1977; assistant professor, pharmacology, Yale University School of Medicine, 1977–1980; assistant professor to professor, University of Wisconsin, Madison, 1980–1987; professor, chemistry, Massachusetts Institute of Technology (MIT), 1987–1992, professor, chemistry and biology, 1992–

Concurrent Positions: National Institutes of Health (NIH) postdoctoral fellow, Brandeis University, 1975–1977

JoAnne Stubbe has made notable contributions to understanding how enzymes catalyze, or cause, chemical reactions. Her research has potential applications for antitumor, antivirus, and antiparasite activity, because inhibiting these enzymes interferes with the biosynthesis of DNA and cell growth. She held a prestigious postdoctoral appointment at UCLA, and was later an NIH fellow at Brandeis University. After having appointments at the Yale University School of Medicine and the University of Wisconsin, Madison, she moved to MIT as professor of chemistry and was appointed a distinguished professor of both chemistry and biology in 1992.

Chemist JoAnne Stubbe is presented with the National Medal of Science by President Barack Obama, 2009. (AP/Wide World Photos)

Stubbe was elected to membership in the National Academy of Sciences in 1992, and received the National Academy of Sciences Award in Chemical Sciences (2008) and a National Medal of Science (2009). She received a career development award from the NIH and the Pfizer Award in enzyme chemistry from the American Chemical Society (1986), given each year to a young scientist (under 40) for outstanding work in the field. She has also received the ICI-Stuart Pharmaceutical Award for excellence in chemistry (1989), a teaching award from MIT (1990), the Arthur C. Cope Scholar Award (1993), and the F. A. Cotton Medal for Excellence in Chemical Research (1998). She is a member of the American Chemical Society, American Society of Biological Chemists, Protein Society, and American Academy of Arts and Sciences.

Further Resources

Massachusetts Institute of Technology. Faculty website. http://web.mit.edu/chemistry/www/faculty/stubbe.html.

Sudarkasa, Niara

b. 1938
Anthropologist

Education: student, Fisk University, 1953–1956; B.A., anthropology and English, Oberlin College, 1957; M.A., anthropology, Columbia University, 1959, Ph.D., anthropology, 1964

Professional Experience: assistant professor, anthropology, New York University, 1964–1967; assistant professor, anthropology, and research associate, Center for Research in Economic Development, University of Michigan, 1967–1970, associate professor, 1970–1976, professor, 1976–1986, director, Center for Afro-American and African Studies, 1981–1984, associate vice president, Academic Affairs, 1984–1986; president, Lincoln University, Pennsylvania, 1987–1998

Concurrent Positions: visiting scholar, Florida Atlantic University; Distinguished Scholar-in-Residence, African-American Research Library and Cultural Center, Ft. Lauderdale, Florida

Niara Sudarkasa is renowned as an authority in the fields of African women, especially Yoruba women traders, West African migration, and the African American and African family. She has also researched higher-education policies for black Americans and other minorities, and she is an advocate for minority access to education at the university level. Born Gloria Marshall, *Sudarkasa* was her first husband's name, and she adopted the African name *Niara* (an adaptation of a Swahili word for "a woman of high purpose") as a result of her studies of the African continent in the 1970s. She studied Yoruba culture and language for her doctoral work, and in 2001, she was honored with the title of "Chief" in the Ife kingdom of the Yoruba of Nigeria, the first African American to hold the title. Sudarkasa has applied her study of West African culture to the African American family structure, with an emphasis on the role of black women within the family and society. Her published works include *Where Women Work: A Study of Yoruba Women in the Marketplace and in the Home* (1973), *Exploring the African-American Experience* (1995), and *The Strength of Our Mothers: African and African-American Women and Families* (1996).

At the age of only 14, she won a Ford Foundation Early Entrant Scholarship to Fisk University. In her junior year, she went to Oberlin College as an exchange student and decided to stay there to receive her undergraduate degree. After completing her undergraduate degree at age 18, she went to Columbia University for graduate study, receiving another Ford Foundation Foreign fellowship to study in Nigeria and at the University of London School of Oriental and African Studies. She also spent two years at the University of Chicago as a fellow with a Carnegie Foundation project on a comparative study of new nations. She began her academic career at New York University and then spent 20 years at the University of Michigan. She directed the Center for Afro-American and African Studies and was a research scientist at the Center for Research in Economic Development. She became politically active while she was at Michigan, advocating on behalf of the students for a black studies program and for increasing the number of black and minority students in the university. She then spent a decade serving as the first female president of Lincoln University in Pennsylvania, one of the oldest black colleges in the United States. In 2000, she returned to her native Ft. Lauderdale as Distinguished Scholar-in-Residence at Florida Atlantic University, where she helped establish an African-American Research Library and Cultural Center.

Sudarkasa is a member of the American Ethnological Society, American Anthropological Association, African Studies Association, American Association for Higher Education, and Council on Foreign Relations. She has been awarded more than a dozen honorary degrees from institutions such as Fisk University, Oberlin College, Sojourner–Douglass College, Franklin and Marshall College, Susquehanna University, the University of Nigeria, and Fort Haiti University in South Africa.

Sullivan, Kathryn D.

b. 1951
Geologist, Astronaut

Education: B.S., earth sciences, University of California, Santa Cruz, 1973; Ph.D., geology, Dalhousie University, Nova Scotia, 1978

Professional Experience: staff member, National Aeronautics and Space Administration (NASA), 1978, astronaut, 1979–1993; chief scientist, National Oceanic and Atmospheric Administration (NOAA), 1992–1996; president and chief

Astronaut Kathryn Sullivan aboard the Space Shuttle Challenger, 1984. (NASA)

executive officer, Center for Science and Industry (COSI), Columbus, Ohio, 1996–2005; director, Battelle Center for Mathematics and Science Education Policy, Ohio State University, 2005–

Concurrent Positions: adjunct professor, Rice University, 1984–1992; captain, U.S. Naval Reserve; volunteer science advisor, COSI, 2005–

Kathryn D. Sullivan was one of the first women trained in the astronaut program in 1978, and she was the first American woman to perform a space walk. The first six women were selected for a training program in scientific, engineering, and medical duties, but none was to be trained in piloting the space shuttle. However, most of the women in the program took flying lessons anyway so they would be prepared to land the shuttle in an emergency. Sullivan passed her training tests and became an astronaut in 1979. Her shuttle assignments included software development, lead chase photography of launches and landings, and orbiter and cargo testing. She was a member of the spacesuit monitoring and extravehicular activity (EVA) crew, and served as capsule communicator in Mission Control for numerous shuttle missions. Her first space mission was as a mission specialist on STS 41-G in 1984; **Sally Ride** was also a member of the crew. Sullivan was the first woman to perform an EVA, with orbiter commander David Leetsma, and the two demonstrated the feasibility of in-flight satellite refueling. On her second mission, STS-31 in 1990, she was a mission specialist when the crew deployed the Hubble Space Telescope (the telescope proved to have a defective mirror, and several years later, another shuttle crew installed a new mirror). On her third mission, STS-45 in 1992, she was a mission specialist and payload commander. Overall, she logged over 500 hours in space in her career as an astronaut.

Prior to completing a doctorate in geology, Sullivan had participated in several oceanographic expeditions under the auspices of the U.S. Geological Survey and the Woods Hole Oceanographic Institution. She resigned from the astronaut corps in 1992 and was selected by President George H. W. Bush to be chief scientist of NOAA, replacing oceanographer **Sylvia Earle** in that position. Always looking for new challenges in her career, Sullivan resigned from NOAA in 1995 to become director of COSI in Columbus, Ohio. In 2005, she became the director of the new Battelle Center for Mathematics and Science Education Policy at the John Glenn School of Public Affairs, Ohio State University. She remains affiliated with COSI as a science advisor.

Sullivan has served on various committees and government commissions related to marine science and ecosystems. She was appointed by President Ronald Reagan to the National Commission on Space in 1985 and participated in preparing guidelines for U.S. space exploration. In 2004, she was appointed to the National Science Board. She also served on the Pew Oceans Commission, which issued a 2003 report entitled "America's Living Oceans: Charting a Course for Sea Change," urging reform in ocean wildlife protection policy.

Sullivan has received a number of awards, including the NASA Exceptional Service Medal (1988 and 1991), National Air and Space Museum Trophy (1985), NASA Space Flight Medals (1984 and 1990), Haley Space Flight Award of the American Institute of Aeronautics and Astronautics (1991), Space Achievement Award of the American Aeronautic Society (1991), NASA Outstanding Leadership Medal (1992), Public Service Award of the National Science Board (2003), Astronaut Hall of Fame (2004), and *Aviation Week & Space Technology*'s Aerospace Legend Award (2005). She is a member of the American Institute of Aeronautics and Astronautics, Geological Society of America, American Geophysical Union, Society of Women Geographers, Explorers Club, Association of Space Explorers, and American Association for the Advancement of Science.

Further Resources

Kevles, Bettyann H. 2003. *Almost Heaven: The Story of Women in Space*. New York: Basic Books.

National Aeronautics and Space Administration. "Kathryn D. Sullivan (Ph.D.)." http://www.jsc.nasa.gov/Bios/htmlbios/sullivan-kd.html.

Sweeney, (Eleanor) Beatrice Marcy

1914–1989
Botanist

Education: A.B., Smith College, 1936; Ph.D., biology, Radcliffe College, 1942

Professional Experience: laboratory assistant, endocrinology, Mayo Clinic, 1942; fellow, University of Minnesota, 1942–1943; junior research biologist, Scripps Institution of Oceanography, California, 1947–1955, assistant research biologist, 1955–1960, associate research biologist, 1960–1961; research staff biologist, Yale University, 1961–1962, lecturer, biology, 1962–1967; lecturer, biology, University of California, Santa Barbara (UCSB), 1967–1969, associate professor, 1969–1971, professor, 1971–1981, associate provost, College of Creative Studies, 1978–1981

Beatrice Sweeney was recognized for her research on circadian rhythms (or biological clocks) in plants, and their effect on plant processes such as bioluminescence, photosynthesis, and cell division. In addition to hundreds of scientific papers, she also published a book, *Rhythmic Phenomena in Plants* (1969; 2nd ed., 1987). She completed her doctoral research in biology at Radcliffe and in 1947 moved to the Scripps Institution of Oceanography in California. She spent six years as a research biologist and lecturer at Yale University before returning to California as a professor at UCSB, where she spent the remainder of her career. Even after formally retiring in 1981, Sweeney remained active as a researcher, reviewer, and visiting lecturer. She was lecturing at the Woods Hole Oceanographic Institution in Massachusetts, when she suffered a stroke and passed away in the summer of 1989.

Sweeney began her scientific studies as an undergraduate botany student at Smith College. She was greatly influenced by her female teachers at Smith, but dismayed that, unlike male professors, most of the career women at that time (in the early 1930s) were not married and did not have families. She resolved to find her own way to balance family life with a scientific career and, throughout her years as a professor and administrator (including as an advisor for the UCSB Women's Studies Program), mentored female students on not giving up on their career plans. Sweeney was herself a powerful role model, as she was the mother of four children.

Sweeney's work was recognized by the Botanical Society of America, which awarded her the Darbaker Award in 1983. She was president of the Western Section of the American Society for Plant Physiology (1977–1978), the American Institute for Biological Sciences (1979–1980), the American Society for Photobiology (1979), the American Association for the Advancement of Science (1980), and the Phycological Society of America (1986), and was also a member of the American Association for the Advancement of Science, the American Society of Plant Physiologists, the Society of General Physiologists, and the Society for the Study of Biological Rhythms. She was awarded honorary doctorates from Umea University in Sweden (1985) and from Knox College in Illinois (1986). UCSB established the Beatrice M. Sweeney Memorial Fund in her name.

Further Resources

University of California. "Eleanor Beatrice March Sweeney, Biological Sciences: Santa Barbara." http://content.cdlib.org/xtf/view?docId=hb4p30063r&doc.view=frames &chunk.id=div00063&toc.depth=1&toc.id=.

T

Talbot, Mignon

1869–1950
Geologist

Education: A.B., Ohio State University, 1892; Ph.D., Yale University, 1904

Professional Experience: high school teacher, physical geography, Columbus, Ohio, 1896–1902; instructor, geology, Mount Holyoke College, 1904–1905, associate professor and chair, 1905–1908, professor and chair, 1908–1935, professor and chair, geography, 1928–1935

Mignon Talbot was among the first women to enter the field of geology and paleontology, and she made an important discovery of a rare dinosaur skeleton, *Podokesaurus holyokensis*, found near Mount Holyoke College in Massachusetts. She spent her entire career at Mount Holyoke, where she headed the geology department and helped build a renowned program there in the early 1900s. Talbot grew up in Iowa, studied geology at Ohio State University, and then traveled abroad before returning to Ohio to teach physical geography to high school students. She did some graduate work and summer research at Ohio State, Harvard, and Cornell Universities before settling at Yale, where she received a Ph.D. in 1904. She joined the geology faculty at Mount Holyoke that same year and, three years later, became head of the department.

In addition to her teaching and research, she built a world-class fossil and mineral collection and science library at Mount Holyoke. In 1910, while on an expedition with her sister, Ellen (a Mount Holyoke philosophy professor), Talbot discovered the most complete dinosaur skeleton to date found in the Northeast; the rare find of a 45-foot-long, 150-million-year-old dinosaur was subsequently reported in the *American Journal of Science* (June 1911). A cast of the skeleton was made and kept at Yale University, but the original fossil was lost in a fire at Mount Holyoke's science hall in 1917. Besides the loss of the fossil, Talbot had to restock the collections of books as well as specimens, rocks, and minerals from scratch. She took her female students on numerous field trips to conduct this work over the next several years, building an even more extensive collection than before. In 1928, she traveled throughout Europe on a sabbatical, collecting materials for use in her teaching. Upon her return, she was made chair and professor of the joint program in geology and geography.

In 1909, Mignon Talbot became the first woman elected to the Paleontological Society (1909); she was elected vice president of the Society in 1926. She was a fellow of the Geological Society of America and a member of the American Association for the Advancement of Science.

Further Resources

"Lost Dinosaur." 1937. In Frances Lester Warner, *On A New England Campus*. Boston, MA: Houghton Mifflin.

Taussig, Helen Brooke

1898–1986
Cardiologist, Endocrinologist

Education: Radcliffe College, 1917–1919; A.B., University of California, 1921; M.D., Johns Hopkins University, 1927

In 1945, physician Helen Taussig developed a surgical technique for treating "blue baby" syndrome in newborn babies with heart defects. (Library of Congress)

Professional Experience: fellow, medicine, Johns Hopkins Hospital, 1927–1928, intern, pediatrics, 1928–1930; physician in charge, cardiac clinic, Harriet Lane Home, 1930–1963; associate professor, pediatrics, Johns Hopkins University, 1946–1959, professor, 1959–1963

Helen Taussig originated the idea for the "blue-baby" operation, first tried in 1945 as the Blalock-Taussig procedure, which involves treating babies with congenital malformations of the heart within the first few days after birth. After she received her M.D. from Johns Hopkins in 1927, and completed her internship, she spent her entire career as physician in charge of the cardiac clinic at the

Harriet Lane Home and as a member of the faculty of Johns Hopkins Medical School from 1946 until retiring in 1963. In 1962, she was the first physician to alert the United States to the dangers of thalidomide, a medicine routinely given to pregnant women to control nausea that was later found to cause deformities in the limbs of numerous newborns. She was also the first to demonstrate that changes in the heart and lungs could be diagnosed by X-ray and fluoroscope. Her colleague, Dr. Blalock, was elected to the National Academy of Sciences in 1946, the year after the two introduced the Blalock-Taussig procedure; however, Helen Taussig was not elected to the National Academy of Sciences until 1973.

In addition to many scientific papers, Taussig was the author of *Congenital Malformations of the Heart* (1947, rev. 1960). She was elected a master of the American College of Physicians and was the first woman president of the American Heart Association (1965). She also was a member of the American Pediatric Society and the Society for Pediatric Research. In recognition for her achievements, she received 20 honorary degrees plus numerous awards, including the Lasker Award (1954), the Gold Heart Award (1963), and the Medal of Freedom (1964).

Taussky-Todd, Olga

1906–1995
Mathematician

Education: Ph.D., mathematics, University of Vienna, 1930; Bryn Mawr College, 1934–1935; M.A., Cambridge University, 1937

Professional Experience: assistant, University of Göttingen, 1931–1932; assistant, University of Vienna, 1932–1934; lecturer, University of London, 1937–1943; scientific officer, Ministry of Aircraft Production, England, 1943–1946; researcher, Department of Scientific and Industrial Research, 1946–1947; mathematician, National Bureau of Standards, 1947–1957; research associate, California Institute of Technology (CalTech), 1957–1971, professor, 1971–1977

Concurrent Positions: member, Institute for Advanced Study, Princeton University, 1948; visiting faculty, Courant Institute for Mathematical Sciences, New York University, 1955; Fulbright visiting professor, University of Vienna, 1965

Olga Taussky-Todd was known for her work in algebraic number theory and matrix theory, which she helped popularize. Born in Austria-Hungary, she enrolled in the University of Vienna, where she first majored in chemistry but quickly dropped that study to concentrate on mathematics, graduating in 1930. She received an appointment as an assistant at the University of Göttingen, where she edited several volumes on number theory. She received a fellowship to study at Bryn Mawr in Pennsylvania and then at Girton College in Cambridge, England. She taught briefly at the University of London, where she met fellow mathematician John "Jack" Todd, and the two were married in 1938. During World War II, she worked for a government agency in England and then moved to the United States again, where they both worked at the National Bureau of Standards. In 1957, the couple was recruited to CalTech, where he was a professor and she was a research associate; she was promoted to professor in 1971, the first female full professor at CalTech. During her tenure there, she mentored many graduate students in matrix theory before retiring in 1977. The couple collaborated for more than 50 years, and she authored or co-authored more than 300 papers. She was founding editor of the journal, *Linear Algebra and Its Applications*, and served as editor of *Linear and Multilinear Algebra, Journal of Number Theory*, and *Advances in Mathematics*.

Taussky-Todd was named a "Woman of the Year" in 1964 by the *Los Angeles Times*. She received the Ford Prize of the Mathematical Association of America (1970) and the Gold Cross of Honor for Science and Art from the Austrian government (1978). She was elected a fellow of the American Association for the Advancement of Science and was a member of the American Mathematical Society. She was also elected to the Austrian Academy of Sciences (1975) and the Bavarian Academy of Sciences (1985). She received an honorary Golden Doctorate from the University of Vienna (1980) and an honorary Doctor of Science degree from the University of Southern California (1988). In 1990, CalTech established the Olga Taussky–John Todd Lecture Program.

Further Resources

Luchins, Edith H. and Mary Ann McLoughlin. "In Memoriam: Olga Taussky-Todd." *Notices of the American Mathematical Society.* 43(8): 838–847. (August 1996). http://www.ams.org/notices/199608/taussky.pdf.

Case, Bettye Anne and Anne M. Leggett. 2005. *Complexities: Women in Mathematics.* Princeton, NJ: Princeton University Press.

Taylor, Kathleen Christine

b. 1942
Chemical Engineer

Education: B.A., chemistry, Douglass College (Rutgers University), 1964; Ph.D., physical chemistry, Northwestern University, 1968

Professional Experience: fellow, University of Edinburgh, 1968–1970; associate senior research chemist, General Motors (GM) Corporation, 1970–1974, senior research chemist, 1974–1975, assistant head, Physical Chemistry Department, 1975–1983, head, Environmental Sciences Department, Research Laboratories, 1983–1985, head, Physics and Physical Chemistry Department, 1985–, chief scientist, GM of Canada, 2000–, director, Materials and Processes Laboratories, GM Research and Planning (retired)

Concurrent Positions: chair, Center for Automotive Materials and Manufacturing, Canada, 2002–2003

Kathleen Taylor is an expert on catalytic converters for automobiles, and her research includes surface chemistry, heterogeneous catalysis, and catalytic control of automobile exhaust emissions. The U.S. Congress passed the Clean Air Act in 1970, demanding that automobile manufacturers begin to significantly reduce auto exhaust emissions of carbon monoxide, hydrocarbons, and nitrogen oxides. That same year, Taylor began working for GM, where her early research led to the development of the catalytic converter, introduced in new vehicles by the mid-1970s. Her group was interested in understanding the catalytic conversion of nitrogen oxides in automobile exhaust, and she published a book on the topic, *Automobile Catalytic Converters* (1984). She spent more than 30 years at GM in a variety of research and administrative positions involving the development of catalysis, surface chemistry, surface coatings, corrosion, combustion, batteries, fuel cells, and chemical processes.

Taylor published dozens of scientific papers on her research, a significant number for a corporate scientist. Even after formally retiring, Taylor remains committed and active on issues related to energy efficiency, reduction of greenhouse gases, and new fuel technologies. She has served on numerous government and industry committees, including the Department of Energy (DOE) Hydrogen and Fuel Cell Technical Advisory Committee, DOE Council on Materials Science and Engineering, DOE Basic Energy Sciences Advisory Committee, Advisory Committee for Columbia University Center for Electron Transport in Molecular Nanostructures, and National Academies Board on Energy and Environmental Systems.

Taylor was elected to membership in the National Academy of Engineering in 1995. She is the recipient of the Garvan Medal of the American Chemical Society (1989) and was nominated by the National Women's History Project as one of their "Women Taking the Lead to Save Our Planet" (2009). She has been a member of the North American Catalysis Society, Materials Research Society (president, 1987), Society of Automotive Engineers (SAE), American Academy of Arts and Sciences, and Indian National Academy of Engineering (elected 2006), and a fellow of SAE International.

Further Resources

"Kathleen C. Taylor." http://www.hydrogen.energy.gov/docs/bio_taylor.doc.

Tesoro, Giuliana (Cavaglieri)

1921–2002
Polymer Chemist

Education: Ph.D., organic chemistry, Yale University, 1943

Professional Experience: research chemist, Calco Chemical Company, 1943–1944; research chemist, Onyx Oil and Chemical Company, 1944–1946, head, organic synthesis department, 1946–1955, assistant director of research, 1955–1957, associate director, 1957–1958; assistant director of organic research, central research laboratory, J. P. Stevens & Company, Inc., 1958–1968; senior chemist, Textile Research Institute, 1968–1969; senior chemist, Burlington Industries, Inc., 1969–1971, director, chemical research, 1971–1972; research professor, Polytechnic Institute, 1982–1996

Concurrent Positions: visiting professor, Massachusetts Institute of Technology, 1972–1976, adjunct professor and senior research scientist, 1976–1982; member, committee on military personnel supplies, National Research Council, 1979–1982, committee on toxic combustion products, 1984–1989

Giuliana Tesoro was an internationally recognized expert on the science and technology of polymers. Her research involved synthesis of pharmaceuticals, textile chemicals, and chemical modification of fibers, and she made important contributions to developments in polymer flammability and flame retardants in her work with several textile companies. After receiving her doctorate in 1943, she worked summers for Calco Chemical Company before accepting a position as research chemist at Onyx Oil and Chemical Company in 1944. She was promoted to head of the organic synthesis department in 1946, assistant director of research in 1955, and associate director in 1957. She was appointed assistant director of organic research for J. P. Stevens & Company, then moved to the Textile Research Institute for two years. She accepted a position as senior chemist at Burlington Industries in 1969 and was appointed director of chemical research in 1971. She was appointed research professor at Polytechnic Institute in 1982.

Tesoro was a member of several committees of the National Academy of Sciences and the National Research Council concerning toxic materials and fire safety. She was president of the Fiber Society in 1974, and has been a member of the American Chemical Society, the American Association of Textile Chemists and Colorists, the American Institute of Chemists, and the American Association for the Advancement of Science.

Tharp, Marie

1920–2006
Geologist

Education: B.A., Ohio University, 1943; M.A., geology, University of Michigan, 1945; B.S., mathematics, University of Tulsa, 1948

Professional Experience: junior geologist, U.S. Geological Survey, 1944; geologist, Stanolind Oil & Gas Company, Oklahoma, 1945–1948; assistant, Lamont-Doherty Geological Observatory, Columbia University, 1949–1952, research geologist, 1952–1960, research scientist, 1961–1963, research associate, 1963–1968; oceanographer, U.S. Naval Oceanographic Office, 1968–1983; owner and consultant, Marie Tharp Maps, 1983–2006

Marie Tharp was a geologist who pioneered charting the ocean floor at a time when little was known about undersea geology. The detailed maps she prepared indicated features that helped other scientists understand the structure and evolution of the bottom of the ocean. Of particular importance was her discovery of the valley that divides the Mid-Atlantic Ridge, which convinced other geologists that the ocean floor was being created at these ridges in various parts of the world and spreading outward. The confirmation of "seafloor spreading" led to the eventual acceptance of the theory of continental drift, or "plate tectonics."

Although a few studies of the Mid-Atlantic Ridge had been done by the 1920s, scientists had not fully explored the seafloor until an earthquake near the Great Banks in the Atlantic Ocean in 1929 broke the transatlantic cables and there was a need to anticipate future earthquakes before laying new cables. Working with geologist Bruce C. Heezen at the Lamont-Doherty Geological Observatory in the 1950s, Tharp began preparing a "physiographic" diagram of the Atlantic Ocean floor. The resulting maps show how the floor would look if all the water were drained away, and her first map showed a deep valley dividing the crest of newly formed rocks making up the ridge. At the time, most scientists believed that the Earth was a shrinking globe, cooling and contracting from its initial hot birth, and that continental drift was impossible. For many years, Tharp herself was not able to participate in recording ocean-floor soundings because women were not permitted on U.S. Navy ships. Beginning in the late 1960s, she went on several research cruises and, in 1977, Heezen and Tharp published the World Ocean Floor Panorama, based on all available geological and geophysical data as well as more than 5 million miles of ocean-floor soundings. They received the Hubbard Medal of the National Geographic Society in 1978, and their work was chronicled in the book by John Noble Wilford, *The Mapmakers*, first published in 1981.

Tharp retired from the observatory and from her later appointment with the U.S. Navy in 1983 and began doing independent oceanography consulting and writing articles about Heezen's life and work. She received the Lamont-Doherty Heritage Award in 2001, and her former institute has established a fellowship in her name to support women in the sciences. Until her death in 2006, she operated a map-distribution business, Marie Tharp Maps, which still sells prints of her ocean floor map.

Further Resources

Marie Tharp Maps. http://www.marietharp.com.

Columbia University, Lamont-Doherty Earth Observatory. "Marie Tharp, Pioneering Mapmaker of the Ocean Floor." http://www.ldeo.columbia.edu/news/2006/08_23_06.htm.

Wilford, John Noble. 2001. *The Mapmakers*. 2nd ed. New York: Random House.

Thomas, Martha Jane (Bergin)

1926–2006
Analytical Chemist, Physical Chemist

Education: B.A., chemistry, Radcliffe College, 1945; M.A., Boston University, 1950, Ph.D., chemistry, 1952; M.B.A., Northeastern University, 1981

Professional Experience: senior engineer, chemical laboratory, General Telephone and Electronics Corporation (GTE), 1945–1959, group leader, lamp material engineering laboratories, Lighting Products Division, 1959–1966, section head, chemical and phosphor laboratory, Sylvania Lighting Center, 1966–1972, manager, technical assistance laboratories, Lighting Products, 1972–1981, technical director, technical service laboratories, 1981–1983, director, technical quality control, 1983–1990

Concurrent Positions: instructor, chemistry, Boston University, 1952–1970; adjunct professor, chemistry, University of Rhode Island, 1974–1993

Martha Thomas is renowned for her work in phosphor chemistry at Sylvania Lighting/GTE. Her research includes phosphors, photoconductors, ion exchange membranes, complex ions, and instrumental analysis. Phosphors are the powdery substances used to coat the inside of fluorescent lighting tubes, and her inventions included developing the phosphors that made possible Sylvania's natural-daylight fluorescent lamps and made mercury lamps 10% brighter. She holds more than 20 patents, including her first for a method of etching the fine tungsten coils that were designed to improve telephone switchboard lights. She went on to establish two pilot plants for the preparation of phosphors—pilot plants are experimental industrial setups in which processes or techniques planned for use in full-scale operations are tested in advance. She also developed a natural white phosphor that allowed fluorescent lamps to impart daylight hues and a phosphor that increased the brightness of mercury lamps.

Thomas studied chemistry at Radcliffe, intending to enter medical school, but instead accepted a job at Sylvania (later GTE), where she remained for 45 years. She attended graduate school at Boston University part-time while working and received her doctorate in 1952. She returned to school again in 1980 to obtain a master's degree in business administration so she could handle her new responsibilities at GTE as a manager. In 1983, she was the first woman to be made a director in her division, and she was one of the few women then working in phosphor chemistry. Although she had a heavy schedule as a researcher, manager, and mother of four, she also taught evening chemistry classes at Boston University and then served as an adjunct professor of chemistry at the University of Rhode Island.

In 1991, Thomas was named New England Inventor of the Year by Boston's Museum of Science, the Inventors Association of New England, and the Boston Patent Law Association. She also received the National Achievement Award of the Society of Women Engineers (1965) and the Gold Plate of the American Academy of Achievement (1966), and was the first woman to receive the New England Award of the Engineering Societies of New England. She was a fellow of the American Institute of Chemists and a member of the American Chemical Society, Electrochemical Society, and Society of Women Engineers.

Thompson, Laura Maud

1905–2000
Anthropologist

Education: B.A., Mills College, 1927; Ph.D., anthropology, University of California, Berkeley, 1933

Professional Experience: assistant ethnologist, Bishop Museum, Honolulu, 1929–1934; social scientist, U.S. Navy, Guam, 1938–1940; social scientist, Community Survey of Education, Territory of Hawaii, 1940–1941; coordinator, Indian education research project, U.S. Office of Indian Affairs, 1941–1947; research associate, Institute for Ethnic Affairs, 1946–1954; professor, anthropology, City

College of New York, 1954–1956; visiting professor, University of North Carolina, 1957–1958; visiting professor, North Carolina State College, 1958–1960; distinguished visiting professor, anthropology, Pennsylvania State University, 1961; professor, anthropology, University of Southern Illinois, 1961–1962; professor, anthropology, San Francisco State College, 1962–1963; consulting anthropologist

Laura Thompson was recognized for her research on Native Americans. Her research has included comparative interdisciplinary research in small communities, especially among Native Americans and Lower Saxons of West Germany; human ecology; and ecosystem approach toward population control.

She conducted field research in Fiji, Germany, Guam, Hawaii, Iceland, and the United States with the Papago, Navajo, Zuni, Sioux, and Hopi people. She also consulted for the Hutterite communities in Pennsylvania in tracing their early history in Germany before they immigrated to the United States. After receiving her doctorate from Berkeley in 1933, she held numerous faculty and visiting positions, including at City College of New York, the University of Southern Illinois, and San Francisco State College. She regularly consulted for educational and government agencies, including the U.S. National Indian Institute in Mexico, the U.S. Office of Indian Affairs, and the Hutterite Socialization Project at Pennsylvania State University.

Thompson was a prolific writer as well, publishing numerous scientific papers and books, including *Archaeology of the Mariana Islands* (1932), *Fijian Frontier* (1940), *Guam and Its People* (1940), *The Hopi Way* (1944; co-author), *Guam and Its People* (1947), *Culture in Crisis: A Study of the Hopi Indians* (1950), *Personality and Government* (1951), *Toward a Science of Mankind* (1961), and *The Secret of Culture* (1969). She held her last formal academic appointment in 1963, but continued to work as a consulting anthropologist, invited speaker, and author after that date. She received grants from the Viking Fund, a Wenner-Gren fellowship to study in New York and Iceland, and Rockefeller Foundation grants in 1951 and 1952. She was the founder of the Society for Applied Anthropology, and was elected a fellow of the American Anthropological Association, New York Academy of Sciences, American Association for the Advancement of Science, and Association for Social Anthropology in Oceania.

Thornton, Kathryn (Cordell)

b. 1952
Physicist, Astronaut

Education: B.S., physics, Auburn University, 1974; M.S., physics, University of Virginia, 1977, Ph.D., physics, 1979

Professional Experience: NATO fellow, Max Planck Institute for Nuclear Physics, Germany, 1979–1980; physicist, U.S. Army Foreign Science and Technical Center, Charlottesville, Virginia, 1980–1984; staff member, National Aeronautics and Space Administration (NASA), 1984, astronaut, 1985–1996; professor and associate dean, School of Engineering and Applied Science, University of Virginia, 1997–

Kathryn Thornton is a physicist who joined the NASA astronaut training program in 1984 and was involved in four shuttle missions: STS-33 (1989), STS-49 (1992), STS-61 (1993), and STS-73 (1995). She was a mission specialist aboard the space shuttle *Discovery* in November 1989. Her second mission was in 1992 aboard the space shuttle *Endeavour* on its maiden flight, and her third was again on the *Endeavour* in 1993, which was a Hubble Space Telescope servicing and repair mission. Her last flight was aboard the space shuttle *Columbia* in 1995, and the mission included conducting scientific experiments on the SpaceLab module. The Columbia flight was a unique experience for Thornton, who had a starring role in footage that was used on the television show *Home Improvement*, the first entertainment footage shot in space. In the scene, Thornton was taped using a screwdriver in the gravity-free environment. Her technical assignments have included flight software verification in the Shuttle Avionics Integration Laboratory, serving as a team member of the Vehicle Integration Test Team at Kennedy Space Center, and serving as a spacecraft communicator.

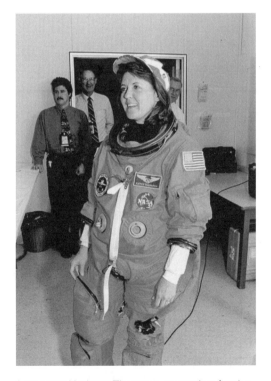

Astronaut Kathryn Thornton preparing for the launch of the Space Shuttle Columbia, 1995. (NASA)

Like the other women astronauts who entered the NASA program since the late 1970s, she has a solid scientific background, with a doctorate in physics and a postdoctoral appointment at the Max Planck Institute for Nuclear Physics in Germany. Following that appointment, she was a physicist with the U.S. Army science and technology center before joining NASA. She retired from NASA in 1996 and remains committed to science education and to encouraging women in science and technology, including space flight. At the University of Virginia, she has been a faculty member and director of the Center for Science, Mathematics, and Engineering Education. She has also been a leading voice urging the U.S. Congress to support a manned mission to Mars.

Thornton received a NASA Distinguished Service Medal. She is a member of the American Association for the Advancement of Science and the American Physical Society.

Further Resources

Kevles, Bettyann H. 2003. *Almost Heaven: The Story of Women in Space*. New York: Basic Books.

National Aeronautics and Space Administration. "Kathryn C. Thornton (Ph.D.)." http://www.jsc.nasa.gov/Bios/htmlbios/thornt-k.html.

"University of Virginia's Kathryn Thornton Urges Congress to Back Manned Space Exploration to Mars . . . and Beyond." *UVA Today*. (3 April 2008). http://www.virginia.edu/uvatoday/newsRelease.php?id=4748.

Tilden, Josephine Elizabeth

1869–1957
Botanist

Education: B.S., University of Minnesota, 1895, M.S., 1897

Professional Experience: instructor, botany, University of Minnesota, 1898–1902, assistant professor, 1902–1910, professor, 1910–1937

Josephine Tilden was an expert on algae and phycology of the Pacific Ocean. She was ahead of her time in realizing the ecological and economic importance of algae as a marine life food source and spoke often on problems of ocean pollution, conservation, and industrial uses of algae. Her primary area of research was the Pacific Rim, and she conducted research on the shores of Japan, Australia, New Zealand, Hawaii, and the South Pacific islands, often accompanied by her elderly mother. She was also a specialist in local freshwater algae, and her book, *Minnesota Algae* (1910), remained a widely used technical reference for many decades. Her later work, *The Algae and Their Life Relations* (1935–1937), was the first effort by an American scientist to summarize the known characteristics of these important marine and freshwater plants.

Tilden spent her entire career in the Department of Botany at the University of Minnesota, where she was the first woman scientist on the faculty. In 1900, she discovered an area of algae, seaweed, and tide pools along a desolate stretch of Canadian coastline, and subsequently used her own money, and donated land, to establish the Minnesota Seaside Station for research on Vancouver Island. Tilden was the subdirector and, along with her mother and the chair of the Minnesota botany department, hosted 25 to 30 professors, students, and lecturers every summer to conduct research on algae, lichen, animals, and the natural environment. The station operated between 1900 and 1906, when the University of Minnesota chose not to continue the program. Throughout her career, Tilden was at odds with the university over funding for her expeditions, as it was unusual for a Midwestern university to dedicate resources to ocean research. She retired in 1937 and took 300 boxes of her own algae specimens with her to Florida, which were later returned to the University of Minnesota after her death. Tilden had built an impressive collection of algae samples from around the world and drawn numerous students to the botany program; 10 years after her death, however, there was no longer a program in marine algae studies at Minnesota.

Born in Davenport, Iowa, Tilden received her undergraduate degree from the University of Minnesota in 1895 and her master's degree in 1897. She was appointed instructor in botany at the school in 1898 and promoted to assistant professor in 1902. Her promotion to full professor in 1910, even though she did not hold a doctorate, was an indication of the recognition she had received for her research and teaching. She was a delegate to the First Pan-Pacific Scientific Congress of 1920 and attended succeeding congresses in 1923 and 1926. She was a member of the American Society for the Advancement of Science, the American Society of Naturalists, the American Geographical Society, the Botanical Society of America, and the Torrey Botanical Club.

Further Resources

Brady, Tim. 2008. "Of Algae and Acrimony." University of Minnesota Alumni Association. (11 January 2008). http://www.minnesotaalumni.org/s/1118/content.aspx?sid =1118&gid=1&pgid=1077&sparam=algae&scontid=0.

Tilghman, Shirley M.

b. 1946
Molecular Biologist

Education: B.Sc., chemistry, Queen's University, Ontario, 1968; Ph.D., biochemistry, Temple University, Pennsylvania, 1975

Professional Experience: teacher, Sierra Leone, West Africa, 1968–1970; postdoctoral fellow, National Institutes of Health (NIH), 1975–1978; assistant professor, Fels Research Institute, Temple University School of Medicine, 1978–1979; adjunct associate professor, human genetics and biochemistry and biophysics, University of Pennsylvania, 1980–1986; Howard A. Prior Professor of Life Sciences, Princeton University, 1986–2001, professor, molecular biology, 1986–, president, Princeton University, 2001–

Concurrent Positions: independent investigator, Institute for Cancer Research, Fox Chase Cancer Center, Philadelphia, 1979–1986; investigator, Howard Hughes Medical Institute, 1988–2001; founding director, Lewis-Sigler Institute for Integrative Genomics, Princeton University, 1998–2003

Shirley Tilghman is a molecular biologist and in 2001 became the first female president of Princeton University in New Jersey. Tilghman specialized in mammalian developmental genetics and was on the faculty at Princeton for 15 years before assuming the presidency. She was a founding member of the council on the Human Genome Project, a project for mapping all human DNA. Born and educated in Canada, she received her undergraduate degree in chemistry from Queen's University in Kingston, Ontario. She went on to earn a doctorate in biochemistry from Temple University in Philadelphia, Pennsylvania, and held a postdoctoral fellowship with the NIH, where she was involved in the first efforts to clone mammalian genes. She was then an investigator for the Institute for Cancer Research in Philadelphia and taught at the University of Pennsylvania before joining the life sciences faculty at Princeton in 1986. In 1998, she founded and became director of the interdisciplinary Institute of Integrative Genomics at Princeton. During this time, she simultaneously held an affiliation as an investigator at Howard Hughes Medical Institute. Tilghman has served on numerous committees and advisory councils for both academic and government organizations, and has been committed to science education and to

Molecular biologist Shirley Tilghman became the first female president of Princeton University in 2001. (AP/Wide World Photos)

careers for women in the sciences. She served as chair of Princeton's Council on Science and Technology for seven years (1993–2000).

Tilghman was elected a foreign associate to the National Academy of Sciences (1996) and a member of the Institute of Medicine (1995), and served on the National Research Council's Commission on Life Sciences (1993–2001). She is a fellow of the American Philosophical Society, American Academy of Arts and Sciences, New York Academy of Sciences, and Royal Society of London, and a member of the American Society for Biochemistry and Molecular Biology, American Society for Cell Biology, and Society for Developmental Biology. In addition to 25 honorary degrees from universities in the United States, Canada, and England, she is the recipient of a Basic Science Award of the Society for the Advancement of Women's Health Research (1997), the Mellon Prize from the University of Pittsburgh (2000), the L'Oréal-UNESCO Women in Science Award (2002), a Lifetime Achievement Award from the Society of Developmental Biology (2003), the Radcliffe Institute Medal (2004), a Genetics Society of America Medal for outstanding contributions to the field (2007), and numerous other awards and honors.

Further Resources

Princeton University. "Office of the President." http://www.princeton.edu/president/.

Tinsley, Beatrice Muriel (Hill)

1941–1981
Astronomer

Education: B.Sc., University of Canterbury, New Zealand, 1961, M.Sc., physics, 1963; Ph.D., astronomy, University of Texas, Austin, 1967

Professional Experience: fellow, University of Texas, Austin, 1967–1968; visiting scientist, physics, University of Texas, Dallas, 1969–1973, assistant professor, astronomy, 1973–1974; associate professor, astronomy, Yale University, 1975–1978, professor, 1978–1981

Beatrice Tinsley was the first person to make a realistic, computer-generated model of how the color and brightness of a galaxy change as the stars that make up the galaxy are born, grow old, and die. Before her research, astronomers treated galaxies as static, unchanging objects. Since galaxies are the milestones that astronomers use to measure the universe as a whole, her evolutionary models of galaxies have had a profound impact on cosmology, the branch of astronomy that deals with the general structure and evolution of the universe. In developing her models of galaxies while working on her doctoral dissertation, she added up the colors and luminosities of the evolving stars to find the total color and luminosity of the entire galaxy as it developed. Her models demonstrated how the results of work in many other areas of astronomy could be synthesized into models of the evolution of galaxies far more accurately than any previous models, and she was largely responsible for establishing the photometric evolution of galaxies as a field of study in astronomy.

Despite her short life, she had an extremely prolific and successful career, although she encountered many obstacles as a female scientist. A native of New Zealand, Tinsley studied physics at Canterbury University, and although she was interested in astronomy and cosmology as an undergraduate, there were no facilities available for her to write a master's thesis on the subject. She moved to the United States and received her doctorate at the University of Texas, only to find there were no job opportunities for her as an astronomer in Dallas, where she lived with her husband, fellow physicist Brian Tinsley, and their two adopted children. She obtained a position as a visiting scientist at the newly formed University of Texas, Dallas and received part-time National Science Foundation funding. She conducted research at Mt. Wilson, Lick, and Mt. Palomar observatories in California, and at the University of Maryland and Cambridge University. After she and her husband divorced, she took a tenure-track faculty position at Yale. In 1978, the same year she was promoted as the first female full professor of astronomy at Yale, she learned that a lesion on her leg was malignant skin cancer. The cancer later spread to her vital organs, and she died in 1981 at only 40 years old.

Among the several awards Tinsley received was the Annie Jump Cannon prize in 1974, named for the Harvard astronomer who specialized in stellar spectra. Tinsley herself is commemorated by a biennial prize awarded by the American Astronomical Society for exceptionally creative or innovative research and by a visiting professorship of astronomy at the University of Texas, Austin. She was a member of the American Astronomical Society, Royal Astronomical Society, and International Astronomical Union.

Tolbert, Margaret Ellen (Mayo)

b. 1943
Biochemist

Education: B.S., chemistry, Tuskegee University, 1967; M.S., analytical chemistry, Wayne State University, 1968; Ph.D., biochemistry, Brown University, 1974

Professional Experience: research technician, biochemistry, Tuskegee University, 1969, instructor, mathematics, 1969–1970; instructor, science and mathematics, Opportunities Industrialization Center, Providence, Rhode Island, 1971–1972; assistant professor, chemistry, Tuskegee University, 1973–1976; associate professor, pharmaceutical chemistry, and associate dean, School of Pharmacy, Florida A&M University, 1977–1979; professor, chemistry, and director, Research and Development, Carver Research Foundation, Tuskegee University, 1979–1988; budgets and control analyst, and senior planner, British Petroleum (BP) America Research Center, 1988–1990; director, Research Improvement in Minority Institutions (RIMI) Program, National Science Foundation, 1990–1993; director, educational programs, Argonne National Laboratory, University of Chicago, 1994–1996; director, New Brunswick Laboratory, U.S. Department of Energy, 1996–2002; senior advisor, Office of Integrative Activities, National Science Foundation, 2002–

Concurrent Positions: instructor, chemistry, summer Transitional Program, Brown University, 1973; visiting associate professor, medical sciences, Brown University, 1979; consulting scientist, Howard Hughes Medical Institute, 1994

Margaret Tolbert has moved successfully from a distinguished research career to academia to science administration in the corporate and government realms. She had already established herself as a noted researcher on the biochemistry of the liver when she changed her career plans in the late 1980s and became an administrator and dean for the Carver Research Foundation, and then took a number of positions in government and industry. In 1990, the National Science Foundation recruited her as program director for the RIMI Program. In 1994, she worked with the Howard Hughes Medical Institute to establish international research programs in Eastern Europe. Dr. Tolbert was the first African American and the first female to serve as Director of the U.S. Department of Energy's New Brunswick facility at Argonne National Laboratories, which brings together researchers in nuclear science.

Tolbert's high school teachers arranged for her to take advanced placement courses in mathematics and science, and she enrolled in Tuskegee University. Her initial goal was to study medicine, but she switched to chemistry for both financial and research reasons. She also had the opportunity for summer internships at Central State College in North Carolina and at Argonne National Laboratories in Illinois, where she was a member of a team that was studying the various chemical combinations made by uranium. In her later, high-level positions at Argonne, she remained committed to creating high school and post–high school programs and opportunities in science education such as she received.

After graduating from Tuskegee in 1967, she went on to Wayne State University in Detroit, Michigan, where she earned a master's degree in analytical chemistry in 1968. She returned as a researcher and mathematics instructor at Tuskegee before being recruited to the doctoral program at Brown University in Rhode Island. While at Brown, she received financial aid from the Southern Fellowship Fund to research biochemical reactions in liver cells. She also taught basic science to nurses and mathematics to welders in night school in Providence, Rhode Island, where adults sought to upgrade their employment skills. In her later positions representing research institutes and government agencies, she has traveled widely for her research and educational mission to increase international communication among scientists and educators in different countries, and to increase the numbers of women and minorities in science, work she continues as a Senior Advisor for the National Science Foundation.

Among her numerous awards and honors, Tolbert has received a Certificate of Distinguished Service from the Federal Reserve System (1987), the Secretary of Energy Pride Award for Community Service (1998), a Chicago-Tuskegee Alumni Club President's Merit Award (1999), Performance Awards from the Chicago Operations Office of the U.S. Department of Energy (1997–2001), the Women of Color in Government and Defense Technology Award in Managerial Leadership (2001), and a Performance Award from the National Science Foundation (2005). In 2007, she received the Dr. George Washington Carver Distinguished Service Award of Tuskegee University. She is a fellow of the American Association for the Advancement of Science and a member of the American Chemical Society, New York Academy of Sciences, Institute of Nuclear Materials Management, and Society for Environmental Toxicology and Chemistry.

Further Resources

National Science Foundation. "Dr. Margaret E. M. Tolbert, Senior Advisor." http://www.nsf.gov/od/oia/staff/tolbert.jsp.

Townsend, Marjorie Rhodes

b. 1930
Aerospace Engineer, Electronics Engineer

Education: B.S., electrical engineering, George Washington University, 1951

Professional Experience: aide, physical science, National Bureau of Standards, 1948–1951; electronics engineer, basic and applied sonar research, Naval Research Laboratory, 1951–1959; section head, design and development of electronic instruments, Goddard Space Flight Center, National Aeronautics and Space Administration (NASA), 1959–1965, technical assistant to chief of systems division, 1965–1966, project manager, small astronomical satellites, 1966–1975, project manager, applied explorer mission, 1975–1976, manager, preliminary systems design group of advanced systems design, 1976–1980; consultant, 1980–1990; director, Space Systems Engineering BDM International, 1990–1993; consultant, 1993–

Marjorie Townsend is renowned for her work in launching the first astronomical satellites in the Small Astronomy Satellite (SAS) program for NASA in the 1970s. She co-invented (and received a patent for) a digital telemetry system, and her research includes advanced space and ground systems design for a large variety of missions in space and terrestrial applications and in the space sciences, new applications for the use of the space shuttle, and improvements in the data system design of space stations. During her years with NASA, she was the only woman to work as a project manager for a satellite program, and as such, she was responsible for the origination, design, construction, and testing of the satellites, as well as for the actual launches of the instruments.

Townsend was the first woman to earn an engineering degree from George Washington University. She joined NASA's Goddard Space Flight Center in 1959, the year after the agency was established. She had been conducting sonar research at the Naval Research Laboratory by developing frequency multiplication systems, an analog logic computer, and new submarine detection and classification techniques; at NASA, her first assignment was to design a ground system for the forerunner of meteorological satellites. In 1966, she was placed in charge of the SAS program, a joint U.S.–Italian project, and she created quite a controversy when she persuaded NASA administrators to use a launch site owned and operated by the Italian government. Her research indicated that the launch site in the Indian Ocean off the coast of Kenya was the best site because it was located in an area where the satellite could be placed in an equatorial orbit, thereby missing the radiation belt and avoiding a significant amount of background noise. The data received from SAS revolutionized the study of x-ray-emitting stars.

Townsend left NASA in 1980 and has continued to work and consult on space systems and satellite programs. She has received numerous awards, including the Exceptional Service Medal (1971), Knight of the Italian Republic Order from Italy (1972), Federal Woman's Award from the U.S. Government (1973), George Washington University Distinguished Alumni Achievement Award (1976), and NASA Outstanding Leadership Medal (1980). She is a fellow of the American Institute of Aeronautics and Astronautics and of the Institute of Electrical and Electronics Engineers (IEEE), and a member of the American Association for the Advancement of Science, American Geophysical Union, and Society of Women Engineers.

Further Resources

"Marjorie Rhodes Townsend." 2009. Interview for *Online Journal of Space Communication*. 15. (Spring 2009). http://satjournal.tcom.ohiou.edu/issue15/townsend.html.

Treisman, Anne

b. 1935
Psychologist

Education: B.A., psychology, Cambridge University, England, 1957; D.Phil., Oxford University, 1962

Professional Experience: research assistant, experimental psychology, University of Oxford, 1961–1963; staff member, M. R. C. Psycholinguistics Research Unit, 1963–1966; visiting research scientist, Behavioral Sciences Department, Bell Telephone Laboratories, New Jersey, 1966–1967; university lecturer, psychology, Oxford University, 1968–1978; professor, psychology, University of British Columbia, 1978–1986; professor, psychology, University of California, Berkeley, 1986–1994; professor, psychology, Princeton University, 1993–, James S. McDonnell Distinguished University Professor of Psychology, 1995–

Concurrent Positions: lecturer, Trinity College, Oxford University, 1961–1977, Somerville College, 1962–1966, St. Anne's College, 1964–1967, fellow, St. Anne's College, 1967–1978; fellow, Center for Advanced Study in Behavioral Sciences, Stanford University, 1977–1978; fellow, Canadian Institute for Advanced Research, 1984–1986; visiting scholar, Russell Sage Foundation, New York, 1991–1992

Anne Treisman is a research psychologist who has created models for testing visual perception and analyzing how the brain combines visual and auditory input in selective attention and memory. Her research combines, and has implications for, work in cognitive psychology, neuropsychology, and neuroscience. Treisman developed a Feature Integration Theory (FIT) to explain how human vision processes color, shape, size, light, motion, and other input by creating and combining separate "feature maps" that correspond to different areas of the brain. She studied patients with behavioral differences, such as attention problems, or with brain damage, to see how their brains combined the various visual stimuli (a process called the "binding problem") to make sense of the whole. Her research revealed that there are neurological as well as behavioral or learned explanations for attention, memory, and perception.

Research psychologist Anne Treisman with her husband, Daniel Kahneman, in 2002. Treisman created models for testing visual perception, and analyzes other brain processes, such as attention and memory. (AP/Wide World Photos)

Treisman holds dual citizenship in the United States and Britain. She was elected to the National Academy of Sciences in 1994 as a foreign associate, but later became a regular member as a U.S. citizen. Among her awards are the Spearman Medal of the British Psychological Society for experimental research (1963), Howard Crosby Warren Medal of the Society of Experimental Psychologists (1990), Distinguished Scientific Contribution Award of the American Psychological Association (1990), and Golden Brain award of the Minerva Foundation (1996), and she was named a William James Fellow of the American Psychological Society (2002). She was elected to the Society of Experimental Psychologists, Royal Society of London, and American Academy of Arts and Sciences, and is a member of the Psychonomic Society, Association for Research in Vision and Ophthalmology, and Cognitive Neuroscience Society. She was awarded an honorary doctorate from the University of British Columbia (2004) and was named an Honorary Professor in the Institute of Psychology by the Chinese Academy of Sciences (2004). Her husband, Daniel Kahneman, is also a psychologist and won the Nobel Prize in Economics in 2002.

Further Resources

Princeton University. Faculty/laboratory website. http://weblamp.princeton.edu/~psych/psychology/research/treisman/index.php.

Turkle, Sherry

b. 1948
Psychologist, Sociologist

Education: B.A., social studies, Harvard University, 1970, M.A., 1973, Ph.D., sociology and psychology, 1976

Professional Experience: clinical intern, psychology, University Health Services, Harvard University, 1974–1975; assistant professor, sociology, Massachusetts Institute of Technology (MIT), 1976–1980, associate professor, 1980–1989, professor, 1991–

Concurrent Positions: licensed clinical psychologist, Commonwealth of Massachusetts, 1978–; founder and director, MIT Initiative on Technology and Self, 2001–

Sherry Turkle is a psychologist and sociologist who has done pioneering research on how humans interact with computers and computer programs, and how computers shape our very identities. Her work on robots and computers began before the age of the Internet, but the intrusion of computers into nearly every aspect of our lives has raised even more questions about the boundaries between computers and humans, and between real life and virtual reality. She has written on computer games, online and digital pets, and Internet chat rooms, and is considered by some to be an anthropologist or ethnographer of computer culture. A licensed clinical psychologist, she has been professor of sociology, technology, and science studies at MIT since 1976, and in 2001 founded an Initiative on Technology and Self research center. Her work has made her a high-profile media figure on the psychology of computer users, and she has been interviewed for popular magazines and television and radio shows.

Turkle entered Radcliffe College in 1965, but dropped out and moved to Paris after her mother died. She returned to Harvard to complete her degrees, earning a bachelor's and a master's, and, in 1976, her doctorate in sociology and psychology on the influence of Freud's psychoanalytic theory in France, the subject of her first book, *Psychoanalytic Politics* (1978). She became interested in computers and computer users when she accepted a position as an assistant professor of sociology at MIT. She noticed students using computer language in their everyday conversations, even when talking about their emotions, speaking of "debugging their relationships" or excusing verbal slips as "information processing errors." In her second book, *The Second Self: Computers and the Human Spirit* (1984; rev. ed., 2005), she theorized that the computer is not just a tool, but an evocative object with which one can have intense, almost intimate, relations.

For her third book, *Life on the Screen: Identity in the Age of the Internet* (1995), she interviewed more than 1,000 people, 300 of them children. The use of computers had increased dramatically, and she expressed her concern about the concept of computer literacy defined as teaching computer skills, but not necessarily critical thinking. She also warned about the abuses of online identities. She is not against the idea of having multiple identities, but warned of adults assuming false identities in order to prey on children. Other books were compiled from seminars and lectures from the Initiative on Technology and Self, including *Evocative Objects: Things We Think With* (2007), *Falling for Science: Objects in Mind* (2008), *The Inner History of Devices* (2008), and *Simulation and Its Discontents* (2009). New issues on which she has written articles and lectured include cellphone use and the effect of an ever-present availability of friends through texting, chatting, and online access, especially among teenagers.

Turkle has been honored by numerous magazines and organizations as an innovator and voice of the computer age. She was named Woman of the Year by *Ms. Magazine* (1984), one of the Computer 200 innovators for the Association of Computing Machinery's Fiftieth Anniversary celebration (1997), one of *Time* magazine's Innovators of the Internet (2000), and one of the Top Ten Wired Women by ABC News (2002). She is a member of the American Psychological Association and the American Sociological Association, and a fellow of the American Association for the Advancement of Science, Boston Psychoanalytic Society and Institute, and World Economic Forum.

Further Resources

Massachusetts Institute of Technology. Faculty website. http://www.mit.edu/~sturkle/.

Tyson, Laura (D'Andrea)

b. 1947
Economist

Education: B.A., economics, Smith College, 1969; Ph.D., economics, Massachusetts Institute of Technology, 1974.

Professional Experience: staff economist, World Bank, 1974; assistant professor, economics, Princeton University, 1974–1977; assistant to associate professor, economics, University of California, Berkeley, 1977–1988, professor, 1988–2001, professor, Haas School of Business, University of California, Berkeley, 1990–2001, dean, 1998–2001; dean, London Business School, 2002–2006; professor, Haas School of Business, University of California, Berkeley, 2007–2008, S. K. and Angela Chan Professor of Global Management, 2008–

Concurrent Positions: Director of Research, Berkeley Roundtable on the International Economy, University of California, Berkeley, 1988–1992; director, Institute of International Studies, University of California, Berkeley, 1990–1992; chair, President's Council of Economic Advisors, 1993–1995; chair, National Economic Council, 1995–1996; principal, Law and Economics Consulting Group, 1997–2001

Laura Tyson is an economist renowned as an authority and advisor on global economic issues, global markets and trade, healthcare reform, government deficits, and the high-tech industry. She has served as economic advisor to two presidents and has been the dean of two prestigious business schools. As the first female chair of the National Economic Council (NEC) under President Clinton in the 1990s, and as a member of President Obama's Economic Advisory Panel to address the national economic crisis in 2009, she has been one of the most influential economists in the nation. In 1992, she published a book that examined the American trade imbalance problem in depth, *Who's Bashing Whom: Trade Conflict in High Technology Industries*, in which she advocated aggressive action against foreign traders who close their markets to imports by blocking U.S. markets to the foreign traders. She is known for her ability to explain complex economic concepts in an understandable and interesting way, whether in the classroom, at a conference, or in the media.

In the 1990s, economist Laura Tyson served as chair of President Bill Clinton's Council of Economic Advisors and chair of the National Economic Council. (Hulton Archive/Getty Images)

As an undergraduate at Smith College in the 1960s, she planned to major in mathematics and psychology but changed her major to economics after taking an introductory course in that field. In her graduate program at the Massachusetts Institute of Technology, she became more interested in the practical applications of economic theory rather than the technical and statistical aspects of economics. After receiving her doctorate, she taught at Princeton before accepting a position at the University of California, Berkeley in 1977. She has been affiliated with the Haas School of Business at Berkeley since 1990, and served as dean of the school for three years before becoming the first female dean of the London School of Business, where she founded the Centre for Women in Business. She spent four years in London before returning to teach at Berkeley. In addition to her academic work and government advisory positions, she has consulted with numerous policy organizations, such as the Brookings Institution and the Center for American Progress, and sat on the boards of companies such as Morgan Stanley, AT&T, and Eastman Kodak, among others.

In addition to numerous reports and dozens of newspaper editorials, Tyson has published several other books dealing with international competition, trade, productivity, and politics. She is a member of the American Economic Association and the Association for Comparative Economic Studies, and a fellow of the American Academy of Arts and Sciences. She has been awarded honorary degrees from Smith College and from American University.

Further Resources

University of California, Berkeley. Faculty website. http://www2.haas.berkeley.edu/Faculty/tyson_laura.aspx.

U

Uhlenbeck, Karen (Keskulla)

b. 1942
Mathematician

Education: B.A., University of Michigan, 1964; M.A., mathematics, Brandeis University, 1966, Ph.D., mathematics, 1968

Professional Experience: instructor, mathematics, Massachusetts Institute of Technology (MIT), 1968–1969; lecturer, University of California, Berkeley, 1969–1971; assistant professor, mathematics, University of Illinois, Champaign-Urbana, 1971–1976; associate professor, University of Illinois, Chicago, 1977–1983; professor, mathematics, University of Chicago, 1983–1988; professor and chair, mathematics, University of Texas, Austin, 1988–

Concurrent Positions: visiting associate professor, Northwestern University, 1976; chancellor's visiting professor, University of California, Berkeley, 1979; Albert Einstein fellow, Institute for Advanced Study, Princeton University, 1979–1980; visiting member, Mathematical Sciences Research Institute, Berkeley, 1982; visiting professor, Harvard University, 1983; visiting professor, Max Planck Institute for Mathematics, Bonn, 1985; visiting professor, University of California, San Diego, 1986; visitor, Institut des Hautes Études Scientifiques, Bures-Sur-Yvette, France, 1987; visiting professor, mathematics, University of Texas, Austin, 1988; visitor, Mathematics Research Centre, Warwick University, England, 1992; member, Institute for Advanced Study, Princeton, 1995, distinguished visiting professor, 1997–1998

Karen Uhlenbeck is renowned for mathematical research on calculus of variations, global analysis, and gauge theories. Her work has had applications in theoretical physics and has contributed to current research on instantons, which are models for the behavior of surfaces in four dimensions. Mathematicians are looking at imaginary spaces that have been constructed by scientists who are examining other problems. For example, physicists who were studying quantum mechanics had predicted the existence of particle-like elements known as instantons. Uhlenbeck and other researchers built a model for understanding the behavior of instanton surfaces in three and four dimensions. She is the co-author of the books *Instantons and 4-Manifold Topology* (1984) and *Geometry and Quantum Field Theory* (1995).

Uhlenbeck had planned to major in physics at the University of Michigan, but switched to mathematics. After graduating, she spent a year at New York University's Courant Institute; she then married and moved to Boston, where her husband was attending Harvard. She received a National Science Foundation graduate fellowship to work at Brandeis University; after receiving her Ph.D., she taught at MIT for a year, then moved to the University of California, Berkeley as a lecturer in mathematics. In 1971, Uhlenbeck and her husband both obtained positions at the University of Illinois. In 1988, she joined the faculty at the University of Texas, Austin, where she began a mentoring program for women in mathematics. Throughout her career, she has held a number of fellowships and visiting professorships at institutions in both the United States and abroad. In 1990, she traveled to Japan as only the second woman to present the keynote lecture at the International Congress of Mathematics.

Uhlenbeck was elected to membership in the National Academy of Sciences in 1986 and received a National Medal of Science in 2000. She has received several honorary doctorates and numerous fellowships and awards, including a five-year MacArthur Fellowship (1983), Alumna of the Year from the University of Michigan (1984), Alumni Achievement Award from Brandeis University (1988), Commonwealth Award for Science and Invention of PNC Bank Corporation (1995), and Steele Prize of the American Mathematical Society (2007). She was named one of America's 100 most important women in 1988 by *Ladies' Home Journal*. She is a member of the American Academy of Arts and Sciences, Mathematical Association of America, Association for Women in Mathematics, National Association of Mathematicians, and American Mathematical Society.

Further Resources

Henrion, Claudia. 1997. *Women in Mathematics: The Addition of Difference*. Bloomington: Indiana University Press.

University of Texas. Faculty website. http://www.ma.utexas.edu/users/uhlen/.

V

Van Rensselaer, Martha

1864–1932
Home Economist

Education: Chamberlain Institute, 1884; A.B., Cornell University, 1909

Professional Experience: public school teacher, 1884–1893; school commissioner, Cattaraugus County, New York, 1893–1899; head, extension program for farm wives, Cornell University, 1900–1903, instructor, home economics, 1903–1907, co-director, School of Home Economics, 1907–1925, professor, 1911–1932, co-director, New York State College of Home Economics, 1925–1932

Martha Van Rensselaer taught some of the first accredited home economics courses in the country at Cornell University in the early 1900s. Her early career followed the pattern of many educated women of her generation in that she first taught in various public and private schools. It was while serving as a country school commissioner in New York State that she became interested in the education of farm women and created a program, through Cornell, to provide reading and other classes to rural women. In its first few years, the program attracted thousands of women, and Van Rensselaer began offering other courses in home economics through the agricultural college. In 1907, a separate School of Home Economics was formed, with Van Rensselaer and **Flora Rose** as co-directors. After completing her own undergraduate degree program in 1909, Van Rensselaer became a professor in the home economics degree program. Along with Rose and another colleague, she co-authored *A Manual of Home-Making*, published in 1919. In 1925, their popular program was upgraded to a separate school, the New York State College of Home Economics.

In addition to her academic research, teaching, and administrative duties, Van Rensselaer served in a variety of government positions, including as a staff member of the U.S. Department of Agriculture (USDA) and special service for the American Relief Commission during World War I, and, later, as assistant director of the White House Conference on Child Health and Protection under President Hoover. She was also active in professional organizations, serving as president of the American Home Economics Association (1914–1916), home-making editor of the journal *Delineator* (1920–1926), and assistant director of

Home economist, Martha Van Rensselaer. (Courtesy of Cornell University)

the White House Conference on Child Health and Protection (1930). She was chair of the home economics section of the Association of Land-Grant Colleges and Universities in 1928 and 1929. She was a member of the American Association of University Women committee to welcome physicist Marie Curie on her visit to New York City in 1921. In 2004, Van Rensselaer was posthumously inducted into the National 4-H Hall of Fame.

Further Resources

Cornell University. "Martha Van Rensselaer." http://rmc.library.cornell.edu/homeEc/bios/marthavanrensselaer.html.

Van Straten, Florence Wilhemina

1913–1992
Meteorologist

Education: B.S., New York University, 1933, M.S., 1937, Ph.D., chemistry, 1939

Professional Experience: assistant instructor, chemistry, New York University, 1933–1942; aerology engineer, U.S. Department of the Navy, 1946–1948, head, technical requirements section, Naval Weather Service, 1948–1962; consultant and writer

Concurrent Positions: U.S. Naval Reserve, 1942–1946

Florence Van Straten was a meteorologist whose research focused on metal-gas catalysis, the upper atmosphere, and atmospheric physics. She taught at New York University as an assistant instructor in chemistry while completing her doctorate in chemistry from that university in 1939. She continued teaching there until she joined the U.S. Navy's Women Accepted for Voluntary Emergency Services (WAVES) and launched her career in meteorology, the study of the physical processes that combine to produce weather. While she was in the WAVES during

World War II, she received a Certified Meteorologists diploma, which provided her entry into what had been solely a male profession. The science of meteorology really did not develop until World War II, when there was a need for accurate information to deploy troops and supplies all over the world. She was in the first group of 25 WAVES selected for training to overcome the shortage of available male meteorologists; 22 of these women completed the course. Her responsibility as a meteorologist, or aerology engineer, was to advise commanders of the Pacific Fleet on weather conditions for planning strategy. She also developed safety techniques using sonar and radar, and contributed other innovations to the field.

Van Straten continued working for the Naval Weather Service as a civilian until 1962, forecasting weather for the launching of long-range missiles. One study she initiated was to investigate the pattern of radioactive fallout in case of an atomic attack on the United States. She received the Navy's Meritorious Civilian Service Award in 1956 after 10 years of civilian service. After she left her civilian job with the Navy, she turned to consulting and writing. She was a member of the American Association for the Advancement of Science, the American Meteorological Society, and the American Geophysical Union.

Further Resources

Williams, Kathleen Broome. 2001. *Improbable Warriors: Women Scientists and the US Navy in World War II*. Annapolis, MD: Naval Institute Press.

Vaughan, Martha

b. 1926
Biochemist

Education: Ph.B., University of Chicago, 1944; M.D., Yale University, 1949

Professional Experience: intern, New Haven Hospital, Yale, 1950–1951; research fellow, University of Pennsylvania, 1951–1952; National Research Council fellow, cellular physiology, National Heart Institute, 1952–1954, member of research staff, 1954–1968, head, Metabolism Section, National Heart and Lung Institute, 1968–1974, acting chief, molecular disease, 1974–1976, chief, cell metabolism laboratory, 1974–1994, deputy chief, pulmonary critical care branch, 1994–; principal investigator

Concurrent Positions: senior assistant surgeon to medical director, U.S. Public Health Service, 1954–1989

Martha Vaughan is renowned for her research at the National Institutes of Health (NIH) on the mechanism of hormone action. She has worked for the same institute during her long career with the NIH, but the name has changed from National Heart Institute, to National Heart and Lung Institute, to National Heart, Lung, and Blood Institute. She and her husband, Jack Orloff, were among the several scientist couples hired by the NIH in the post–World War II era. Serving on the research staff, she was appointed head of the Metabolism Section in 1968, acting chief of the molecular disease branch in 1974, chief of the cell metabolism laboratory in 1974, and deputy chief of the pulmonary and critical care medical branch in 1994. She also was senior assistant surgeon to the medical director of the U.S. Public Health Service from 1954 to 1989. She is a co-editor of the book *ADP-Ribosylating Toxins and G Proteins: Insights into Signal Transduction* (1990), published by the American Society for Microbiology, and has written many scientific papers.

Vaughan was elected to membership in the National Academy of Sciences in 1985. She has received numerous awards, including the Harvey Society Lecturer award (1982); the G. Burroughs Mider Lecturer award, National Institutes of Health (1979); and the Meritorious Service Medal (1974), Distinguished Service Medal (1979), Command Officer Award (1982), and Superior Service Award (1993), all of the U.S. Public Health Service. She is a member of the American Society of Biochemistry and Molecular Biology, American Society of Clinical Investigation, Association of American Physicians, American Academy of Arts and Sciences, and American Society of Biological Chemists.

Vennesland, Birgit

1913–2001
Enzymologist, Plant Biologist

Education: B.S., biochemistry, University of Chicago, 1934, Ph.D., biochemistry, 1938

Professional Experience: assistant biochemist, University of Chicago, 1938–1939; research fellow, Harvard University medical school, 1939–1941; instructor, University of Chicago, 1941–1944, assistant to associate professor, 1944–1957, professor, 1957–1968; director, Max Planck Institute of Cell Biology, 1968–1970, director, Vennesland Research Institute, Max Planck Society, 1970–1981

Concurrent Positions: civilian consultant, Office of Scientific Research and Development, 1944; adjunct professor, biochemistry and biophysics, University of Hawaii, 1987–

Birgit Vennesland was a biochemist and plant biologist whose research focused on carboxylation reactions in animals and plants, mechanisms of hydrogen transfer in pyridine nucleotide dehydrogenases, and the enzymology and mechanism of plant photosynthesis. She was one of the first chemists to use radioactive carbon 11 to study carbohydrate metabolism. She served on several study teams for the National Science Foundation and the Public Health Service. After receiving her doctorate in biochemistry from the University of Chicago in 1938, she received a fellowship to study in Paris, but World War II interfered with those plans, and she went to Harvard University medical school instead. After working as a research fellow at Harvard for two years, she returned to Chicago as an instructor in 1941 and rose through the ranks to full professor by 1957. She left Chicago in 1968 for a position at another prestigious institute in Germany, being appointed a director at the Max Planck Institute of Cell Biology in 1968 and then, in 1970, director of another institute of the Max Planck Society that became known as the Vennesland Research Institute. She retired in 1981 and moved to Hawaii with her twin sister, a retired medical doctor. She remained affiliated as an adjunct professor at the University of Hawaii.

Vennesland received the Hales Award of the American Society of Plant Physiologists (1950) and the Garvan Medal of the American Chemical Society (1964), as well as an honorary degree from Mount Holyoke College (1960). She was elected a fellow of the New York Academy of Sciences and the American Association for the Advancement of Science, and was a member of the American Chemical Society, American Society of Biological Chemists, and American Society of Plant Physiologists.

Further Resources

Conn, Eric E. and Larry P. Solomonson. "Birgit Vennesland." Women Pioneers in Plant Biology. American Society of Plant Biologists.http://www.aspb.org/committees/women/pioneers.cfm#Birgit%20Vennesland.

Villa-Komaroff, Lydia

b. 1947
Molecular Biologist, Neurobiologist

Education: B.A., biology, Goucher College, 1970; Ph.D., cell biology, Massachusetts Institute of Technology, 1975

Professional Experience: research fellow, biology, Harvard University, 1975–1978; assistant professor, molecular genetics and microbiology, University of Massachusetts Medical Center, 1978–1982, associate professor, 1982–1985; senior research associate, Division of Neuroscience, Children's Hospital, Boston, 1985–1996, acting head, 1988–1994, associate director, 1995; associate professor, neuropathology (genetics), Harvard Medical School, 1985–1988, associate professor, neurology (neuroscience and genetics), 1988–1996; associate vice president, Research, Northwestern University, 1996–1997, vice president, 1998–2002, and professor, neurology, Northwestern University, 1996–2002; vice president of research and chief operating officer, Whitehead Institute for Biomedical Research, Massachusetts Institute of Technology (MIT), 2003–2005; chief scientific officer, Cytonome, Inc., 2005–, chief executive officer, 2006–

Concurrent Positions: visiting postdoctoral fellow, Cold Spring Harbor Laboratories, 1976–1977; director, Laboratory of Molecular Genetics, Mental Retardation Research Center, Children's Hospital, Boston, 1985–1996, associate director, 1987–1994, director, Transgenic Mouse Facility, 1990–1994

Lydia Villa-Komaroff is renowned for her theory of brain development. Her research includes growth factors in brain development, structure and function of insulin-like growth factors in brain development, and the structure and function of genes expressed in central and peripheral nervous systems. Her particular focus is the flow of information in the cell from DNA to RNA to protein. She has held a variety of research and teaching positions at Harvard University, the University of Massachusetts, Northwestern University, and MIT. In 2005, she became Chief Scientific Officer of Cytonome, Inc., a Boston-based biotechnology company that sells a device for cell sorting.

Both of Villa-Komaroff's parents were the first in their respective families to attend college, and she became one of the first generation of Mexican Americans to receive a doctorate in this country. She became interested in a scientific career after taking part in a National Science Foundation summer program during high school. She began her undergraduate studies at the University of Washington in Seattle, but transferred to Goucher College in Baltimore, Maryland, to follow her husband's job. She received her degree in biology and gained experience working for the National Institutes of Health (NIH) during the summers. She went on to graduate school at MIT, receiving her Ph.D. in cell biology in 1975. During her post-doctoral work at Harvard, she first worked on making proteins in bacteria. In 1976, a national controversy arose over recombinant DNA technology. Some people feared that taking the genes from one organism, such as a human, and putting them into bacteria, might somehow create a supergerm or a new disease. The Cambridge city council temporarily banned certain experiments, and the Harvard research team had to move to a laboratory in Cold Spring Harbor on Long Island for a year. Villa-Komaroff's research team was the first to produce insulin from bacteria, a patented process that is now responsible for most insulin used by diabetics.

Villa-Komaroff has received several honorary degrees, and has been recognized with numerous awards and honors, including the Hispanic Engineer National Achievement Award (1992), the first Catalyst Award from the Science Club for Girls (2008), and MOSI's (Museum of Science and Industry) National Hispanic Scientist of the Year (2008). She was a member of the mammalian genetics study section of the NIH (1982–1984) and member of the Neurological Disease Program Project Review Committee (1989–1994). She is a member of the American Society for Microbiology, American Society of Hematology, American Society of Cell Biology, American Association for the Advancement of Science, International Society for Cellular Therapy, American Society for Blood and Marrow Transplantation, Society for Neuroscience, and Association for Women in Science, and is a founding member of the Society for the Advancement of Chicanos and Native Americans in Science (SACNAS). She has served on the National Academy of Sciences Committee on Women in Science and the National Research Council Committee on Underrepresented Groups.

Further Resources

Whitehead Institute for Biomedical Research. "Lydia Villa-Komaroff among 100 Most Influential Hispanics in America." (16 October 2003). http://www.wi.mit.edu/news/archives/2003/wi_1016.html.

Vitetta, Ellen Shapiro

b. ca. 1942
Microbiologist, Immunologist

Education: B.A., Connecticut College, 1964; M.S., New York University, 1966, Ph.D., immunology, and M.D., 1968

Professional Experience: research assistant, biology, New York University, 1964–1968; postdoctoral fellow, New York University School of Medicine, 1968–1970, assistant research scientist, 1970–1971, assistant professor, microbiology, and associate research scientist, Department of Medicine, 1971–1974; associate professor, microbiology, University of Texas, Southwestern Medical Center, 1974–1976, professor, 1976–, director, Cancer Immunobiology Center, 1988–

Concurrent Positions: member, Medical Research Council, Cambridge, England, 1986

Ellen S. Vitetta is a renowned microbiologist whose most recent research has led to the development of a vaccine against ricin, a highly toxic compound made from castor beans that could be used as a biological weapon. After receiving her master's, doctorate, and M.D. degrees from New York University, she conducted research in the Medical School and Department of Medicine at that university for more than 10 years. She then moved to the University of Texas Southwestern Medical Center, where, in addition to being director of the Cancer Immunobiology Center, she holds the S. S. Patigan Distinguished Chair in Cancer Immunobiology. Her work in immunotoxicology also has implications for the treatment of cancer and of AIDS. In the late 1990s, she and her research team first discovered that a specific form of antibodies, chemically altered with ricin, could kill cancer cells. They applied their findings to target HIV cells as well and then to a vaccine against ricin known as RiVax. One of Vitetta's former students, **Linda Buck**, went on to win the Nobel Prize in Physiology or Medicine.

Vitetta has received a merit grant award from the National Institutes of Health (NIH) every year since 1987. She is a member of many distinguished committees and commissions, including the science board of the Ludwig Institute, the Task Force on Immunology of the National Institute of Allergy and Infectious Diseases, the science advisory board of the Howard Hughes Medical Institute, and the National Cancer Institute's Cancer Treatment Board, and is a consultant for pharmaceutical and biotech companies such as Eli Lilly, Abbott, and Genetics Institute. She has also served on the editorial boards of numerous journals in the field.

Vitetta was elected to membership in the National Academy of Sciences in 1994 and the Institute of Medicine in 2006. She is a recipient of the Taittinger Breast Cancer Research Award from the Komen Foundation (1983), NIH Merit Award (1987), Pierce Immunotoxin Award (1988), Women's Excellence in Science Award from the Federation of American Societies in Experimental Biology (FASEB) (1991), Abbott Award of the American Society of Microbiologists (1992), Rosenthal Award (1995) and Charlotte Friend Award (2002), both of the American Association of Cancer Research, and Mentoring Award (2002) and Lifetime Achievement Award (2007) of the American Association of Immunologists. In 1994, she served as president of the American Association of Immunologists, and in 2006, she was elected to the Texas Women's Hall of Fame.

Further Resources

University of Texas. Faculty website. http://www8.utsouthwestern.edu/findfac/professional/0,2356,17609,00.html.

Waelsch, Salome Gluecksohn

1907–2007
Geneticist

Education: Ph.D., zoology, University of Freiburg, 1932

Professional Experience: assistant, department of experimental cell research, University of Berlin, 1932–1933; research associate and lecturer, zoology, Columbia University, 1936–1953, lecturer, College of Physicians and Surgeons, 1953–1955; associate professor to professor, anatomy, Albert Einstein College of Medicine, 1955–1963, professor, molecular genetics, 1958–1978

Salome Waelsch was a mammalian geneticist whose research on the role of genes in normal and abnormal cell differentiation and on genetically controlled congenital abnormalities helped establish the field of developmental genetics. Her research focused on genetic mutations of mice spines and tails, and she later researched the hereditary nature of blood cells and chromosomal defects that affect liver function. After receiving her doctorate from the University of Freiburg in 1932, she worked briefly at the University of Berlin before fleeing to the United States in 1933. She and her husband (a Freiburg-trained biochemist who also obtained a position at Columbia in New York) were among the many Jewish scientists and academics forced to leave Nazi Germany. Salome did not have a job for three years but finally accepted an initially nonpaying research-associate position at Columbia University in 1936, where she also lectured in zoology. Waelsch joined the faculty of Albert Einstein College of Medicine in 1955, where she taught anatomy and molecular genetics for more than 20 years. She also served as chair of the genetics department between 1963 and 1976. Although she formally retired in 1978, she remained active in her research well into her eighties and nineties.

Waelsch was elected to membership in the National Academy of Sciences in 1979 and was awarded the National Medal of Science in 1993. She was awarded the Thomas Hunt Morgan Medal of the Genetics Society of America and the first Lifetime Achievement Award of the American Cancer Society. In 1982, Freiburg University awarded her an honorary degree, which she was hesitant to accept due to the circumstances that led her to have to leave Germany 50 years earlier. She was a fellow of the Royal Society and a member of the American Academy of Arts and Sciences, New York Academy of Science, American Association of Anatomists, Genetics Society of America, Society for Developmental Biology, and American Society of Zoologists.

Further Resources

Solter, Davor. 2008. "In Memoriam: Salome Gluecksohn-Waelsch (1907–2007)." *Developmental Cell*. 14(1): 22–24. (January 2008). http://www.sciencedirect.com/science ?_ob=ArticleURL&_udi=B6WW3-4RKDVWC-7&_user=10&_rdoc=1&_fmt=&_orig =search&_sort=d&_docanchor=&view=c&_acct=C000050221&_version=1&_urlVersion =0&_userid=10&md5=a49e98dbcf1507db2038cc06a0ca4ac6.

Walbot, Virginia Elizabeth

b. 1945
Biologist, Plant Geneticist

Education: B.A., biology, Stanford University, 1967; M.Phil., biology, Yale University, 1969, Ph.D., biology, 1972

Professional Experience: National Institutes of Health postdoctoral fellow, biochemistry, University of Georgia, 1972–1975; assistant to associate professor, biology, Washington University, St. Louis, 1975–1980; associate professor to professor, biological sciences, Stanford University, 1981–

Concurrent Positions: Guggenheim fellow and visiting scientist, C.S.I.R.O., Australia, 1987; adjunct associate professor, agronomy, University of Missouri, 1979–1990

Virginia Walbot is a plant geneticist whose research focus is corn genetics. Her research combines interests in plant molecular biology and development, genetics, and botany. She and other scientists have found corn to be the ideal organism for studying fundamental questions about genetics and development. The plant geneticist **Barbara McClintock** received the Nobel Prize in Physiology or Medicine in 1983 for her fundamental research on corn, and Walbot was able to confer with her while McClintock was still active in research in the late 1970s, and worked with her in her laboratory at Cold Spring Harbor. Transposable genetic elements, or mobile DNA, discovered by McClintock more than 50 years ago, figure prominently in Walbot's research.

Walbot is particularly interested in developmental timing, as plants have continuous development—that is, they are continuously making organs from scratch. For example, if one places a plant with dark leaves in sunlight, the dark leaves will fall off to be replaced by light-colored leaves to filter the sunlight. Scientists are using recombinant DNA methods to manipulate plant genomes to breed for resistance to disease, while ensuring that there is a diversity of varieties that have any new trait. If only a few genetic variants are developed, it means that the food sources are more susceptible to a new disease or environmental conditions that are fatal to that one type. While she was on the faculty of Washington University, she developed, in cooperation with a team of University of Missouri researchers and commercial corn breeders, a corn that is genetically incapable of losing its sweetness and turning starchy. In addition to Walbot's numerous scientific publications, she is co-author of a textbook, *Developmental Biology* (1986), and co-editor of *The Maize Handbook* (1993), a compendium of the standard procedures and protocols for maize research.

Biologist and plant geneticist, Virginia Walbot. (Courtesy of the Stanford University News Service Library)

Walbot has consulted on numerous scientific, government, and industry advisory boards. She is the recipient of the Eppley Foundation Award (1993) and a National Geographic Society Explorer Award (1998). She is a member of the American Association for the Advancement of Science, Botanical Society of America, Society for Developmental Biology, and American Society of Plant Physiologists, and a Corresponding Member of the Mexican Academy of Sciences.

Further Resources

Stanford University. Faculty website. http://med.stanford.edu/profiles/Virginia_Walbot/.

Wallace, Phyllis Ann

1920–1993
Economist

Education: B.A., New York University, 1943; M.A., Yale University, 1944, Ph.D., economics, 1948

Professional Experience: economist and statistician, National Bureau of Economic Research, 1948–1952; associate professor, economics, Atlanta University, 1953–1957; senior economist, U.S. government, 1957–1965; chief of technical studies, U.S. Equal Employment Opportunity Commission, 1966–1969; vice president for research, Metropolitan Applied Research Center, New York City, 1969–1972; visiting professor, Sloan School of Management, Massachusetts Institute of Technology (MIT), 1972–1975, professor, 1975–1986

Concurrent Positions: lecturer, City College of New York, 1948–1951

Phyllis Wallace was a pioneer in research on the economics of racial and sexual discrimination in the workplace. She was the first black woman to receive a doctorate in economics from Yale University and the first black woman on the faculty to be tenured at MIT. She was also the first African American and first woman president of the Industrial Relations Research Association. Early in her career, she concentrated her research on issues dealing with international trade. She had written her dissertation on commodity trade relationships, concentrating on international sugar agreements. She first had a non-tenure-track position at New York University and, at the same time, did research for the National Bureau of Economic Research. She then moved to Atlanta University, where she was an associate professor, before working as a senior economist for an unnamed government agency that was later revealed to be the Central Intelligence Agency (CIA).

Wallace became chief of technical operations for the Equal Employment Opportunity Commission (EEOC) a few months after it started operations in 1965. She worked to coordinate hearings for the EEOC about racial employment patterns in many industries, and her research focused on the status of African Americans in urban poverty neighborhoods. At the Metropolitan Applied Research Center, she worked on issues affecting urban youth in labor markets and on issues affecting young black women, which had not been explored at that point. After joining the faculty of MIT, Wallace consulted on a federal lawsuit against communications company AT&T for discrimination against women and minority men. She wrote about this case in her 1976 book, *Equal Employment Opportunity and the AT&T Case*. Her other books included *Pathways to Work: Employment among Black Teenage Females* (1974) and *Women, Minorities and Employment Discrimination* (1977); in 1980, she published a study on *Black Women in the Labor Force*, in which she concluded that young black women have the highest unemployment rate and the lowest economic status of any group.

Even after her retirement in 1986, Wallace continued to work on issues related to discrimination and consulted with the Sloan School at MIT on sexual harassment issues and policies. The Sloan School established two funds for black students and scholars in her name. Wallace received numerous honorary degrees, and was a member of the American Economic Association and the Industrial Relations Research Association.

Further Resources

Massachusetts Institute of Technology. "Professor Phyllis A. Wallace Dies." http://web .mit.edu/newsoffice/1993/wallace-0113.html.

Warga, Mary Elizabeth

1904–1991
Physicist

Education: B.S., University of Pittsburgh, 1926, M.S., 1928, Ph.D., spectroscopy, 1937

Professional Experience: industrial assistant, Mellon Institute for Industrial Research, 1928–1930, industrial fellow, 1930–1933; industrial fellow, University of Pittsburgh, 1934–1936, instructor to professor of physics and director of spectroscopy laboratory, 1936–1962, adjunct professor, physics, 1962–1972, emeritus

Concurrent Positions: executive secretary, Optical Society of America (OSA), 1959–1972

Mary Warga was a physicist whose research involved ultraviolet, visible, and infrared optical emission spectroscopy; optical absorption; and upper atmosphere spectroscopy. She worked at the Allegheny Observatory between 1926 and 1928. After receiving her master's degree from the University of Pittsburgh in 1928, she received several fellowships before being appointed instructor of physics in 1936. She rose through the ranks to professor of physics and director of the spectroscopy laboratory after receiving her doctorate in spectroscopy in 1937. After a distinguished career in teaching and research, Warga served as the first executive secretary of the OSA, having previously served four years on the Board of Directors. During her first few years in this position, which was headquartered in Washington, D.C., she still directed the spectroscopy laboratory at the University of Pittsburgh, but in 1962, she reduced her teaching load to become adjunct professor of physics. The laser had been invented in 1960, and the field of optics was an exciting new area of research. In her role as executive secretary of the OSA during this time, Warga brought together many top scientists working in this field by encouraging society membership, organizing conferences, and writing a monthly news column. She retired from teaching and from the OSA in 1972, although she remained involved in professional activities for several more years.

Warga was named a Distinguished Daughter of Pennsylvania (1954) and Woman of the Year in Science Research (1959), and received a District Service Award in Applied Spectroscopy (1962). Upon her retirement, she was honored with a Distinguished Service Award from the OSA (1973). She was a member of the governing board of the American Institute of Physics beginning in 1960 and served as secretary of the Joint Council on Quantum Electronics. She was elected a fellow of the OSA, American Physics Society, Physics Society of London, and American Association for the Advancement of Science. She was a member of the U.S. and International Commission for Optics, American Association of Physics Teachers, American Chemical Society, and Society for Applied Spectroscopy.

Further Resources

Howard, John N. 2002. "An Executive Secretary for OSA." *Optics & Photonics News.* 13(6): 14–15. http://www.opticsinfobase.org/abstract.cfm?URI=OPN-13-6-14.

Washburn, Margaret Floy

1871–1939
Psychologist

Education: A.B., Vassar College, 1891, A.M., 1893; Ph.D., Cornell University, 1894

Professional Experience: professor, psychology and ethics, Wells College, 1894–1900; dean, Sage College, Cornell University, 1900–1902, lecturer, psychology, 1901–1902; assistant professor and head, psychology, University of Cincinnati, 1902–1903; associate professor, philosophy and psychology, Vassar College, 1903–1908, professor, psychology, 1908–1937

Margaret Washburn was recognized in the new field of experimental psychology, in particular for her research on a motor theory of consciousness, or the idea that all thoughts and perceptions produce some type of physical reaction. She merged her interests in science and philosophy in her work on social consciousness, emotions, animal psychology, and comparative psychology. She authored or co-authored (with her Vassar students) hundreds of scientific papers; her most important books were *The Animal Mind* (1908) and *Movement and Mental Imagery* (1916), which presented her theory of consciousness and linked different schools of psychological thought at the time. Her name was included in a study of 50 eminent American psychologists in 1903.

Washburn earned a bachelor's and master's degree at Vassar, then applied to the doctoral program in psychology at Columbia University. Columbia would not admit a woman in the graduate program, however, so she went to Cornell University instead and received her Ph.D. in 1894. She spent six years as a professor of psychology and ethics at Wells College before returning to Cornell as a lecturer in psychology. She spent one year as an assistant professor and head of the psychology department at the University of Cincinnati before returning to her alma mater at Vassar College as associate professor of philosophy and psychology in 1903; she was promoted to full professor in 1908 and remained at Vassar until her retirement in 1937. Between 1925 and 1935, she served as co-editor of the *American Journal of Psychology*.

In 1931, Washburn was only the second woman (after **Florence Sabin**, 1925) to be elected to the National Academy of Sciences. She served as vice president of the American Association for the Advancement of Science and, in 1921, became president of the American Psychological Association. In 1929, she became a member of the Society of Experimental Psychologists, which had previously barred women from membership; two years later, the Society met at Vassar, a women's college, at Washburn's invitation.

Watson, Patty Jo (Andersen)

b. 1932
Anthropologist, Archaeologist

Education: B.A., University of Chicago, M.A., 1956, Ph.D., anthropology, 1959

Professional Experience: field assistant, Oriental Institute Iraq-Jarmo project, University of Chicago, 1954–1955; National Science Foundation (NSF) fellow, University of Michigan, 1957–1958; NSF fellow, University of Minnesota, 1958–1959; archaeologist and ethnographer, Oriental Institute, University of Chicago, 1959–1960, research associate, archaeology, 1964, 1967; assistant to associate professor, anthropology, Washington University, St. Louis, 1969–1973, professor, 1973–1993, distinguished university professor, 1993–

Concurrent Positions: instructor, anthropology, University of Southern California and Los Angeles State College, 1961; summer lecturer, anthropology, University of Michigan, Ann Arbor, 1962–1963; project associate, anthropology curriculum study project, American Anthropological Association, 1965–1967

Patty Jo Watson is a distinguished anthropologist and archaeologist who pioneered the field of ethnoarcheaology. Her research interests have ranged from the prehistory of Iran, to the archaeology of the Mammoth Cave area in Kentucky, to method and theory in shipwreck archaeology. Early in her career, she was a field assistant for the Oriental Institute of the University of Chicago, through which she was involved as an archaeologist and ethnographer on an Iranian prehistory project and directed the excavation of an ancient site in Turkey on behalf of the Istanbul-Chicago Joint Prehistoric Project. Her dissertation project was an investigation of early village farming in the Levant.

Married to an avid caver, she became interested in cave archaeology in North America and has conducted research in Kentucky, New Mexico, and Tennessee; this work has been supported by grants from the National Endowment for the Humanities and National Geographic Society. She is author of *Archaeology of the Mammoth Cave Area* (1974) and co-editor of the book, *Of Caves and Shell Mounds* (1996). She also co-authored *Archaeological Explanation: The Scientific Method in Archaeology* (1984) and co-edited *The Origins of Agriculture: An International Perspective* (1992), a collection of papers from a symposium of the American Association for the Advancement of Science at which all of the speakers were experts in crop evolution or the archaeological record for early plant cultivation. She has served on the editorial board of the journals *Anthropology Today, American Anthropologist*, and *American Antiquity.*

Watson was elected to membership in the National Academy of Sciences in 1988. She received the Fryxell Award for interdisciplinary research given by the Society for American Archaeology at a symposium on interdisciplinary research held in her honor in 1990, and in 1996 received a Distinguished Service Award of the American Association of Anthropology. She is a fellow of the American Anthropological Association and a member of the American Association for the Advancement of Science, Society for American Archaeology, Middle East Studies Association of North America, Cave Research Foundation, and the St. Louis Society, a branch of the Archaeological Institute of America. In 1995, she was featured in a three-part public television miniseries on women scientists called "Discovering Women."

Wattleton, (Alyce) Faye

b. 1943
Nurse-Midwife

Education: B.S., nursing, Ohio State University, 1964; M.S., maternal and infant healthcare, and certificate, nurse-midwifery, Columbia University, 1967

Professional Experience: instructor, Miami Valley School of Nursing, 1964–1966; assistant director, nursing, Dayton Public Health Nursing Association, 1967–1970; executive director, Planned Parenthood Association of Miami Valley, Dayton, Ohio, 1970–1978; president, Planned Parenthood Federation of America (PPFA), 1978–1992; host, syndicated television show, Chicago, 1992; president and founder, Center for Gender Equality, 1995; president, Center for the Advancement of Women, 1995–2010

Faye Wattleton was the first African American woman to serve as president of the PPFA. She led the nation's oldest and largest voluntary family-planning organization in a crusade to guarantee every person's right to decide if and when to have a child. With a background in nursing,

Faye Wattleton, a former president of Planned Parenthood, has been a world leader in the struggle to safeguard women's reproductive rights. (Getty Images)

she became president of Planned Parenthood in 1978, believing that family planning is the best solution to a host of problems that are intensified by the high rate of unintended pregnancies. These problems include child abuse, teenage pregnancy, and sexually transmitted diseases, as well as poverty, hunger, and death and injury from unsafe abortions. She was inspired to work with Planned Parenthood on the local level in Dayton, Ohio, after seeing the number of girls and women who suffered or died from illegal or self-induced abortions. In the early twentieth century, Margaret Sanger founded the American birth-control movement and created an organization that was the forerunner of the PPFA. Continuing Sanger's vision of medical services as well as education and information, Planned Parenthood offers pregnancy diagnosis, prenatal care, infertility counseling, AIDS testing, and contraceptive services, as well as information on sexual health, not only in the United States but through international efforts as well.

In the 1960s and early 1970s, some black activists criticized Planned Parenthood as a white-managed agency whose mission was to reduce the black birthrate through population control. The selection of Wattleton as president expanded the vision and operation of Planned Parenthood. New controversies arose in the 1980s and 1990s with an active anti-abortion movement's attacks on patients and clinics, and even death threats sent to Wattleton personally. The courts continued to address the abortion issue, and setbacks to the pro-choice movement came in the form of decreased funding and efforts to limit abortion rights through parental notification or waiting periods. Throughout her presidency, Wattleton (like her successors) emphasized Planned Parenthood's message of education and choice. She resigned as president in 1992 but remained active as a public figure through hosting a television show and in her work with various organizations on a range of women's issues.

Wattleton has received numerous awards, including the American Humanist Association's Humanist of the Year (1986), Claude Pepper Humanitarian Award (1990), Boy Scouts of America Award (1990), Spirit of Achievement Award of the Albert Einstein College of Medicine, Yeshiva University (1991), Margaret Sanger Award (1992), Jefferson Public Service Award (1991), and Dean's Distinguished Service Award of the Columbia School of Public Health (1992). In addition to her scientific papers, she has written a book, *How to Talk to Your Child about Sexuality* (1986), and her autobiography, *Life on the Line* (1996).

Further Resources

Faye Wattleton. http://www.fayewattleton.com/.

Way, Katharine

1903–1995
Physicist

Education: B.S., Columbia University, 1932; Ph.D., physics, University of North Carolina, 1938

Professional Experience: research fellow, Bryn Mawr College, 1938–1939; instructor, physics, University of Tennessee, 1939–1941, assistant professor, 1941–1942; physicist, Naval Ordnance Laboratory, 1942; physicist, Oak Ridge National Laboratory, 1942–1948; physicist, National Bureau of Standards, 1949–1953; director, nuclear data project, National Research Council (NRC), 1953–1963; director, Oak Ridge National Laboratory, 1964–1968; editor, *Nuclear Data Tables*, 1965–1973; editor, *Atomic Data*, 1969–1973; editor, *Atomic Data and Nuclear Data Tables*, 1973–1982; director, surgery and bioengineering, National Institutes of Health (NIH) study section, 1981–1985

Concurrent Positions: adjunct professor, physics, Duke University, 1968–1988

Katharine Way was a physicist whose research included nuclear fission, radiation shielding, and nuclear constants. One of her most notable contributions to science, however, was the vast project of compiling and editing the *Atomic Data and Nuclear Data Tables*, a journal of regularly updated research information for experimental and theoretical physicists that is now available online. Way was in on the ground floor of the entire project during World War II, when she was involved in the Manhattan Project at the Naval Ordnance Laboratory, the Oak Ridge National Laboratory, and the National Bureau of Standards. She had been a research fellow at Bryn Mawr College and then on the faculty at the University of Tennessee. She moved to Washington, D.C., during the war and worked at various laboratories on nuclear physics. She worked with Eugene Wigner on what became known as the Way-Wigner formula on the decay of nuclear fission products; Wigner was later awarded the Nobel Prize in Physics.

Way's work in nuclear physics led to her concern about the ethical uses and threat of the atomic bomb. In 1946, she co-edited a book, *One World or None: A Report to the Public on the Full Meaning of the Atomic Bomb*, which included essays by top scientists of the era (including Albert Einstein) and became a *New York Times* bestseller; the book has been reprinted numerous times, most recently in 2007. After the war, she worked for the National Bureau of Standards and then as director of the nuclear data project for the NRC, where she served as editor of the new publications for collecting and organizing research data. She and other physicists began compiling the "Nuclear Data Sheets" in 1964 and, after the project moved to Oak Ridge National Laboratory, they published the first issue of the new journal. The project culminated in the combined *Atomic Data and Nuclear Data Tables*, which Way worked on from 1965 to 1982, leaving to accept a directorship at the NIH. During this time, she also spent 20 years as adjunct professor of physics at Duke University. Way was an elected fellow of the American Physical Society and the American Association for the Advancement of Science.

Weertman, Julia (Randall)

b. 1926
Solid-State Physicist, Metallurgist

Education: B.S., Carnegie Institute of Technology, 1946, M.S., 1947, D.Sc., physics, 1951

Professional Experience: Rotary International fellow, École Normale Supérieure, University of Paris, 1951–1952; physicist, U.S. Naval Research Laboratory, 1952–1958; visiting assistant professor, Northwestern University, Illinois, 1972–1973, assistant to associate professor, materials science, 1973–1981, professor, 1982–1987, director, Material Science and Engineering Department, 1987–1992

Julia Weertman is renowned for her research on high-temperature metal failure and the nanocrystalline structures of metals. Her research includes dislocation theory, high-temperature fatigue, small-angle neutron scattering, and nanocrystalline material. She has also contributed to the understanding of the basic characteristics of different materials in her research on small-angle neutron scattering. She received all of her degrees at the Carnegie Institute of Technology (now part of Carnegie-Mellon University). After completing postdoctoral studies at the École Normale Supérieure, she was appointed to a position as a physicist at the U.S. Naval Research Laboratory. Her work there centered on ferromagnetic spin resonance and the study of the basic concepts of magnetism. She interrupted her formal research to accompany her husband to London, where he worked for the Naval Research Laboratory. She and her husband, Johannes Weertman, collaborated on a textbook during this period, *Elementary Dislocation Theory* (1964). When they returned to the United States, her husband accepted a position at Northwestern University, and she took several years off from her career to raise children.

Weertman returned to research formally when she joined Northwestern in 1972 as a visiting assistant professor, then rose through the ranks to full professor, director of the materials science and engineering program, and a distinguished professorship by the time she retired in 1992. She has been an advisor to several government agencies, including the National Science Foundation, Department of Energy, National Bureau of Standard and Technology, and Argonne and Oak Ridge National Laboratories.

Weertman was elected to membership in the National Academy of Engineering in 1988. In addition to her scientific papers, she has been co-author of six books. She was a member of the Evanston (Illinois) Environmental Control Board (1972–1979) and a member of the National Research Council's National Materials Advisory Board (1999–2005). She has received a number of awards, including the Creativity Award of the National Science Foundation (1981 and 1986), a Guggenheim fellowship (1986), the Distinguished Engineering Educator Award of the Society of Women Engineers (SWE) (1989), an Achievement Award of SWE (1991), the Leadership Award of the Minerals, Metals, and Materials Society (TMS) (1996), the Von Hipple Award of Materials Research Society (2003), and the Gold Medal of American Society for Metals (ASM) International (2005). She is a fellow of the American Academy of Arts and Science and a member of the American Institute of Physics, American Crystallographic Association, American Society for Testing and Materials, Materials Research Society, and American Physical Society.

Further Resources

Northwestern University. Faculty website. http://www.matsci.northwestern.edu/faculty/jrw.html.

Weisburger, Elizabeth Amy (Kreiser)

b. 1924
Biochemist, Toxicologist

Education: B.S., chemistry, Lebanon Valley College, 1944; Ph.D., organic chemistry, University of Cincinnati, 1947

Professional Experience: research associate, University of Cincinnati, 1947–1949; postdoctoral research fellow, National Cancer Institute, National Institutes of Health, 1949–1951, researcher, Biochemistry Laboratory, 1951–1961, Carcinogen Screening Section, Experimental Pathology, 1961–1973, chief, Carcinogen Metabolism and Toxicology, Division of Cancer Cause and Prevention, 1973–1981, assistant director, Chemical Carcinogenesis, Division of Cancer Etiology, 1981–1988; consultant, 1989–

Elizabeth Weisburger had a distinguished career as a toxicologist with the National Cancer Institute (NCI), where she conducted pioneering research on the carcinogenic effects of chemicals, pharmaceuticals, food additives, and environmental pollutants. Her research has aided in providing insight at the molecular level of carcinogenesis, which is vital in developing methods for the treatment and prevention of cancer. Among the compounds that she has studied are fluorenes, nitrosamines, aromatic amines, halogenated hydrocarbons, fumigants, and food preservatives. She has also investigated the relationship between mutagens and cancers, and emphasized developing improved test systems for evaluating carcinogenic risk. She was among the first scientists to test some of the drugs used in clinical cancer chemotherapy and to point out their potential dangers.

Weisburger was originally interested in biology as an undergraduate, but changed her major to chemistry, and studied mathematics and physics as well. During World War II, graduate assistantships in chemistry were readily available to women because so many men were in the military service. Weisburger received an assistantship at the University of Cincinnati, Ohio, where she began work in cancer research and continued working as a research associate after graduation. She and her husband, medical researcher John H. Weisburger, then joined the NCI, where they collaborated until their divorce in 1974; he later became research director of the American Health Foundation. Elizabeth Weisburger published more than 200 papers on cancer-causing chemicals, nutrition, and other topics, and for many years was editor of the *Journal of the National Cancer Institute*. She retired from the NCI in 1988 and became a consultant, including as an advisor on project funding for the American Institute of Cancer Research, and as a senior associate with Mandava Associates, a consulting firm that advises biotechnology, pharmaceutical, and related companies on compliance with government safety, environmental, and other regulations.

Weisburger has been a member of the Chemical Substances Committee of the American Conference of Government Industrial Hygienists since 1978. She has received honorary degrees from the University of Cincinnati (1981) and from Lebanon Valley College (1989). Her numerous awards include the Meritorious Service Medal (1973) and Distinguished Service Medal (1985) of the U.S. Public Health Service, the Garvan Medal of the American Chemical Society (1981), the Hillebrand Prize of the Chemical Society of Washington (1981), and the Herbert E. Stokinger Award of the American Conference of Governmental Industrial Hygienists (1996). She is a member of the American Association for the Advancement of Science, American Association for Cancer Research, American Chemical Society, Society of Toxicology, American Society of Biochemistry and Molecular Biology, and Royal Society of Chemistry.

Weisstein, Naomi

b. 1939
Psychologist

Education: B.A., Wellesley College, 1961; Ph.D., psychology, Harvard University, 1964

Professional Experience: lecturer, University of Chicago, 1965; assistant to associate professor, Loyola University, Chicago, 1966–1973; professor, psychology, State University of New York at Buffalo, 1973–emerita

Concurrent Positions: consultant, Xerox Corporation, 1973–1974

Naomi Weisstein is an experimental psychologist known for her research in vision, perception, and cognition. She is also known for her activity in civil rights and feminist causes starting in the 1960s, first as a graduate student at Harvard and then as a postdoctoral lecturer at the University of Chicago. She was one of the first women to speak out about employment practices in academia and, specifically, on the difficulties of women entering the scientific professions. Weisstein attended the Bronx High School of Science and went on to Wellesley College as an undergraduate, where she took for granted the dedicated female professors who had received degrees from first-class universities, not realizing many of them were teaching at a women's college because they were unable to secure positions at other universities. Such faculty members often had heavy teaching loads and little time, equipment, or funds to conduct scientific research. Weisstein went on to Harvard, where the department chair told first-year graduate students that women did not belong in graduate school and restricted women's use of lab equipment. She attended Yale University briefly to use their equipment and transferred her credits and work back to Harvard in order to receive her doctorate in psychology in 1964.

Weisstein was promised a position at the University of Chicago, but about 10 days before classes started, the department invoked an unwritten anti-nepotism rule to deny her a position because her husband, Jesse Lemisch, was a faculty member. She was hired instead as a lecturer to teach in areas outside her research and, after one year, was notified that her contract would not be renewed. She then obtained a tenure-track faculty position at Loyola University in Chicago, where she taught for seven years before both she and her husband relocated for faculty positions at the State University of New York at Buffalo.

One of Weisstein's research interests was the discipline of psychology itself, and the way it describes male and female personalities differently in ways that disadvantage women. She has also questioned the effectiveness of psychotherapy and the clinical definitions of schizophrenia, homosexuality, and even heterosexuality. In a landmark 1968 paper, "Psychology Constructs the Female," she argued that psychology provides little insight into woman's true "nature," instead defining women according to sexist ideas of desired social roles as wives and mothers, and finding them psychologically unstable when they do not fulfill those roles. She went on to write dozens of scientific papers, but by the early 1980s, Weisstein was struck with physical health problems and forced into early retirement. Although bedridden, she continued for some years to write, collaborate, consult, and sit on the editorial board of scientific journals.

In 1970, Weisstein, along with Phyllis Chesler and others, helped found American Women in Psychology. She also founded the Women's Caucus of the Psychonomic Society (1972) and Women in Eye Research (1980), a caucus of the Association for Research in Vision and Ophthalmology. She has been a member of the Optical Society of America, American Association for the Advancement of Science, and American Psychological Association.

Further Resources

Lemisch, Jesse and Naomi Weisstein. 1997. "Remarks on Naomi Weisstein." http://www.uic.edu/orgs/cwluherstory/CWLUMemoir/weisstein.html.

Weisstein, Naomi. 2003. "Adventures of a Woman in Science." In *Autobiographical Writings Across the Disciplines*, edited by Diane P. Freedman and Olivia Frey, 397–413. Durham, NC: Duke University Press.

Westcott, Cynthia

1898–1983
Plant Pathologist

Education: B.A., Wellesley College, 1920; Ph.D., plant pathology, Cornell University, 1932

Professional Experience: science teacher, Northboro High School, Massachusetts, 1920–1921; assistant in plant pathology, Cornell University, 1921–1923, instructor, 1923–1925, research assistant, 1925–1931; assistant horticulturist, seed laboratory, New Jersey Experiment Station, 1931–1933; independent horticulturist, The Plant Doctor, New Jersey, 1931–1961; independent writer and lecturer, New York, 1961–1983

Concurrent Positions: plant pathologist, U.S. Department of Agriculture (USDA), 1943–1945

Cynthia Westcott was a plant pathologist whose research focused on rose diseases, diseases of ornamental trees and flowers, and garden diseases and pests. She established a private practice as the "Plant Doctor" when she was unable to find a full-time professional position, and worked briefly for the USDA. As an independent plant consultant, she maintained the gardens of her wealthy customers, lectured at women's clubs, and published several books: *The Plant Doctor: The How, Why and When of Disease and Insect Control in Your Garden* (1937), *The Gardener's Bug Book* (1946), *Anyone Can Grow Roses* (1952), *Garden Enemies* (1953), and *Are You Your Garden's Worst Pest?* (1961). Many of her books went through numerous editions, including *Westcott's Plant Disease Handbook*, originally published in 1950 and released in a seventh edition in 2008. She also wrote an autobiography, *Plant Doctoring Is Fun* (1957), and wrote articles for popular magazines and newspapers, as well as leaflets on pesticides for the Manufacturing Chemists Association.

Westcott received her Ph.D. in plant pathology from Cornell University in 1932. She had previously taught high school science courses and spent 10 years working on her doctorate at Cornell, supporting herself by working as a research assistant and instructor. She worked in the seed laboratory at the New Jersey Experiment Station for three years before setting up her own business. During World War II, she worked as a plant pathologist for the USDA in order to earn money to obtain the supplies she needed for her business.

Westcott's work was honored with a citation from the American Horticultural Council (1955), a Gold Medal from the American Rose Society (1960), a Gold Medal from the Garden Club of New Jersey, and a Garden Writers Award from the American Association of Nurserymen (1963). She was active in professional scientific and gardening organizations, and was the first president of the North Jersey Rose Society (1954–1956) and director of the American Rose Society (1954–1960), and served as committee chair for the American Rose Foundation and the National Council of State Garden Clubs. She was a fellow of the American Association for the Advancement of Science and a member of the American Phytopathological Society, American Association of Economic Entomologists, American Entomological Society, American Horticultural Society, and Garden Writers Association of America, and an honorary life member of the Garden Club of New Jersey.

Further Resources

Horst, R. K. 1984. "Pioneer Leaders in Plant Pathology: Cynthia Westcott, Plant Doctor." *Annual Review of Phytopathology*. 22: 21–26. (September 1984). http://arjournals .annualreviews.org/doi/abs/10.1146/annurev.py.22.090184.000321?cookieSet=1 &journalCode=phyto.

West-Eberhard, Mary Jane

b. 1941
Entomologist

Education: B.A., University of Michigan, 1963, M.S., 1964, Ph.D., zoology, 1967

Professional Experience: teaching fellow, zoology, University of Michigan, 1963–1965; fellow, biology, Harvard University, 1967–1969; associate entomologist, Smithsonian Tropical Research Institute, 1973–1975, entomologist, 1975–

Concurrent Positions: staff member, biology, University of Valle, Colombia, 1972–1978; distinguished visiting scientist, Museum of Zoology, University of Michigan, 1982

Mary Jane West-Eberhard is a renowned entomologist who has studied the evolution of social behavior in insects of all types, primarily in Central and South America. She has published papers on insects' chemical communication, scent trails, social behavior, and diversity. She has theorized that evolved traits such as cyclic reproductive behavior, aggressiveness, and group life presumably reflect the genetic makeup of the individuals performing them. It seems that even caste determination, according to which some individuals end up as helpers and others as egg-laying queens, depends to some degree on heritable differences in aggressiveness, for example, especially in relatively simple societies in which there is no extensive manipulation of the brood, which can overwhelm heritable variation. She is the co-editor of *Natural History and Evolution of Paper-Wasps* (1996), which is based on a workshop held in Italy in 1993 to celebrate the fiftieth anniversary of Leo Pardi's original description of dominance hierarchies. In her chapter in the volume, she discusses how the differentiation of paper-wasp behavior and physiology may provide an illuminating model for some of the largest questions concerning the interface between development and evolution.

West-Eberhard was elected to membership in the National Academy of Sciences in 1988. Her 2003 book, *Developmental Plasticity and Evolution*, won the R. R. Hawkins Award of the American Association of Publishers for Outstanding Professional, Reference or Scholarly Work. Soon after the book's publication, she was awarded the 2003 Sewall Wright Award of the American Society of Naturalists. In 2005, she received a prestigious international honor when she was elected to Italy's Accademia Nazionale dei Lincei, the oldest scientific society in the world. She has been a member of several distinguished committees and commissions, including the International Committee for the International Union for the Study of Social Insects, the Organization for Tropical Studies, and the advisory committee of the Monteverde Conservation League Committee on Human Rights of the National Academy of Sciences. She is a member of the American Society of Naturalists, and Society for the Study of Evolution (president, 1992).

Further Resources

Smithsonian Tropical Research Institute. "Mary Jane West-Eberhard." http://www.stri.org/english/scientific_staff/staff_scientist/scientist.php?id=35.

Westheimer, (Karola) Ruth (Siegel)

b. 1928
Psychologist

Education: degree, psychology, University of Paris, Sorbonne; M.S., sociology, New School for Social Research, 1959; Ed.D., Columbia University, 1970

Professional Experience: research assistant, Columbia University School of Public Health, 1967–1970; associate professor, Department of Sex Counseling, Lehman College, 1970–1977; radio talk show host, television show host, author, private practice in psychology, 1980–1997, independent author and lecturer, 1997–

Concurrent Positions: adjunct professor, New York University; associate fellow, Calhoun College, Yale University; fellow, Butler College, Princeton University

Ruth Westheimer is popularly known as "Dr. Ruth," a psychologist and sex therapist who has appeared on hundreds of television and radio shows, and who has written numerous books for the general public. She is a trained counselor and also

American psychologist and sex therapist, Ruth Westhiemer, 2007. Dr. Ruth has appeared on hundreds of television and radio shows, and written numerous books for the general public. (AP/Wide World Photos)

maintained a private practice for a number of years. She pioneered the call-in radio and television era of media psychology shows with her show, *Sexually Speaking*, which began in 1980. For more than 20 years, she has engaged the public on controversial subjects related to sexuality, sexual dysfunction, marriage, and relationships. She has joked that her German accent allows her, a diminutive older woman, to call body parts and functions by their proper name and to advise both men and women, and that the American public would not have accepted her if she had an English or American accent. Although formally retired in 1997, she still travels, gives talks, teaches university courses on the family and sexuality, makes television and documentary appearances, and maintains a website.

Karola Ruth Siegel was born in Frankfurt, Germany in 1928. When she was 10 years old, her parents decided to flee Germany, but sent Ruth to a children's refuge in Switzerland. Her father was arrested before she left, and she presumes all of her family died in the concentration camps. Since the school considered her a welfare case, she was trained only as a maid. After the war, she emigrated to Palestine and joined Haganah, an underground movement fighting for the creation of a Jewish homeland. She dreamed of becoming a physician, but without family support and money, it was an impossible dream. She married and accompanied her first husband to Paris, where he studied medicine and she received a degree in psychology from the Sorbonne. She was divorced and remarried a Frenchman, whom she accompanied to New York in 1956 with their young daughter. After divorcing again, she attended evening classes for a master's degree at the New School for Social Research and went on for a doctorate of education from Columbia University Teachers College. In 1961, she married her third husband, Manfred Westheimer, and had another child. She had a position at Lehman College in the Department of Sex Counseling for a time and then worked for Brooklyn College and a few other schools.

After giving a lecture to a group of New York broadcasters about the need for sex education programming, she was invited in 1980 to tape a 15-minute radio show, *Sexually Speaking*. She was immediately popular and went on to nationally and internationally syndicated newspaper columns and award-winning television shows such as *The Dr. Ruth Show, Ask Dr. Ruth*, and *What's Up, Dr. Ruth?*, a show for teens. Her career as an author began with *Dr. Ruth's Guide to Good Sex* (1983). She went on to publish more than 30 titles, including *Dr. Ruth's Guide to Safer Sex* (1992), *Dr. Ruth's Encyclopedia of Sex* (1994; now available completely online), *Sex for Dummies* (1996; 3rd ed., 2006), *Rekindling Romance for Dummies* (2001), *Human Sexuality: A Psychosocial Perspective* (2002, co-authored with Sanford Lopater), and *Dr. Ruth's Guide to Talking about Herpes* (2004, co-authored with Pierre A. Lehu). Her autobiography is *All in a Lifetime* (1987).

Westheimer is a fellow of the New York Academy of Medicine and has received honorary doctorates from Hebrew Union College (2000), City University of

New York, Lehman College (2001), Trinity College (2004), and Westfield State College (2008). Her most recent awards include the Ellis Island Medal of Honor and the Leo Baeck Medal of the International Society for Sexual and Impotence Research, both in 2002, and a 2006 Medal for Distinguished Services from Columbia University Teachers College. In 2009, *Playboy* magazine named her one of the "55 Most Important People in Sex" of the past 55 years.

Further Resources

Dr. Ruth. http://www.drruth.com.

Wexler, Nancy Sabin

b. 1945
Neuropsychologist

Education: B.A., Radcliffe College, 1967; Ph.D., clinical psychology, University of Michigan, 1974

Professional Experience: intern and teaching fellow, University of Michigan, Ann Arbor, 1968–1974; assistant professor, psychology, New School of Social Research, New York City, 1974–1976; executive director, Congressional Commission for Control of Huntington's Disease, National Institute of Neurology, National Institutes of Health (NIH), 1976–1978, health science administrator, 1978–1983; associate professor, clinical neuropsychology, College of Physicians and Surgeons, Columbia University, 1985–1992, professor, 1992–

Concurrent Positions: psychologist, private practice, 1974–1976; president, Hereditary Disease Foundation, Santa Monica, California, 1983–

Nancy Wexler is renowned as one of the primary leaders in the fight to discover the cause of and cure for the hereditary Huntington's disease, named for George Huntington, a physician who identified the disease in 1872. The disease appears in middle age and slowly kills nerve cells in the brain, causing dementia and rapid, uncontrollable movements of the joints and limbs. Patients live an average of 15 years after the symptoms first appear. In 1968, when she was in graduate school at the University of Michigan, Wexler learned that her mother had developed symptoms of the disease, which had killed her grandfather and three of her uncles. Her mother's illness meant that both Nancy and her sister had a 50% chance of having inherited the defective gene that causes the disease—and that they might pass it on if they ever had children. Her father, a psychoanalyst, founded the Hereditary Disease Foundation in order to support research; Nancy assumed the presidency of the Foundation in 1983.

Wexler received her doctorate in 1974 and wrote her dissertation on the neuro-psychological and emotional consequences of being at risk for Huntington's disease. She was executive director of the Congressional Commission for the Control of Huntington's Disease through the NIH. She also helped to organize the Huntington's Disease Collaborative Research Group in 1984, an international consortium of scientists whose mandate was to track down the gene. The gene was isolated in 1993, but unfortunately, there is not yet a treatment for the disease. She has been a member of government committees concerned with ethical, legal, and social issues in medicine, and advisor to several groups related to the Human Genome Project, an international effort to map and identify the approximately 25,000 genes in the human body.

In 1997, Wexler was elected to the Institute of Medicine of the National Academies of Science. She has received several honorary medical doctorates and was awarded the Albert Lasker Public Service Award (1993) for her efforts connected with finding a cure for Huntington's disease. A partial listing of her numerous other awards includes an Alumnae Athena Award from the University of Michigan (1989), Venezuelan Presidential Award (1991), Distinguished Service Award of the National Association of Biology Teachers (1993), National Medical Research Award of the National Health Council (1993), J. Allyn Taylor International Prize in Medicine (1994), Public Advocacy Award of the Society for Neuroscience (2003), Distinguished Investigator Award of NARSAD (National Alliance for Research on Schizophrenia and Depression) (2006), and Benjamin Franklin Medal in Life Science (2007). She is a fellow of the Royal College of Physicians, American Academy of Arts and Sciences, and New York Academy of Sciences, and has been a member of the American Psychological Association, American Society for Human Genetics, American Neurological Association, American Society of Law and Medicine, Society for Neuroscience, and World Federation of Neurology.

Further Resources

Hereditary Disease Foundation. "Meet Nancy Wexler." http://www.hdfoundation.org/bios/nancyw.php.

Wheeler, Anna Johnson Pell

1883–1966
Mathematician

Education: A.B., University of South Dakota, 1903; M.A., University of Iowa, 1904; A.M., Radcliffe College, 1905; Ph.D., mathematics, University of Chicago, 1910

Professional Experience: instructor, mathematics, Mount Holyoke College, 1911–1914, associate professor, 1914–1918; associate professor, Bryn Mawr College, 1918–1925, department chair, 1924–1948, professor, 1925–1927, non-research professor, 1929–1932, professor, 1932–1948

Anna Pell Wheeler was a distinguished research mathematician whose work was primarily in the area of linear algebra of infinitely many variables and integral equations. She was only the second woman to receive a doctorate in mathematics at the University of Chicago, and was one of the few professional women mathematicians recognized in the early twentieth century. She graduated from the University of South Dakota in 1903 and went on to obtain master's degrees from both the University of Iowa and Radcliffe. She received a one-year fellowship to study at the University of Göttingen (1906–1907), then returned to South Dakota, where she married one of her former mathematics professors and taught some classes. The couple moved to Chicago, where she completed her doctorate in mathematics in 1910. When her husband suffered a paralytic stroke, she substituted for him at the Armour Institute of Technology, but she was unable to obtain a position there. She accepted a position at Mount Holyoke in 1911 in order to support them, but she did not have time for her research. She moved to Bryn Mawr as an associate professor in 1918, served as head of the department, and was promoted to professor in 1925, the same year she married her second husband, Arthur Wheeler, a classics professor. They moved to Princeton, New Jersey, but she continued to teach at Bryn Mawr on a part-time basis. When her husband died in 1932, she returned to full-time work at Bryn Mawr, retiring in 1948.

During her 30-year affiliation with Bryn Mawr, Wheeler encouraged female students to pursue mathematics and advised several who went on to earn doctoral degrees. In addition to her teaching and research, Wheeler was active in professional mathematical associations and was an editor of the *Annals of Mathematics* for almost 20 years. She received honorary degrees from the New Jersey College for Women (1932) and Mount Holyoke College (1937). In 1927, she was the first woman invited to give the American Mathematical Society Colloquium Lecture; the next female lecturer was not until 1980. Wheeler was a member of the American Mathematical Society, the Mathematical Association of America, and the American Association for the Advancement of Science.

Further Resources

Agnes Scott College. "Anna Johnson Pell Wheeler." Biographies of Women Mathematicians. http://www.agnesscott.edu/lriddle/women/wheeler.htm.

Wheeler, Mary F.

b. 1931
Mathematician, Engineer

Education: B.S., social sciences and math, University of Texas, Austin, 1960, M.A., mathematics, 1963; Ph.D., mathematics, Rice University, 1971

Professional Experience: instructor, mathematics, Rice University, 1971–1973, assistant to associate professor, mathematical sciences, 1973–1980, professor, 1980–1988; M.D. Anderson Professor of Mathematics, University of Houston, 1988–1990; Noah Harding Professor of Computational and Applied Mathematics, Rice University, 1988–1995; Ernest and Virginia Cockrell Chair in Engineering, University of Texas, Austin, 1995–, professor, mathematics, aerospace engineering, and petroleum and geosystems engineering, 1995–

Concurrent Positions: adjunct professor, University of Texas M. D. Anderson Cancer Center; affiliated senior scientist, University of Houston, 1990–

Mary Wheeler is a mathematician whose research links theory and application in a focus on numerical solutions of partial differential equations, parallel computation, and modeling subsurface and surface flows. Specifically, her work has had industry-related applications to projects in oil recovery, reservoir engineering, and solutions for reducing pollutants in groundwater, bays, and estuaries. Born in Texas, she received her doctorate in mathematics from Rice University in 1971 and has taught at several Texas institutions, including Rice and the University of Houston, and, since 1995, has been a faculty member at the University of Texas, Austin. She is also the director of the Center for Subsurface Modeling in the Texas Institute for Computational and Applied Mathematics (TICAM). She is the author of hundreds of scientific technical papers and has edited or co-edited several books. She has also served on the editorial board of several professional journals, including *Computational Geosciences.*

Wheeler began her college career with interests in pharmacology, or government and law. But her passion was in math, and she held a double major in social sciences and mathematics while an undergraduate at the University of Texas. She went on to study math at the graduate level and became interested in physical and engineering applications rather than theory and economics.

She was invited to give the prestigious Emmy Noether Lecture in 1989 and has been an invited lecturer at universities and organizations around the world. She has served on committees on science policy, industrial mathematics, and science education, and mathematical sciences and research review committees for government organizations such as the National Science Foundation, Argonne and Oak Ridge National Laboratories, and U.S. Department of Energy.

Wheeler was elected to the National Academy of Engineering in 1998. Among her numerous awards and honors is an Educator Award from American Women in Aerospace (1997), Distinguished Alumna Award from Rice University (2000), USACM Computational Fluid Mechanics Award (2003), Joe J. King Award from the University of Texas at Austin (2006), and several IBM Faculty Recognition Awards (2006, 2007, 2008). She has also received honorary doctorates from Technische Universiteit Eindhoven (2006) and the Colorado School of Mines (2008). She is a member of the Mathematical Association of America, American Geophysical Union, Society of Industrial and Applied Mathematics (SIAM), Society of Petroleum Engineers (SPE), and American Women in Mathematics, and a fellow of the International Association for Computational Mechanics.

Further Resources

Agnes Scott College. "Mary F. Wheeler." Biographies of Women Mathematicians. http://www.agnesscott.edu/lriddle/women/mwheeler.htm.

University of Texas. Faculty website. http://users.ices.utexas.edu/~mfw/.

Whitman, Marina (von Neumann)

b. 1935
Economist

Education: B.A., government, Radcliffe College, 1956; M.A., economics, Columbia University, 1959, Ph.D., economics, 1962

Professional Experience: administrative assistant, Educational Testing Service, 1956–1957; consultant, Pittsburgh Regional Planning Association, 1961, staff economist, Economic Study of the Pittsburgh Region, 1962; lecturer, economics, University of Pittsburgh, 1962–1964, assistant to associate professor, 1964–1971, professor, 1971–1973, Distinguished Public Service Professor of Economics, 1973–1979; vice president and chief economist, General Motors (GM) Corporation, New York, 1979–1985, vice president and group executive, public affairs and marketing staff, 1985–1992; distinguished visiting professor, business administration and public policy, University of Michigan, Ann Arbor, 1992–1994, professor, 1994–

Concurrent Positions: fellow, Center for Advanced Study in the Behavioral Sciences, Stanford University, 1978–1979

Marina v. N. Whitman is a renowned international economist who has worked in business, education, and government. She has served as a senior staff economist for the Council of Economic Advisers, its first woman member, and she was the only woman member of the National Price Commission. She earned her doctorate in economics from Columbia University in 1962 and subsequently joined the faculty at the University of Pittsburgh. She rose quickly through the ranks to full professor, but left academia in 1979 to join GM as vice president and chief economist in charge of economic and environmental policy and industry–government relations. She left GM in 1992 to return to teaching at the University of Michigan, Ann Arbor. In all of her work, in teaching, business, or as a government advisor, Whitman has advocated for a greater global vision and international economic role for the United States.

As the only daughter of the eminent mathematician John von Neumann, she grew up in an atmosphere of stimulating people. Many famous people visited her family home, and she had tremendous intellectual drive and intense pressure to achieve as an undergraduate at Radcliffe. She married after graduation and, in order to be near her husband's job at Princeton, worked as an administrative assistant for Educational Testing Service, a nonprofit organization specializing in educational measurement and research. She then enrolled in Columbia University, planning to receive a master's degree in economics and journalism, and to pursue a career in financial writing. Instead, she concentrated on economic theory and, as part of her graduate studies, prepared an economic development plan for the Pittsburgh Regional Planning Association. She then accepted an appointment as a lecturer in economics at the University of Pittsburgh. In 1970, she was selected as a member of the prestigious Council of Economic Advisers under President Nixon. In this role, Whitman made a special report on women in the American economy, stating that, despite 10 years of civil rights legislation, women had made little progress toward job equality with men; in 1971, the average female worker earned only 59.5 cents on the male dollar for comparable work. She later served on President Carter's Economic Advisory Committee.

Whitman has published numerous articles and books, including *Government Risk-Sharing in Foreign Investment* (1965), *Reflections of Interdependence: Issues for Economic Theory and U.S. Policy* (1979), *New World, New Rules: The Changing Role of the American Corporation* (1999), and *American Capitalism and Global Convergence* (2003). She has examined the effect of global markets on American corporations and society, and advocated for an open market economy. She has been a member of several government committees, including the President's Council of Economic Advisors (1970–1973), National Price Commission (1971–1972), Economic Advisory Committee of the U.S. Department of Commerce (1979–1980), Commission on Security and Economic Assistance (1983–1984), President's Export

Council (1986–1987), and President's Advisory Committee on Trade Policy and Negotiations (1987–1993). She has also been a board member or trustee of numerous national, international, and academic advisory committees.

Whitman has been awarded honorary doctorates from more than 20 universities and is the recipient of a Columbia University Medal for Excellence (1973 and 1984), the George Washington Award of the American Hungarian Foundation (1975), the Catalyst Award for women in business (1976), a Women's Equity Action League Achievement Award (1979), and the William F. Butler Memorial Award of the New York Association of Business Economists (1988). She is a member of the American Economic Association, National Association of Business Economists, and Council on Foreign Relations, and a fellow of the American Academy of Arts and Sciences.

Further Resources

University of Michigan. Faculty website. http://www.bus.umich.edu/FacultyBios/
    FacultyBio.asp?id=000119718.

Whitson, Peggy A.

b. 1960
Astronaut

Education: B.S., biology and chemistry, Iowa Wesleyan College, 1981; Ph.D., biochemistry, Rice University, 1985

Professional Experience: Robert A. Welch postdoctoral fellow, Rice University, 1986; National Research Council Resident Research Associate, National Aeronautics and Space Administration (NASA) Johnson Space Center, Houston, Texas, 1986–1988; supervisor, Biochemistry Research Group, KRUG International, 1988–1989; research biochemist, Biomedical Operations and Research Branch, NASA Johnson Space Center, 1989–1993, deputy division chief, Medical Sciences Division, 1993–1996, astronaut, 1996–, deputy chief, Astronaut Office, 2003–2005, chief, Station Operations Branch, Astronaut Office, 2005

Concurrent Positions: adjunct assistant professor, Departments of Internal Medicine and Human Biological Chemistry and Genetics, University of Texas Medical Branch, Galveston, 1991–1997; adjunct assistant professor, Maybee Laboratory for Biochemical and Genetic Engineering, Rice University, 1997–

Peggy Whitson is a biochemist and astronaut who was the first woman commander of the International Space Station (ISS). She logged two long-term stays at the

Astronaut Peggy Whitson preparing for the launch of the Space Shuttle *Endeavour* on a mission to the International Space Station, 2002. (NASA)

ISS, in 2002 and 2007, and has accumulated more than 377 days in space and almost 40 hours of space walks, more than any other female astronaut. Whitson completed her doctorate in biochemistry at Rice University in Houston, Texas, and was a research associate at NASA before working briefly for KRUG International, a NASA-contracted medical sciences company. She returned to NASA in 1989 in Biomedical Operations and Research, and became a member of the U.S.–USSR Joint Working Group in Space Medicine and Biology, training astronauts in both the United States and Russia. Between 1992 and 1995, she was a project scientist on the Shuttle-Mir Program. She applied to the astronaut training program several years before being accepted in 1996. In 2002, she flew aboard the *Endeavour* for the Expedition-5 mission to dock with the ISS. She spent 6 months (nearly 185 days) with only two other astronauts on the ISS as NASA Science Officer, conducting research on human biology and microgravity conditions. It was unusual for a first-time astronaut to be assigned such an extended mission, but her science research background, and her 10 years of NASA training on the ground, had prepared her well. In 2005, she began training as a backup ISS Commander and flew as Commander of the ISS for a second long-term stay (more than 191 days) with Expedition-16 in the fall of 2007.

Besides setting records for women in space, Whitson has been acknowledged for her numerous achievements at NASA, including but not limited to the following awards: Sustained Superior Performance Award (1990), Certificate of Commendation (1994), Exceptional Service Medal (1995, 2003, 2006), Silver Snoopy Award (1995), Space Act Board Award (1995, 1998), Group Achievement Award for Shuttle-Mir Program (1996), Space Flight Medal (2002), and Outstanding

Leadership Medal (2006). She was also awarded the Randolph Lovelace Award of the American Astronautical Society (1995).

Further Resources

National Aeronautics and Space Administration. "Peggy A. Whitson (Ph.D.)." http://www .jsc.nasa.gov/Bios/htmlbios/whitson.html.

Widnall, Sheila (Evans)

b. 1938
Aeronautical Engineer

Education: B.S., Massachusetts Institute of Technology, 1960, M.S., 1961, D.Sc., aeronautical engineering, 1964

Professional Experience: staff, Boeing, summers 1947–1959, 1961; staff, Aeronautical Research Institute of Sweden, summer 1960; research staff engineer, aerodynamics, Massachusetts Institute of Technology (MIT), 1961–1962, research assistant, 1962–1964, assistant to associate professor, aeronautics, 1964–1974, professor, 1974–1986, Abby Rockefeller Mauzè Professor and Chair, 1986–1993; Secretary of U.S. Air Force, 1993–1997; Institute Professor, aeronautics and astronautics, MIT, 1998–

Concurrent Positions: director, university research, U.S. Department of Transportation, 1974–1975; associate provost, MIT, 1992–1993; vice president, National Academy of Engineering, 1998–2006

Sheila Widnall is an aeronautical engineer whose research interests include unsteady aerodynamics, aeroelasticity, aerodynamic noise, turbulence, applied mathematics, vortex flows, numerical analysis, aerospace, transportation, aerodynamics and fluid mechanics, acoustics, and noise and vibration. She has been a professor of aeronautics and astronautics at MIT for more than 40 years, where her research has centered particularly on problems associated with fluid dynamics and air turbulence. Another area of research is the vortices of aircraft that make vertical, short takeoffs and landings (V/STOL), and the noise associated with them. One of her projects was to design an anechoic wind tunnel at MIT to study V/STOL aircraft—a wind tunnel that has a low degree of reverberation and is echo-free. In 1993, she became the first woman to head a branch of the U.S. military when she was selected to be Secretary of the U.S. Air Force. In this position, Widnall was responsible for recruiting, organizing, training, administration, logistical support, maintenance, and welfare of personnel, as well as overseeing research and development projects outlined by the president or the Secretary of Defense. She co-chaired the Department of Defense Task Force on Sexual Harassment and Discrimination. She left the Air Force in 1997 to return to her faculty position at MIT.

As a young woman, she was encouraged by teachers and parents to pursue a career in science. Still, there were only 20 women in her class of about 900 at MIT. As president of the American Association for the Advancement of Science (1988), she was committed to encouraging more women to pursue careers in science and engineering, and outlined the problems they face in attaining their degrees and achieving professional goals. She has been an advisor for numerous government and industry projects and scientific agencies, including for the Carnegie Corporation, Sloan Foundation, Institute for Defense Analysis, Smithsonian Institution of Washington, Boston Museum of Science, GenCorp Inc., Chemfab Inc., Space and Aeronautics Board of the National Research Council, and National Science Foundation, to name just a few.

Widnall was elected to membership in the National Academy of Engineering in 1985 and was vice president of the National Academy of Engineering from 1998 to 2006. Her numerous awards and honors include the Lawrence Sperry Award of the American Institute of Aeronautics and Astronautics (1972), Outstanding Achievement Award of the Society of Women Engineers (1975), Washburn Award of the Boston Museum of Science (1986), Distinguished Service Award of the National Academy of Engineering (1993), Medal of Distinction from Barnard College (1994), W. Stuart Symington Award (1995) and Maxwell A. Kriendler Memorial Award (1995), both from the Air Force Association, Applied Mechanics Award of American Society of Mechanical Engineers (ASME) (1996), Distinguished Civilian Service Medals from both the Army and Navy (1997), Reed Aeronautics Award from American Institute of Aeronautics and Astronautics (AIAA) (2000), and Spirit of St. Louis Medal from ASME (2001). In 1996, she was inducted into the Women in Aviation Pioneer Hall of Fame. She is a fellow of the AIAA (president, 1999–2000), American Physical Society, Royal Aeronautical Society, and American Association for the Advancement of Science. She is a member of the Society of Women Engineers, American Society of Mechanical Engineers, Puerto Rican Academy of Sciences, International Academy of Astronautics, Institute of Electrical and Electronics Engineers (IEEE), and American Philosophical Society.

Further Resources

Massachusetts Institute of Technology. Faculty website. http://web.mit.edu/aeroastro/www/people/widnall/.

Wilhelmi, Jane Anne Russell

1911–1967
Endocrinologist

Education: B.A., University of California, Berkeley, 1932, Ph.D., biochemistry, 1937

Professional Experience: technical assistant in biochemistry, University of California, Berkeley, 1932–1933, assistant, institute of experimental biology, 1934–1937; research associate, pharmacology, Washington University, St. Louis, 1936; National Research Council (NRC) Fellow, School of Medicine, Yale University, 1938–1939, fellow, 1939–1941, instructor, physiological chemistry, 1941–1950; assistant professor, biochemistry, Emory University, 1950–1953, associate professor, 1953–1965, professor, 1965–1967

Jane Russell Wilhelmi was an endocrinologist whose research interests included endocrine control of intermediate metabolism; adrenal cortex, anterior pituitary, growth hormone, and insulin in carbohydrate and protein metabolism; the metabolic aspects of shock; and the use of isotopic tracers in metabolism. After receiving her undergraduate degree at the University of California, Berkeley, she worked as a technical assistant while she completed her doctorate in biochemistry in 1937. She spent a year in 1936 as a pharmacology research associate working on carbohydrate metabolism with Carl and **Gerty Cori** at Washington University in St. Louis. She was appointed a research fellow at Yale School of Medicine in 1938, a fellow in 1939, and an instructor in physiological chemistry in 1941. In 1940, she married her colleague, Alfred Ellis Wilhelmi, with whom she collaborated on research and co-authored dozens of scientific papers on metabolism and the role of growth hormones in breaking down proteins.

Jane Wilhelmi received outside recognition for her pioneering research, and consulted on committees of the National Institutes of Health, NRC, National Science Foundation, and National Science Board. Her research was acknowledged and supported with a California Fellowship in Biochemistry, a Rosenberg Fellowship, and the American Physiological Society's Porter Fellowship. Despite these honors, she did not advance at Yale, remaining at the rank of instructor before accepting a position as assistant professor of biochemistry at Emory University in Atlanta, Georgia, in 1950. She finally reached the rank of full professor in 1965, just two years before her death. Up until the time of her death, she remained active in research, writing, and working as an editor for the *American Journal of Physiology*. She received the Ciba

Award in 1946, and she shared the Upjohn Award of the Endocrine Society with her husband in 1961. She was also named Atlanta's Woman of the Year in 1961.

Williams, Anna Wessels

1863–1954
Bacteriologist

Education: diploma, New Jersey State Normal School, Trenton, 1883; M.D., Women's Medical College, New York, 1891

Professional Experience: public school teacher, 1883–1885; instructor, pathology, Women's Medical College of the New York Infirmary, 1891–1893, assistant to department chair, pathology and hygiene, 1891–1895; assistant bacteriologist, diagnostic laboratory, New York City Department of Health, 1895–1905, assistant director, 1905–1934

Concurrent Positions: consulting pathologist, Women's Medical College of the New York Infirmary, 1902–1905

Anna Williams was a pioneering bacteriologist who gained national recognition for her work on infectious diseases. At the diagnostic laboratory of the New York City Department of Health, she made significant contributions on effective immunization for diphtheria, streptococcal (strep throat) and pneumococcal (pneumonia) infections, scarlet fever, and rabies. In the first year of her research, she isolated a strain of the diphtheria bacillus that made possible the widespread immunization of children and almost complete eradication of the disease that, at that time, was one of the primary causes of death among young children. She played a significant role in building the New York laboratory into a nationally known center as the first municipal laboratory to apply bacteriology to the problems of public health. After receiving her diploma from the New Jersey State Normal School, she taught public school for several years to earn funds to obtain her M.D. in 1891 from the Women's Medical College of New York. She had convinced her family to allow her to become a physician after a sister almost died due to complications of childbirth.

After working as a pathologist for the Women's Medical College for several years, she was initially a volunteer with the diagnostic laboratory before joining the staff of the New York City Department of Health in 1895. She spent a year at the Pasteur Institute in Paris, unsuccessfully researching an antitoxin for scarlet fever, but her work did lead to the development of a rabies vaccine by 1898 and a new, faster method for identifying rabies in animals. She later served as chair of a new rabies committee for the American Public Health Association and, during World War I, worked on government programs related to diagnosing influenza and meningitis. She was appointed assistant director of the diagnostic laboratory of the New York Department of Health in 1905, a position she held until forced into mandatory retirement in 1934 at the age of 71.

Williams was co-author of a book for the general public entitled *Who's Who among the Microbes* (1929). In addition to her scientific papers, she was also co-author of *Pathogenic Microorganisms Including Bacteria and Protozoa: A Practical Manual for Students, Physicians and Health Officers* (1905) and author of *Streptococci in Relation to Man in Health and Disease* (1932). She was a member of the American Public Health Association and the New York Women's Medical Association (president, 1915).

Further Resources

National Institutes of Health. "Dr. Anna Wessels Williams." Changing the Face of Medicine: Celebrating America's Women Physicians. National Library of Medicine, National Institutes of Health. http://www.nlm.nih.gov/changingthefaceofmedicine/physicians/biography_331.html.

Williams, Roberta

b. 1952
Computer Games Designer

Education: high school

Professional Experience: part-time programmer; designer of computer action games, 1979–1980; co-founder and chief game designer, Sierra On-Line, Inc., 1980–1999

Roberta Williams is considered a pioneer of the graphic adventure multimedia computer game. Williams did not attend college, but had some technical training and experience with mainframe computers in the 1970s. She became intrigued by text-based video games after purchasing an Apple computer. Her career started when her husband, programmer Kenneth Williams, brought home a computer game that she found was too easy to solve, so she was challenged to create more difficult games. The couple founded their own company, On-Line Systems (later known as Sierra On-Line) in 1980. Their first game, "Mystery House," debuted

Roberta Williams poses with a copy of her computer game, *Phantasmagoria*, 1995. (AP/Wide World Photos)

in 1980 and became part of a six-part series of Apple games that included the first bestselling games with colored graphics. Their second game, "The Wizard and the Princess," was programmed on a disk rather than on a cassette, a format that revolutionized the microcomputer game industry by making possible much longer games. They also designed a computer game based on the Jim Henson movie *The Dark Crystal*, released at the same time, in 1992. Roberta also advised on some of the layouts for the movie. She was one of the first designers to use a female protagonist in an adventure game. She has also designed a range of other computer software products for home use.

By 1983, Sierra On-Line was earning $10 million a year in sales, and by 1991, annual sales were $43 million and the company employed some 500 people. When the Williamses sold the company to CUC International, Inc. in 1996 for about $1 billion, Roberta stayed on briefly as chief designer. In 1997, Sierra On-Line, Inc. released the "Roberta Williams Anthology," a collection of 15 of her games. Most of the early ones are primitive by today's standards, but the anthology is a compact history of the form. She began all of her games by drawing them out on large sheets of paper, but the later games eventually involved the work of more than 100 people, including animators, programmers, musicians, and composers. Williams has won numerous awards and honors for her games. She retired from Sierra On-Line in 1999 to travel with Ken, and the couple maintain a website and message boards for gaming enthusiasts.

Further Resources

MobyGames. "Roberta Williams: Developer Bio." http://www.mobygames.com/developer/sheet/view/developerId,60/.

Sierra Gamers. http://www.sierragamers.com.

Witkin, Evelyn Maisel

b. 1921
Geneticist

Education: B.A., zoology, New York University, 1941; M.A., Columbia University, 1943, Ph.D., zoology, 1947

Professional Experience: research associate, bacterial genetics, Cold Spring Harbor Laboratories, Carnegie Institution, 1945–1955; associate professor, medicine, Downstate Medical Center, State University of New York, 1955–1969, professor, 1969–1971; professor, biology, Douglass College (Rutgers University), 1971–1983, Barbara McClintock Professor of Genetics, Rutgers, 1979, professor, Waksman Institute of Microbiology, 1983–1991, emerita

Concurrent Positions: postdoctoral fellow, American Cancer Society, 1947–1949

Evelyn Witkin is a geneticist who has been recognized for her work on mutation in bacteria. Her research has involved mechanism of spontaneous and induced mutation in bacteria, genetic effects of radiation, and enzymatic repair of DNA damage. While completing her doctorate in zoology from Columbia University, she spent a summer as a research associate in bacterial genetics at Cold Spring Harbor Laboratories where she first isolated a radiation-resistant strain of *E. coli*. She became a regular staff member at Cold Spring Harbor in 1945 and remained there for 10 years. She was appointed an associate professor of medicine at the State University of New York in 1955 and promoted to full professor in 1969. She moved to Douglass College, the women's campus at Rutgers University in New Jersey, in 1971 and joined the Waksman Institute of Microbiology at Rutgers in 1983. In the early 1970s, she made a breakthrough discovery on bacterial response to genetic damage and repair. She retired in 1991 as the Barbara McClintock Professor Emerita of Genetics.

Witkin was elected a member of the National Academy of Sciences in 1977 and awarded a National Medal of Science by President George W. Bush in 2003. She is also the recipient of honorary doctorates from New York Medical College (1978), Rutgers University (1995), and Clark University (2006). Among her other honors are the Lindback Award (1979), the American Women of Science Award for Outstanding Research (1982), the Thomas Hunt Morgan Medal of the Genetics Society of America (2000), and the Distinguished Research Award of the New Jersey Association for Biomedical Research (2004). She is a fellow of the American Academy of Arts and Sciences and American Society of Microbiology, and a member of the Genetics Society of America, American Society of Naturalists, and Radiation Research Society.

Further Resources

Rutgers University, Office of Media Relations. "President Bush Names Rutgers' Evelyn Witkin for Nation's Highest Science Honor." (22 October 2003). http://ur.rutgers.edu/ medrel/viewArticle.html?ArticleID=3545.

Wood, Elizabeth Armstrong

1912–2006
Crystallographer

Education: B.A., Barnard College, 1933; M.A., Bryn Mawr College, 1934, Ph.D., geology, 1939

Professional Experience: instructor, geology, Bryn Mawr College, 1934–1935, 1937–1938; assistant, Barnard College, 1935–1937, lecturer, geology and mineralogy, 1938–1941; research assistant, Columbia University, 1941–1942; technical staff, crystal research, Bell Telephone Laboratories, AT&T, 1942–1967

Elizabeth Wood was recognized for her research on x-ray crystallography and the physical properties of crystals. She also studied the geology and petrology of igneous and metamorphic rocks, and optical mineralogy. Wood received her undergraduate degree from Barnard in 1933. After earning her master's degree at Bryn Mawr in 1934, she was employed there as a demonstrator in geology while she did further graduate work. In 1938, she returned to Barnard as a lecturer, and in 1939, she received her doctorate in geology from Bryn Mawr. She was promoted to research assistant at Barnard in 1941 before joining the technical staff in crystallographic research at Bell Telephone Labs in 1942—the first woman scientist in the physical research department. Her career coincided with the beginning of the discipline of solid-state physics, and Bell Labs was one of the first developers of lasers and other solid-state devices that required crystals. Wood became an acknowledged authority at Bell and was even called upon to receive the first call on a "picture-phone," made from First Lady Johnson from the White House to Wood in New York in 1964. She spent 25 years at Bell/AT&T, retiring in 1967.

Wood was also committed to science education and published several textbooks and guides, including *Rewarding Careers for Women in Physics* (1962) and *Pressing Needs in School Sciences* (1969), both published by the American Institute of Physics, and *Crystal Orientation Manual* (1963) and *Crystals and Light: An Introduction to Optical Crystallography* (1964), which remain classics in the field. In the 1960s, she also published (through Bell Labs) a high school curriculum, *Experiments with Crystals and Light*, and a general-interest book, *Science for the Airplane Passenger*, which was sold through airport bookstores. The American Crystallographic Association (ACA) established the Elizabeth A. Wood Science Writing Award in her honor.

Wood was active in professional scientific organizations, serving as secretary of the American Society for X-Ray and Electron Diffraction (ASXRED) in 1947, and was the first female president of the American Crystallographic Association in 1957 (**Isabella Karle** became the second, in 1976). She received honorary doctorates from Wheaton College (1963), Western College, Ohio (1965), and Worcester Polytechnic (1970). She was a fellow of the American Academy of Arts and Sciences, American Physical Society, International Union of Crystallography, and Mineralogical Society of America.

Further Resources

Abrahams, S. C. "Death Notice. Elizabeth A. Wood. 19 October 1912–23 March 2006." *Physics Today*. (12 May 2006). http://www.physicstoday.org/obits/notice_060.shtml.

Woods, Geraldine (Pittman)

1921–1999
Embryologist, Science Consultant

Education: B.S., biology, Howard University, 1942; M.A., Radcliffe College, 1943, Ph.D., neuroembryology, 1945

Professional Experience: instructor, biology, Howard University, 1945–1946; special consultant, National Institute of General Medical Sciences, National Institutes of Health (NIH), 1969–1987

Geraldine Woods was an embryologist who was primarily known for her efforts to improve access to higher education for minorities. In addition to her volunteer work, she served as a consultant to the National Institute of General Medical Sciences in implementing various programs. She was one of the earliest black women to hold a Ph.D. in the biological sciences, and her doctoral research involved the early development of nerves in the spinal cord, studying whether the nerve specialization process was governed by the cell's heredity or by stimulation from nearby cells. While attending Talladega College in Alabama, her mother became seriously ill. The physicians recommended she take treatments at Johns Hopkins University, so Geraldine transferred to nearby Howard University in Washington,

D.C. An embryology professor at Howard encouraged her to continue her studies at Harvard University. At that time, the women enrolled in Radcliffe College took all of their science classes at Harvard, and she earned two graduate degrees in three years.

After receiving her doctorate, Woods taught biology at Howard before moving to California, where her husband set up his dental practice. She raised three children and began volunteering with social services projects and civil rights efforts, first locally, in Los Angeles, and then statewide. She served four years (1963–1967) as president of Delta Sigma Theta, a national public-service sorority of black, college-educated women. It was through this group that she helped establish several Head Start preschools in the Los Angeles area. Her work attracted national attention when Lady Bird Johnson, wife of President Lyndon Johnson, invited her to the White House in 1965 to help launch Project Head Start, a federal program to help children from low-income families attend preschool. In 1968, President Johnson appointed her chair of the Defense Advisory Committee on Women in the Services.

In 1969, Woods was appointed as a special consultant to the National Institute of General Medical Sciences of the NIH, where she addressed problems of minority students and institutions gaining access to grants and other funding, their overall lack of adequate equipment for scientific research, and educational opportunities for minority students in the sciences. The NIH installed two programs under her guidance: the Minority Biomedical Support (MBS) program to guide researchers through the grant application process, and Minority Access to Research Careers (MARC), which provided counseling and scholarships for students and faculty members in science careers.

Woods was a member of the American Association for the Advancement of Science and the New York Academy of Sciences. Among her awards and honors are several biomedical scholarships given in her name, and the Mary Church Terrell Award of Delta Sigma Theta (1979), the Scroll of Merit of the National Medical Association (1979), the Howard University Achievement Award (1980), and a Distinguished Leadership Achievement Award from the National Association for Equal Opportunity in Higher Education (1987). She received honorary degrees from several institutions, including Benedict College (1977), Talladega College (1980), Fisk University (1991), Bennett College (1993), Meharry Medical College (1988), and Howard University (1989).

Further Resources

Giddings, Paula A. 1994. *In Search of Sisterhood: Delta Sigma Theta and the Challenge of the Black Sorority Movement*, 2nd ed. New York: William Morrow.

Warren, Wini. 1999. *Black Women Scientists in the United States*. Bloomington: Indiana University Press.

Woolley, Helen Bradford Thompson

1874–1947
Psychologist

Education: B.A., University of Chicago, 1897, Ph.D., neurology, 1900

Professional Experience: instructor and professor, psychology, Mount Holyoke College, 1901–1905; experimental psychologist, Bureau of Education, Philippine Islands, 1905–1906; health inspector, serum laboratory, Bangkok, Thailand, 1907–1908; instructor, philosophy, University of Cincinnati, 1910–1912; director, Bureau for the Investigation of Working Children, Cincinnati public schools, 1911–1921; psychologist and assistant director, Merrill-Palmer School, Michigan, 1921–1926; professor of education and director, bureau of child development, Teachers College, Columbia University, 1926–1930

Helen Woolley was a pioneer in the study of child development and of gender differences. Her research involved the psychology of adolescence and of young childhood, mental development, testing, and educational methods, and exposing what she termed the "inconsistencies, contradictions, and lack of data behind the conventional wisdom on sex differences." Woolley (then Thompson) challenged beliefs about women's "natural" roles and interests, and used scientific data to support women's participation in academia and the workplace. For her doctoral research at the University of Chicago, she created a series of tests of male and female students' physical and mental processes. In her thesis, *Psychological Norms in Men and Women* (published in 1903 as *The Mental Traits of Sex*), she concluded that there were few biological or psychological differences between men and women, and that social and environmental factors accounted for most differences. Not surprisingly, Woolley became an advocate of both civil rights and women's rights, becoming a member and chair of the Ohio Woman Suffrage Association.

After she received her doctorate from the University of Chicago in 1900, she undertook further studies at the Universities of Berlin and Paris before she accepted a position in the psychology department at Mount Holyoke in 1901. When she was married in 1905, she and her husband spent several years in Southeast Asia, where she worked in the Philippine Bureau of Education and at a laboratory run by her physician husband in Thailand. The couple returned to the United States, where she taught at the University of Cincinnati for three years and became involved in child welfare reforms and child psychology. After Ohio passed a child labor law in 1910, she served as director of a program to compare the development of working children with those who stayed in school, and her work contributed to educational reforms, such as compulsory attendance laws. She accepted a position as assistant director and psychologist at the Merrill-Palmer School, a child development institute in Detroit, in 1921, and she helped develop a teacher-training program and design educational tests, such as the Merrill-Palmer Scale of Mental Tests. In 1926, Woolley took a position at Teachers College, Columbia University, as professor of education and director of the bureau of child development. She was forced to retire in 1930 due to health issues.

Woolley contributed a chapter on "The Psychologist" for a 1920 guide to *Careers for Women* (edited by Catherine Filene). She was elected president of the National Vocational Guidance Association in 1921. She also was a member of the American Psychological Association and the American Association for the Advancement of Science.

Further Resources

Morse, Jane Fowler. 2002. "Ignored but Not Forgotten: The Work of Helen Thompson Bradford Woolley." *NWSA Journal*. 14(2): 121–147. (Summer 2002).

Scarborough, Elizabeth and Laurel Furumoto. 1987. *Untold Lives: The First Generation of American Women Psychologists*. New York: Columbia University Press.

Wright, Margaret H.

b. 1944
Computer Scientist, Mathematician

Education: B.S., mathematics, Stanford University, M.S., computer science, Ph.D., computer science, 1976

Professional Experience: research associate, Systems Optimization Laboratory, Operations Research, Stanford University, 1976–1981, senior research associate, 1981–1988; technical staff, Bell Laboratories, AT&T, 1988–1993, Distinguished Member of Technical Staff, 1993–2001, head, Scientific Computing Research Department, 1997–2000; Silver Professor of Computer Science and

Chair, Department of Computer Science, Courant Institute of Mathematical Sciences, New York University, 2001–

Margaret Wright is a computer scientist and applied mathematician whose research interests include optimization, linear algebra, numerical analysis, scientific computing, and scientific and engineering applications. She builds mathematical and computer models for problem solving in a variety of practical applications. She earned degrees from Stanford University and spent more than 20 years in Scientific Computing Research at AT&T's Bell Laboratories (now Lucent Technologies) before entering academia in 2001 as professor and chair of computer sciences at the Courant Institute of Mathematical Sciences at New York University. She has co-authored two books on optimization and has published dozens of scientific papers, articles, and technical reports.

Wright was elected to the National Academy of Engineering in 1997 and the National Academy of Sciences in 2005. She has been a distinguished lecturer and committee member for numerous academic and government scientific organizations, including the National Science Foundation, National Research Council, and U.S. Department of Energy. She has received an honorary doctorate from the University of Waterloo (2003) and was the Emmy Noether Lecturer of the Association for Women in Mathematics (2000). Her other awards and honors include a Special Award for Distinguished Service to the Profession from the Society for Industrial and Applied Mathematics (SIAM) (2000) and an Award for Distinguished Public Service from the American Mathematical Society (2001). She served as president of SIAM in 1995–1996. She is a fellow of the Institute for Operations Research and the Management Sciences and of the American Academy of Arts and Sciences, and a member of the Mathematical Programming Society.

Further Resources

Agnes Scott College. "Margaret Wright." Biographies of Women Mathematicians. http://www.agnesscott.edu/lriddle/women/wright.htm.

New York University. Faculty website. http://cs.nyu.edu/mhw/.

Wrinch, Dorothy Maud

1894–1976
Biochemist, Mathematician

Education: mathematics and philosophy, Girton College, Cambridge; M.Sc., University of London, 1920, D.Sc., 1922; M.A., Oxford University, 1924, D.Sc., 1929

Biochemist and mathematician, Dorothy Maud Wrinch, right, shows physicist Katharine Blodgett of General Electric her protein molecule model, 1938. (AP/Wide World Photos)

Professional Experience: lecturer, pure mathematics, University College, University of London, 1918–1920; lecturer, mathematics and director, studies for women, member, faculty of physical sciences, Oxford University, 1923–1939; research fellow, Somerville College, Oxford, 1939–1941; lecturer, chemistry, Johns Hopkins University, 1939–1941; visiting professor, natural sciences, Amherst College and Mount Holyoke College, 1941–1942; lecturer, physics, Smith College, 1941–1954, visiting professor, 1954–1971

Dorothy Wrinch was a biochemist and mathematician whose interests spanned mathematical physics, molecular biology, chemistry, genetics, the philosophy of science, and sociology. In 1929, she was the first woman to receive a doctorate in science from Oxford University. In the mid-1930s, she developed an important contribution to science—the first theory of protein structure, or the "cyclol theory" of amino acids holding the keys to the genetic code. She received a Rockefeller Foundation grant for this groundbreaking work applying mathematics to molecular biology, but her funding and her reputation were damaged when prominent scientists, notably Linus Pauling, publicly rejected her theory. Although her theory was proven incorrect (as was Pauling's early theory), it later applied to other aspects of chemical bonds in alkaloids and thus contributed to scientific advances. Her argument with Pauling began in the late 1930s, but she published her research in two books, *Chemical Aspects of the Structure of Small Peptides: An Introduction* (1960) and *Chemical Aspects of Polypeptide Chain Structure: An Introduction* (1960). Wrinch held a wide range of scientific interests and engaged in collaborative work with other scientists on topics related to theoretical physics and philosophy, and published nearly 200 articles and papers.

In addition to the degree from Oxford, Wrinch also received a doctorate from the University of London and spent many years at a student at Cambridge and at the Universities of Vienna and Paris. She alternated between teaching at London and Oxford before coming to the United States with her daughter after her marriage ended in 1938. Wrinch accepted a position as a lecturer in chemistry at Johns Hopkins and went on to hold lectureships and fellowships at Amherst, Mount Holyoke, and at Smith College, where she spent 30 years as a teacher but never secured a permanent faculty appointment.

Wrinch was elected a fellow of the American Physical Society and the London Royal Society, and was a member of the American Chemical Society and the American Crystallographic Association. In 1930, she published a book, *The Retreat from Parenthood*, under a pseudonym (Jean Ayling), in which she addressed the choice many educated women had to make between careers and family life, and advocated for greater childcare services.

Further Resources

Abir-Am, Pnina G. and Dorinda Outram. 1987. *Uneasy Careers and Intimate Lives: Women in Science, 1789–1979*. New Brunswick, NJ: Rutgers University Press.

Agnes Scott College. "Dorothy Maud Wrinch." Biographies of Women Mathematicians. http://www.agnesscott.edu/lriddle/women/wrinch.htm.

Wu, Chien-Shiung

1912–1997
Nuclear Physicist

Education: B.S., physics, National Central University, China, 1934; Ph.D., physics, University of California, Berkeley, 1940

Professional Experience: lecturer, University of California, Berkeley, 1940–1942; assistant professor, Smith College, 1942–1943; instructor, Princeton University,

Physicist Chien-Shiung Wu with a particle accelerator at Columbia University, 1963. (Robert W. Kelley/Time Life Pictures/Getty Images)

1943–1944; senior scientist, Columbia University, 1944–1947, associate, 1947–1952, associate professor to professor, physics, 1952–1972, professor, physics, 1972–1981

Concurrent Positions: member, advisory committee to director, National Institutes of Health (NIH), 1975–1982

Chien-Shiung Wu was one of the top women in elementary particle physics in the world in the mid-twentieth century, and her work contributed to the research that earned two of her Columbia University colleagues, Drs. Tsung Dao Lee and Ning Yang, the Nobel Prize in Physics in 1957. She researched the separation of uranium isotopes and experimentally established nonconservation of parity in beta decay and conservation of vector current in beta decay. At the time she received her doctorate from Berkeley, in 1940, not one of the nation's top research universities had a female physics professor. She was hired as an instructor at Princeton due to the shortage of male scientists during World War II. In 1944, she was appointed a senior scientist at Columbia, where she helped develop sensitive radiation detectors for the atomic bomb project. After the war ended and the Manhattan Project was completed, she was asked to remain at Columbia, where she spent the remainder of her career as a physics professor. Wu's research focused on radiation detection equipment and, as she moved through the faculty ranks, she conducted experiments to test the theories of Lee and Yang. The two scientists who shared the Nobel Prize acknowledged Wu's role in the success of proving their theory; Lee later said of Wu that she "was one of the giants of physics."

Born in Shanghai, Wu was the daughter of an elementary school principal who founded a women's vocational school and impressed upon her the importance of education. She studied English and science in high school and graduated with a physics degree from the National Central University in Nanking. She did graduate-level study and worked as a research assistant at Zhejiang University and at the Institute of Physics of the Academia Sinica but, wishing to take her education further, Wu moved to the United States to study at the University of California, Berkeley, where she worked with professor Ernest Lawrence, who won the Nobel Prize in Physics in 1939, while Wu was a student there. She worked as Lawrence's research assistant and, after receiving her Ph.D. in 1940, continued as a lecturer at Berkeley, then taught at Smith College and Princeton before moving to Columbia in New York.

Wu was elected a member of the National Academy of Sciences in 1958. She received honorary degrees from several universities, including Princeton, where, also in 1958, she was the first woman to receive an honorary doctorate in science. She was the recipient of numerous awards, both in China and in the United States, including an Achievement Award for the American Association of University Women (1960), the Comstock Award of the National Academy of Sciences (1964), an Achievement Award from the Chi-Tsin Culture Foundation of Taiwan (1965), the Scientist of the Year Award from *Industrial Research Magazine* (1974), the Bonner Prize of the American Physical Society (1975), the National Medal of Science (1975), and the Wolf Prize in Physics in Israel (1978). She was the first living scientist with an asteroid named after her (1990). She was a fellow of the American Association for the Advancement of Science, an honorary fellow of the Royal Society of Edinburgh, and a member of the American Physical Society (president, 1975), the American Academy of Arts and Sciences, and Academia Sinica, the Academy of Sciences in China.

Further Resources

McGrayne, Sharon Bertsch. 1993. *Nobel Prize Women in Science: Their Lives, Struggles, and Momentous Discoveries*. Secaucus, NJ: Carol Pub. Group.

Byers, Nina and Gary A. Williams. 2006. *Out of the Shadows: Contributions of Twentieth-Century Women to Physics*. New York: Cambridge University Press, 2006.

Wu, Ying-Chu (Lin) Susan

b. 1932
Aerospace Engineer

Education: B.S., mechanical engineering, National Taiwan University, 1955; M.S., aerospace engineering, Ohio State University, 1959; Ph.D., aeronautics, California Institute of Technology, 1963

Professional Experience: engineer, Taiwan Highway Bureau, 1955–1956; senior engineer, Electro-Optical Systems, 1963–1965; assistant professor, University of Tennessee, 1965–1967, associate professor, 1967–1973, professor, aerospace engineering, University of Tennessee Space Institute (UTSI), 1973–1988; president and chief executive officer, Engineering Research Consulting, Inc., 1988–

Concurrent Positions: laboratory manager, research and development laboratory, University of Tennessee, 1977–1981, administrator, Energy Conversion Research and Development Program, University of Tennessee, 1981–1988

Susan Wu is an aerospace engineer renowned for her research on the potential for cleaner and more efficient methods of coal-fired power generation in the United States through the use of magnetohydrodynamics (MHD), which produce electric power without the use of rotating machinery by passing a plasma through a magnetic field. This method of power generation is cleaner and more efficient than the traditional power plant, and MHD generation is also used as a power source for aircraft. Wu's field of research is an important one primarily because of increasing mandates from the federal government to reduce emissions from coal-fired power plants and to reduce the use of fossil fuels, such as coal, to preserve them for future generations. After a productive career as an engineer and then aerospace engineering professor, Wu founded her own company in 1988, Engineering Research Consulting, Inc. Wu still serves as the company chairman, and her oldest son, Dr. Ernie Wu, is the president and chief executive officer.

After she received her undergraduate degree in 1955, Wu found that engineering jobs for women were scarce in China. She moved to the United States, where she received graduate degrees in aerospace engineering and aeronautics. She became a professor at the University of Tennessee Space Institute (UTSI), but left academia after 23 years to found her own aerospace and energy research consulting firm, ERC, Inc., now headquartered in Huntsville, Alabama. ERC consults for such agencies as the National Aeronautics and Space Administration, the Department of Energy, and the Argonne National Laboratory, and for corporations such as Boeing and McDonnell Douglas.

Wu has been a member of the advisory board of the National Air and Space Museum of the Smithsonian Institution since 1993 and has received several awards, including the University of Tennessee's Chancellor's Research Scholar Award (1978), Outstanding Educators of America Award (1973 and 1975), Society of Women Engineers Achievement Award (1985), and Plasmadynamics and Lasers Award of the American Institute of Aeronautics and Astronautics (1994). She is a three-time recipient of the Amelia Earhart Fellowship (1958, 1959, 1962) from the women's advocacy organization, Zonta International, for women in aerospace science and engineering. She is a fellow of the American Institute of Aeronautics and Astronautics and the American Society of Mechanical Engineers, and a member of the Society of Women Engineers.

Further Resources

ERC Incorporated. http://erc-incorporated.com/comphistory.aspx.

Yalow, Rosalyn Sussman

b. 1921
Medical Physicist

Education: A.B., Hunter College, 1941; M.S., University of Illinois, 1942, Ph.D., physics, 1945

Professional Experience: assistant, physics, University of Illinois, 1941–1943, instructor, 1944; assistant engineer, Federal Telecommunications Laboratory, 1945–1946; lecturer and assistant professor, physics, Hunter College, 1946–1950; physicist, assistant chief, chief, radioimmunoassay service, Veterans Administration (VA) Hospital, Bronx, New York, 1950–1970, nuclear medical service, 1970–1980; chair, clinical science, Montefiore Hospital and Medical Center, 1980–1985

Concurrent Positions: consultant, radioisotope unit, Veterans Administration Hospital, 1947–1950; research professor, Mount Sinai School of Medicine, 1968–1974, distinguished service professor, 1974–1979; distinguished professor at large, Albert Einstein College of Medicine, Yeshiva University, 1980–1985; Solomon A. Berson distinguished professor at large, Mount Sinai School of Medicine, 1986–

Rosalyn Yalow was a co-recipient of the Nobel Prize in Physiology or Medicine in 1977, the second woman to win in that category (**Gerty Cori** had been the first, in 1947). She and her collaborators were pioneers in the new science of neuroendocrinology, a discipline that enables doctors to diagnose conditions caused by hormonal changes. Yalow's work combined immunology, isotope research, mathematics, and physics, and established the field of modern biomedical physics. She set up one of the first radioisotope labs in the United States when she was hired in 1947 at the VA Hospital in the Bronx. The initial plan was that radioisotopes would be a cheap alternative to radium for cancer treatment. With her engineering experience, she was able to design her own equipment, as no commercial instrumentation existed at the time.

As a graduate student in physics at the University of Illinois, Yalow was assigned to teach only pre-med students, as no female faculty taught male engineering and science students. This changed, however, as more men were called to war and women were called to fill teaching positions. After completing her Ph.D., she became the first woman engineer at the Federal Telecommunications

Physicist Rosalyn Yalow was co-recipient of the 1977 Nobel Prize in Physiology or Medicine for her development of the radioimmunoassay (RIA) technique. (National Library of Medicine)

Laboratory for a year before returning to her alma mater, Hunter College, to teach. In 1947, she began her long tenure with the VA Hospital and a fruitful collaboration with physician Solomon Berson. Together, they invented radioimmunoassay (RIA), or the method of using radioactively tagged substances to measure antibodies produced by the immune system. By accident, they discovered that the insulin obtained from animal sources had minor but important differences from human insulin, namely that human insulin contains antibodies created by the immune system. The result of their research was that manufactured insulin could be genetically engineered to be precisely the same as human insulin. She and Berson did not patent their discovery, and commercial laboratories have realized enormous profits from performing RIA.

Yalow and Berson published numerous papers together, always alternating first authorship, and earned numerous awards for their work. Although Berson accepted a position at Mount Sinai School of Medicine in 1968, they continued their work together until he died in 1972. It was already rumored at that time that the two were candidates for a shared Nobel Prize, but Berson's premature death in 1972 removed his name from consideration, as the prize is not awarded post-humously. Yalow continued her research and was finally recognized with the Nobel Prize in Physiology or Medicine in 1977. She went on to teach at Mount Sinai, at Montefiore Hospital and Medical Center, and at Albert Einstein College of Medicine. She helped establish and direct the Solomon A. Berson Research Laboratory at the Bronx VA Hospital and held the Berson Distinguished Professorship at Mount Sinai School of Medicine. She retired from full-time research in the 1980s, but retained positions as affiliated faculty at several schools and continued to use her office at the VA Hospital until 2002.

Yalow was elected to the National Academy of Sciences in 1975. She was the first woman and first nuclear physicist to win the Albert Lasker Medical Research Award (1976), and is also the recipient of a National Medal of Science (1988). She was elected president of the Endocrine Society (1978–1979) and fellow of the New York Academy of Sciences. She has been a member of the American Academy of Arts and Sciences, the Radiation Research Society, the American College of Radiology, the Biophysical Society, the American Diabetes Association, and the American Physiological Society.

Further Resources

Byers, Nina and Gary A. Williams. 2006. *Out of the Shadows: Contributions of Twentieth-Century Women to Physics*. New York: Cambridge University Press.

McGrayne, Sharon Bertsch. 1998. *Nobel Prize Women in Science: Their Lives, Struggles, and Momentous Discoveries*. Secaucus, NJ: Birch Lane Press.

Straus, Eugene. 1998. *Rosalyn Yalow, Nobel Laureate: Her Life and Work in Medicine*. New York: Basic Books.

Young, Anne Sewell

1871–1961
Astronomer

Education: B.L., Carleton College, 1892, M.S., 1897; University of Chicago, 1898, 1902; Ph.D., astronomy, Columbia University, 1906

Professional Experience: instructor, mathematics, Whitman College, 1892–1893, professor, 1893–1895; high school principal, 1897–1899; instructor to professor, astronomy, and director, John Payne Williston Observatory, Mount Holyoke College, 1899–1936

Anne Young was an astronomer recognized for her research on observations of variable stars, measurement of astronomical photographs, and reduction of occultation observations. She conducted an active program at Mount Holyoke on sunspot observations, asteroid positions, comet orbits, and variable stars. Young had an early interest in astronomy, and her uncle, Charles Young, was a renowned professor of astronomy at Princeton University. After receiving her undergraduate degree from Carleton College in 1892, she was an instructor and then professor of mathematics at Whitman College for four years. She returned to Carleton for her master's degree and was a high school principal for a year. She took additional studies at the University of Chicago before receiving her doctorate in astronomy from Columbia in 1906. Her doctoral research was based on the photographic measurements of stars within the constellation of Perseus.

Young was appointed an instructor at Mount Holyoke in 1899 and rose through the ranks to professor, retiring in 1936. Throughout her tenure at Mount Holyoke, she was also director of the Williston Observatory. She published numerous papers in astronomical journals, and in 1900, she started a program of daily sunspot observations at Mount Holyoke that led to a worldwide cooperative research project. One of her contributions to the profession was that she promoted popular interest in astronomy by writing a monthly column on astronomy for a local paper, the *Springfield Republican*, and by providing a series of open nights at the observatory for the public. She was beloved as a teacher and, in 1925, took an entire class of students from Mount Holyoke in Massachusetts to Connecticut by train to see the total eclipse of the sun that year.

Young was elected a fellow of the American Astronomical Society, the Royal Astronomical Society, and the American Association for the Advancement of Science, and was elected president of the American Association of Variable Star Observers (1923).

Young, Roger Arliner

1899–1964
Zoologist

Education: B.S., Howard University, 1923; M.S., University of Chicago, 1926; Ph.D., zoology, University of Pennsylvania, 1940

Professional Experience: instructor and interim department head, zoology, Howard University, Washington, D.C., 1923–1936; researcher, Marine Biological Laboratory, Woods Hole, Massachusetts; assistant professor, biology, North Carolina College for Negroes; instructor, biology, Shaw University, Raleigh, North Carolina; instructor, Jackson State College, Mississippi; instructor, Paul Quinn College, Texas; lecturer, biology, Southern University, Louisiana

Roger Arliner Young was a zoologist and marine biologist who was the first African American woman to receive a doctorate in zoology. Her research focused on the effects of radiation and ultraviolet light on sea urchin eggs, and hydration and salt concentration in other organisms. She published an article in *Science*, "On the Excretory Apparatus in Paramecium," before even receiving her master's degree. She published several other scientific papers in the 1930s.

Young enrolled at Howard University in Washington, D.C., in 1916. Her family was poor, and she was responsible for the care of her invalid mother, causing her grades in school to suffer. For these reasons, it took seven years for Young to earn her bachelor's degree from Howard. She originally intended to study music, but took her first science course in 1921 with biology and zoology professor Ernest Everett Just, who became an important mentor for Young and encouraged her to pursue graduate work in the sciences. After receiving her degree from Howard in 1923, she went on to attend the University of Chicago part-time. She received her master's degree in 1926 and was invited by Just to work with him at the Woods Hole Marine Biological Laboratory in Massachusetts during the summers. Young began her work on marine embryology, fertilization, and the processes of hydration and dehydration. Just asked her to stand in for him as head of the zoology department at Howard on several occasions when he made trips to Europe to seek research funding.

Young returned to the University of Chicago in 1929 to pursue a doctorate with another professor she had met at Woods Hole. She did not pass her qualifying exams, however, and returned to teach at Howard for several more years. In 1936, she was fired by Everett Ernest Just for reasons that seemed to be both political (pressures from the dean) and personal (a rift with Just over rumors about the nature of their relationship). She left Howard and moved to the University of Pennsylvania to resume work toward a doctorate, which she finally received in 1940 with a dissertation on "The Indirect Effects of Roentgen Rays on Certain Marine Eggs." After 1940, she taught at colleges in North Carolina, Mississippi, Louisiana, and Texas. She continued to care for her mother until her mother's death in 1953 and lived on the brink of poverty, unable to retain a teaching position very long. Her research using ultraviolet light had damaged her eyesight and, at one point, she was admitted to the Mississippi State Mental Asylum due to poor mental health. She died in New Orleans in 1964.

Further Resources

Manning, Kenneth R. 1983. *Black Apollo of Science: The Life of Ernest Everett Just*. New York: Oxford University Press.

Warren, Wini. 1999. *Black Women Scientists in the United States*. Bloomington: Indiana University Press.

Z

Zoback, Mary Lou

b. 1952
Geophysicist

Education: B.S. geophysics, Stanford University, 1974, M.S. 1975, Ph.D., 1978

Professional Experience: National Research Council postdoctoral fellow, Heat Flow Studies, U.S. Geological Survey (USGS), 1978–1979, research scientist, Earthquake Hazards Team, USGS, 1979–1999, chief scientist, Western Earthquake Hazards Team, 1999–2003, senior research scientist and program coordinator, Northern California Earthquake Hazards Program, 2003–2006; vice president, Earthquake Risk Applications, Risk Management Solutions (RMS), 2006–

Concurrent Positions: visiting scholar, Geophysical Institute, Karlsruhe, Germany, 1990–1991

Mary Lou Zoback is an internationally recognized geophysicist who specializes in plate tectonics and earthquakes. She has researched and mapped plate stresses, in particular focused on the San Andreas fault system which runs through California. She led the World Stress Map Project (1986–1992) of the International Lithosphere Program, a coalition of scientists from 30 countries who compiled geologic data on worldwide active tectonics and stress for environmental scientists and government risk assessments. Zoback earned three degrees in geophysics from Stanford University and has spent nearly 25 years at the USGS Office of Earthquake Studies. She left the USGS in 2006 to become vice president of Earthquake Risk Applications at RMS in Newark, California. At RMS, she provides scientific data for purposes of assessing earthquake risk, risk-reduction plans, insurance needs, and disaster management and response. In 2006, she helped found the 1906 Earthquake Centennial Alliance to commemorate the San Francisco Earthquake of 1906 and raise public awareness about earthquake safety. Zoback is also committed to science education and has been involved with Expanding Your Horizons, a national program to encourage young girls in science, math, and technology careers.

Zoback has served on numerous scientific and government committees, including for the National Science Foundation, National Research Council, National Aeronautics and Space Administration (NASA), and several universities. She was elected to the National Academy of Sciences in 1995. She is a fellow of the Geological Society of America (GSA) (president, 1999–2000) and American Geophysical Union, and a member of the American Geological Institute, Seismological Society of America, and Earthquake Engineering Research Institute. She has received the Macelwane Award of the AGU (1987), the USGS Gilbert Fellowship Award for a visiting scholarship in Germany (1990–1991), Meritorious Service Award of the Department of Interior (2002), Bownocker Medal of Ohio State University (2003), Innovation and Exemplary Practice in Earthquake Risk Reduction Award from the Earthquake Engineering Research Institute (2006), and Arthur L. Day Medal (2007) and Public Service Award (2007) of the GSA.

Further Resources

National Academy of Sciences. 2003. "InterViews: Mary Lou Zoback, Geophysics." http://www.nasonline.org/site/PageServer?pagename=INTERVIEWS_Mary_Lou _Zoback.

Women Nobel Prize Winners
in the Sciences

Physics

1903   Marie Curie
1963   Maria Goeppert-Mayer

Chemistry

1911   Marie Curie
1935   Irène Joliot-Curie
1964   Dorothy Crowfoot Hodgkin
2009   Ada E. Yonath

Physiology or Medicine

1947   Gerty Cori
1977   Rosalyn Yalow
1983   Barbara McClintock
1986   Rita Levi-Montalcini
1988   Gertrude B. Elion
1995   Christiane Nüsslein-Volhard
2004   Linda B. Buck
2008   Françoise Barré-Sinoussi
2009   Elizabeth H. Blackburn
2009   Carol W. Greider

Economic Sciences

2009   Elinor Ostrom

Scientists by Discipline

Aerospace & Astronautics

Berger, Marsha J.

Brill, Yvonne (Claeys)

Cleave, Mary L.

Cobb, Geraldyne M.

Collins, Eileen

Cowings, Patricia Suzanne

Darden, Christine V. Mann

Dunbar, Bonnie J.

Fisher, Anna L.

Flugge Lotz, Irmgard

Hamilton, Margaret

Jemison, Mae Carol

Johnson, Barbara Crawford

Johnston, Mary Helen

Kurtzig, Sandra L. (Brody)

Leveson, Nancy G.

Long, Irene (Duhart)

Lucid, Shannon (Wells)

Ocampo, Adriana C.

Ochoa, Ellen

Resnik, Judith A.

Ride, Sally Kristen

Seddon, Margaret Rhea

Simon, Dorothy Martin

Stoll, Alice Mary

Sullivan, Kathryn D.

Thornton, Kathryn (Cordell)

Townsend, Marjorie Rhodes

Whitson, Peggy A.

Widnall, Sheila (Evans)

Wu, Ying-Chu (Lin) Susan

Animal Sciences

Altmann, Jeanne

Altmann, Margaret

Fossey, Dian

Grandin, Temple

Moss, Cynthia Jane

Poole, Joyce

Saif, Linda

Anthropology & Archaeology

Aberle, Sophie Bledsoe

Archambault, JoAllyn

Bateson, Mary Catherine

Beall, Cynthia

Benedict, Ruth Fulton

Bricker, Victoria (Reifler)

Buikstra, Jane Ellen

Cole, Johnnetta (Betsch)

Colson, Elizabeth Florence

De Laguna, Frederica Annis

Ellis, Florence May Hawley

Haas, Mary Rosamond

Harrison, Faye Venetia

Hawkes, Kristen

Helm, June

Hrdy, Sarah C. (Blaffer)

Leacock, Eleanor (Burke)

Linares, Olga Frances

Lubic, Ruth (Watson)

Lurie, Nancy (Oestreich)

Marcus, Joyce

Martin, Emily

Mead, Margaret

Medicine, Beatrice A.

Parsons, Elsie Worthington Clews

Reichard, Gladys Amanda

Semple, Ellen Churchill

Shipman, Pat

Slye, Maud Caroline

Sudarkasa, Niara

Thompson, Laura Maud

Watson, Patty Jo (Andersen)

Astronomy & Astrophysics

Bahcall, Neta

Burbidge, (Eleanor) Margaret

Cannon, Annie Jump

Cordova, France Anne-Dominic

Elmegreen, Debra Meloy

Faber, Sandra (Moore)

Furness, Caroline Ellen

Geller, Margaret Joan

Gill, Jocelyn Ruth

Hammel, Heidi

Hoffleit, (Ellen) Dorrit

Intriligator, Devrie (Shapiro)

Leavitt, Henrietta Swan

Lippincott, Sarah Lee

Makemson, Maud Worcester

Maury, Antonia Caetana de Paiva Pereira

McFadden, Lucy-Ann Adams

Meinel, Marjorie Pettit

Payne Gaposchkin, Celelia Helena

Prince, Helen Walter Dodson

Prinz, Dianne Kasnic

Roemer, Elizabeth

Roman, Nancy Grace

Rubin, Vera (Cooper)

Shoemaker, Carolyn (Spellmann)

Sitterly, Charlotte Emma Moore

Small, Meredith F.

Smith, Elske (van Panhuys)

Tinsley, Beatrice Muriel (Hill)

Young, Anne Sewell

Biochemistry

Banfield, Jillian F.

Benesch, Ruth Erica (Leroi)

Blackburn, Elizabeth

Briscoe, Anne M.

Brown, Barbara B.

Brown, Rachel Fuller

Chilton, Mary-Dell (Matchett)

Cohn, Mildred

Cori, Gerty Theresa Radnitz

Daly, Marie Maynard

Delmer, Deborah

Edwards, Cecile Hoover

Elion, Gertrude Belle

Emerson, Gladys Anderson

Fink, Kathryn Ferguson

Fuchs, Elaine V.

Greider, Carol W.

Gross, Elizabeth Louise

Guttman, Helene Augusta (Nathan)

Hamilton, Alice

Haschemeyer, Audrey E. V.

Hay, Elizabeth Dexter

Hollinshead, Ariel Cahill

Horning, Marjorie G.

Hubbard, Ruth (Hoffman)

Jones, Mary Ellen

Kaufman, Joyce (Jacobson)

Klinman, Judith (Pollock)

Macy-Hoobler, Icie Gertrude

Maling, Harriet Mylander

Miller, Elizabeth Cavert

Morgan, Agnes Fay

Neufeld, Elizabeth (Fondal)

Osborn, Mary Jane (Merten)

Petermann, Mary Locke

Ratner, Sarah

Richardson, Jane S.

Rolf, Ida P.

Seibert, Florence Barbara

Shockley, Dolores Cooper

Shotwell, Odette Louise

Simmonds, Sofia

Singer, Maxine (Frank)

Stadtman, Thressa Campbell

Stanley, Louise

Stearns, Genevieve

Steitz, Joan (Argetsinger)

Stubbe, JoAnne

Tilghman, Shirley M.

Tolbert, Margaret Ellen (Mayo)

Vaughan, Martha

Vennesland, Birgit

Weisburger, Elizabeth Amy (Kreiser)

Whitson, Peggy A.

Wrinch, Dorothy Maud

Biomedical Sciences

Avery, Mary Ellen

Baetjer, Anna Medora

Bartoshuk, Linda

Benesch, Ruth Erica (Leroi)

Blackburn, Elizabeth

Bliss, Eleanor Albert

Briscoe, Anne M.

Broome, Claire Veronica

Brown, Rachel Fuller

Brugge, Joan S.

Buck, Linda B.

Cobb, Jewel Plummer

Cohn, Mildred

Colwell, Rita (Rossi)

Cori, Gerty Theresa Radnitz

Cowings, Patricia Suzanne

Daly, Marie Maynard

Dick, Gladys Rowena Henry

Dunbar, Bonnie J.

Elion, Gertrude Belle

Estrin, Thelma Austern

Evans, Alice Catherine

Farquhar, Marilyn (Gist)

Ferguson, Angela Dorothea

Fink, Kathryn Ferguson

Free, Helen (Murray)

Friend, Charlotte

Fuchs, Elaine V.

Gayle, Helene Doris

Giblett, Eloise Rosalie

Glusker, Jenny (Pickworth)

Gordon, Ruth Evelyn

Greider, Carol W.

Griffin, Diane Edmund

Gross, Carol A. (Polinsky)

Guthrie, Mary Jane

Guttman, Helene Augusta (Nathan)

Harris, Mary (Styles)

Hay, Elizabeth Dexter

Hazen, Elizabeth Lee

Hockfield, Susan

Hollinshead, Ariel Cahill

Horning, Marjorie G.

Horstmann, Dorothy Millicent

Huang, Alice Shih-Hou

Jones, Mary Ellen

Kaufman, Joyce (Jacobson)

Kenyon, Cynthia J.

King, Mary-Claire

Koshland, Marian Elliott

Krim, Mathilde (Galland)

Lancaster, Cleo

Lancefield, Rebecca Craighill

Leeman, Susan (Epstein)

Lesh-Laurie, Georgia Elizabeth

L'Esperance, Elise Depew Strang

Levi-Montalcini, Rita

Lewis, Margaret Adaline Reed

Lucid, Shannon (Wells)

Maling, Harriet Mylander

Marrack, Philippa Charlotte

McSherry, Diana Hartridge

Mendenhall, Dorothy Reed

Micheli-Tzanakou, Evangelia

Mielczarek, Eugenie Vorburger

Miller, Elizabeth Cavert

Mintz, Beatrice

Murray, Sandra Ann

Neufeld, Elizabeth (Fondal)

New, Maria (Iandolo)

Osborn, Mary Jane (Merten)

Pearce, Louise

Pert, Candace Dorinda (Bebe)

Petermann, Mary Locke

Pitelka, Dorothy Riggs

Pittman, Margaret

Pool, Judith Graham

Profet, Margie

Quimby, Edith Hinkley

Ramaley, Judith (Aitken)

Ramey, Estelle Rosemary White

Ranney, Helen Margaret

Ratner, Sarah

Rowley, Janet Davison

Sabin, Florence Rena

Sager, Ruth

Saif, Linda

Scharrer, Berta Vogel

Sedlak, Bonnie Joy

Seibert, Florence Barbara

Shapiro, Lucille (Cohen)

Shaw, Jane E.

Shockley, Dolores Cooper

Simmonds, Sofia

Sinkford, Jeanne Frances (Craig)

Stoll, Alice Mary

Stroud-Lee, F. Agnes Naranjo

Taussig, Helen Brooke

Vaughan, Martha

Villa-Komaroff, Lydia

Vitetta, Ellen Shapiro

Waelsch, Salome Gluecksohn

Weisburger, Elizabeth Amy (Kreiser)

Wexler, Nancy Sabin

Wilhelmi, Jane Anne Russell

Williams, Anna Wessels

Witkin, Evelyn Maisel

Woods, Geraldine (Pittman)

Yalow, Rosalyn Sussman

Botany (Plant Sciences)

Bennett, Joan Wennstrom

Berenbaum, May Roberta

Braun, (Emma) Lucy

Britton, Elizabeth Knight

Charles, Vera Katherine

Chase, (Mary) Agnes Meara

Chilton, Mary-Dell (Matchett)

Chory, Joanne

Davis, Margaret Bryan

Delmer, Deborah

Earle, Sylvia Alice

Eastwood, Alice

Esau, Katherine

Farr, Wanda Kirkbride

Fedoroff, Nina Vsevolod

Ferguson, Margaret Clay

Gantt, Elisabeth

Gerry, Eloise B.

Goldring, Winifred

Gross, Elizabeth Louise

Hart, Helen

Leopold, Estella Bergere

Long, Sharon (Rugel)

Mathias, Mildred Esther

McClintock, Barbara

McCoy, Elizabeth Florence

Moore, Emmeline

Patrick, Ruth

Patterson, Flora Wambaugh

Rissler, Jane Francina

Roberts, Edith Adelaide

Shields, Lora Mangum

Sommer, Anna Louise

Sweeney, (Eleanor) Beatrice Marcy

Tilden, Josephine Elizabeth

Vennesland, Birgit

Walbot, Virginia Elizabeth

Westcott, Cynthia

Chemistry

Anderson, Gloria (Long)

Benerito, Ruth Rogan

Berkowitz, Joan B.

Brill, Yvonne (Claeys)

Carr, Emma Perry

Caserio, Marjorie Constance (Beckett)

Cox, Geraldine Anne (Vang)

Dicciani, Nance Katherine

Drake, Elisabeth (Mertz)

Fitzroy, Nancy (Deloye)

Flanigen, Edith Marie

Fox, Marye Anne (Payne)

Free, Helen (Murray)

Gast, Alice P.

Glusker, Jenny (Pickworth)

Good, Mary (Lowe)

Grasselli (Brown), Jeanette

Greer, Sandra Charlene

Hahn, Dorothy Anna

Harrison, Anna Jane

Hoffman, Darleane (Christian)

Jeanes, Allene Rosalind

Karle, Isabella Helen Lugoski

Kaufman, Joyce (Jacobson)

Kwolek, Stephanie Louise

Libby, Leona Woods Marshall

MacLeod, Grace

Marlatt, Abby Lillian

Michel, Helen (Vaughn)

Mitchell, Helen Swift

Nightingale, Dorothy Virginia

Patrick, Jennie R.

Payne, Nellie Maria de Cottrell

Pennington, Mary Engle

Prichard, Diana (Garcia)

Reichmanis, Elsa

Rose, Mary Davies Swartz

Savitz, Maxine (Lazarus)

Schwan, Judith A.

Sherman, Patsy O'Connell

Shotwell, Odette Louise

Shreeve, Jean'ne Marie

Simon, Dorothy Martin

Solomon, Susan

Stiebeling, Hazel Katherine

Stubbe, JoAnne

Taylor, Kathleen Christine

Tesoro, Giuliana (Cavaglieri)

Thomas, Martha Jane (Bergin)

Computer Science & Information Technology

Bell, Gwen (Dru'yor)

Berezin, Evelyn

Berger, Marsha J.

Butler, Margaret K.

Conway, Lynn Ann

Davis, Ruth Margaret

Estrin, Thelma Austern

Goldberg, Adele

Goldwasser, Shafrira

Graham, Susan Lois

Granville, Evelyn (Boyd)

Greibach, Sheila Adele

Hamilton, Margaret

Hopper, Grace Murray

Hutchins, Sandra Elaine

Irwin, Mary Jane

Jones, Anita Katherine

Kempf, Martine

Kurtzig, Sandra L. (Brody)

Leveson, Nancy G.

Liskov, Barbara Huberman

McSherry, Diana Hartridge

Mitchell, Joan L.

Pour-El, Marian Boykan

Reichmanis, Elsa

Sammet, Jean Elaine

Shaw, Mary M.

Turkle, Sherry

Williams, Roberta

Wright, Margaret H.

Crystallography

Donnay, Gabrielle (Hamburger)

Glusker, Jenny (Pickworth)

Karle, Isabella Helen Lugoski

Richardson, Jane S.

Wood, Elizabeth Armstrong

Economics

Adelman, Irma Glicman

Hewlett, Sylvia Ann

Kanter, Rosabeth (Moss)

Kreps, Juanita (Morris)

Krueger, Anne (Osborn)

Ostrom, Elinor

Paté-Cornell, (Marie) Elisabeth
Lucienne

Rivlin, Alice (Mitchell)

Stern, Frances

Stokey, Nancy

Tyson, Laura (D'Andrea)

Wallace, Phyllis Ann

Whitman, Marina (von Neumann)

Engineering

Abriola, Linda M.

Agogino, Alice M.

Baranescu, Rodica

Benmark, Leslie Ann (Freeman)

Berger, Marsha J.

Brill, Yvonne (Claeys)

Clarke, Edith

Cleave, Mary L.

Colmenares, Margarita H.

Conway, Lynn Ann

Conwell, Esther Marly

Darden, Christine V. Mann

Davis, Ruth Margaret

De Planque, E. Gail

Dicciani, Nance Katherine

Drake, Elisabeth (Mertz)

Dresselhaus, Mildred (Spiewak)

Edwards, Helen Thom

Estrin, Thelma Austern

Fitzroy, Nancy (Deloye)

Flugge Lotz, Irmgard

Garmire, Elsa (Meints)

Gast, Alice P.

Gilbreth, Lillian E. Moller

Goldwasser, Shafrira

Good, Mary (Lowe)

Graham, Susan Lois

Hamilton, Margaret

Hicks, Beatrice Alice

Hutchins, Sandra Elaine

Hwang, Jennie S.

Irwin, Mary Jane

Jackson, Shirley Ann

Johnson, Barbara Crawford

Johnston, Mary Helen

Jones, Anita Katherine

Kempf, Martine

Kuhlmann-Wilsdorf, Doris

Levelt-Sengers, Johanna Maria Henrica

Liskov, Barbara Huberman

Matthews, Alva T.

Mitchell, Mildred Bessie

Napadensky, Hyla Sarane (Siegel)

Nichols, Roberta J.

Ochoa, Ellen

Paté-Cornell, (Marie) Elisabeth
Lucienne

Patrick, Jennie R.

Peden, Irene (Carswell)

Pressman, Ada Irene

Rand, (Marie) Gertrude

Resnik, Judith A.

Roy, Della Martin

Savitz, Maxine (Lazarus)

Schwan, Judith A.

Shaw, Mary M.

Taylor, Kathleen Christine

Townsend, Marjorie Rhodes

Widnall, Sheila (Evans)

Wu, Ying-Chu (Lin) Susan

Environmental Sciences & Ecology

Abriola, Linda M.

Ancker-Johnson, Betsy

Anderson, Mary P.

Baetjer, Anna Medora

Beattie, Mollie Hanna

Berenbaum, May Roberta

Berkowitz, Joan B.

Bonta, Marcia (Myers)

Braun, (Emma) Lucy

Braun, Annette Frances

Cady, Bertha Louise Chapman

Caldicott, Helen Mary (Broinowski)

Carson, Rachel Louise

Cleave, Mary L.

Colmenares, Margarita H.

Cox, Geraldine Anne (Vang)

Davis, Margaret Bryan

DeFries, Ruth

Drake, Elisabeth (Mertz)

Ehrlich, Anne (Fitzhugh) Howland

Grasselli (Brown), Jeanette

Hamerstrom, Frances (Flint)

Hamilton, Alice

Haschemeyer, Audrey E. V.

LaBastille, Anne

Leopold, Estella Bergere

Libby, Leona Woods Marshall

Lubchenco, Jane

Margulis, Lynn (Alexander)

Matson, Pamela Anne

McCammon, Helen Mary (Choman)

McCoy, Elizabeth Florence

McWhinnie, Mary Alice

Moore, Emmeline

Morgan, Ann Haven

Nichols, Roberta J.

Patch, Edith Marion

Patrick, Ruth

Ray, (Marguerite) Dixy Lee

Rissler, Jane Francina

Roberts, Edith Adelaide

Savitz, Maxine (Lazarus)

Scott, Juanita (Simons)

Shields, Lora Mangum

Shotwell, Odette Louise

Solomon, Susan

Stickel, Lucille Farrier

Taylor, Kathleen Christine

Tilden, Josephine Elizabeth

Weisburger, Elizabeth Amy (Kreiser)

Genetics

Adams, (Amy) Elizabeth

Altmann, Margaret

Bennett, Joan Wennstrom

Blackburn, Elizabeth

Carothers, (Estrella) Eleanor

Chilton, Mary-Dell (Matchett)

Fausto-Sterling, Anne

Fedoroff, Nina Vsevolod

Fuchs, Elaine V.

Fuchs, Elaine V.

Giblett, Eloise Rosalie

Greider, Carol W.

Harris, Mary (Styles)

Hoy, Marjorie Ann (Wolf)

Huang, Alice Shih-Hou

Hubbard, Ruth (Hoffman)

Jones, Mary Ellen

Kidwell, Margaret Gale

Kimble, Judith

King, Helen Dean

King, Mary-Claire

Krim, Mathilde (Galland)

Long, Sharon (Rugel)

Macklin, Madge Thurlow

Margulis, Lynn (Alexander)

McClintock, Barbara

Mintz, Beatrice

Nelkin, Dorothy (Wolfers)

Neufeld, Elizabeth (Fondal)

Pardue, Mary Lou

Rissler, Jane Francina

Rowley, Janet Davison

Russell, Elizabeth Shull

Sager, Ruth

Shapiro, Lucille (Cohen)

Singer, Maxine (Frank)

Steitz, Joan (Argetsinger)

Stroud-Lee, F. Agnes Naranjo

Tilghman, Shirley M.

Villa-Komaroff, Lydia

Waelsch, Salome Gluecksohn

Walbot, Virginia Elizabeth

Wexler, Nancy Sabin

Witkin, Evelyn Maisel

Geography

Baber, Mary Arizona "Zonia"

Bell, Gwen (Dru'yor)

Boyd, Louise Arner

DeFries, Ruth

Fischer, Irene (Kaminka)

Semple, Ellen Churchill

Tharp, Marie

Geology

Anderson, Mary P.

Banfield, Jillian F.

Bascom, Florence

Bunce, Elizabeth Thompson

Davis, Margaret Bryan

Donnay, Gabrielle (Hamburger)

Dreschhoff, Gisela Auguste-Marie

Fowler Billings, Katharine Stevens

Gardner, Julia Anna

Goldring, Winifred

Herzenberg, Caroline Stuart
(Littlejohn)

Kieffer, Susan Werner

Knopf, Eleanora Frances Bliss

Lochman Balk, Christina

Loeblich, Helen Nina Tappan

Marvin, Ursula Bailey

Maury, Carlotta Joaquina

McCammon, Helen Mary (Choman)

McNutt, Marcia Kemper

Michel, Helen (Vaughn)

Navrotsky, Alexandra A. S.

Ocampo, Adriana C.

Ogilvie, Ida Helen

Owens, Joan Murrell

Palmer, Katherine Hilton Van Winkle

Peden, Irene (Carswell)

Romanowicz, Barbara

Roy, Della Martin

Schwarzer, Theresa Flynn

Sullivan, Kathryn D.

Talbot, Mignon

Tharp, Marie

Wood, Elizabeth Armstrong

Zoback, Mary Lou

Mathematics

Bates, Grace Elizabeth

Berger, Marsha J.

Bertell, Rosalie

Butler, Margaret K.

Clarke, Edith

Cox, Gertrude Mary

Daubechies, Ingrid

Davis, Ruth Margaret

Fischer, Irene (Kaminka)

Flugge Lotz, Irmgard

Geiringer (Von Mises), Hilda

Granville, Evelyn (Boyd)

Greibach, Sheila Adele

Hazlett, Olive Clio

Hopper, Grace Murray

Karp, Carol Ruth (Vander Velde)

Kopell, Nancy J.

Luchins, Edith Hirsch

Menken, Jane Ava (Golubitsky)

Morawetz, Cathleen (Synge)

Partee, Barbara (Hall)

Pour-El, Marian Boykan

Rees, Mina Spiegel

Robinson, Julia Bowman

Rosenblatt, Joan (Raup)

Rudin, Mary Ellen (Estill)

Sammet, Jean Elaine

Taussky-Todd, Olga

Uhlenbeck, Karen (Keskulla)

Wheeler, Anna Johnson Pell

Wheeler, Mary F.

Wright, Margaret H.

Wrinch, Dorothy Maud

Medicine

Apgar, Virginia

Avery, Mary Ellen

Broome, Claire Veronica

Caldicott, Helen Mary (Broinowski)

Crosby, Elizabeth Caroline

Delgado, Jane L.

Densen-Gerber, Judianne

Dick, Gladys Rowena Henry

Elders, (Minnie) Joycelyn (Jones)

Ferguson, Angela Dorothea

Fisher, Anna L.

Gayle, Helene Doris

Graham, Frances (Keesler)

Harris, Jean Louise

Harrison-Ross, Phyllis Ann

Healy, Bernadine Patricia

Horstmann, Dorothy Millicent

Hyde, Ida Henrietta

Jackson, Jacquelyne Mary (Johnson)

Jemison, Mae Carol

Kübler-Ross, Elisabeth

Lancaster, Cleo

Leeman, Susan (Epstein)

L'Esperance, Elise Depew Strang

Long, Irene (Duhart)

Love, Susan M.

Lubic, Ruth (Watson)

Macklin, Madge Thurlow

McSherry, Diana Hartridge

Mendenhall, Dorothy Reed

New, Maria (Iandolo)

Nielsen, Jerri Lin

Northrup, Christiane

Novello, Antonia (Coello)

Pearce, Louise

Pert, Candace Dorinda (Bebe)

Pool, Judith Graham

Profet, Margie

Ramaley, Judith (Aitken)

Ramey, Estelle Rosemary White

Rand, (Marie) Gertrude

Ranney, Helen Margaret

Rolf, Ida P.

Rowley, Janet Davison

Sabin, Florence Rena

Seddon, Margaret Rhea

Shalala, Donna Edna

Sinkford, Jeanne Frances (Craig)

Spurlock, Jeanne

Taussig, Helen Brooke

Wattleton, (Alyce) Faye

Wilhelmi, Jane Anne Russell

Williams, Anna Wessels

Meteorology

Ackerman, Bernice

Austin, Pauline Morrow

Kalnay, Eugenia

Ledley, Tamara (Shapiro)

LeMone, Margaret Anne

Simpson, Joanne Malkus (Gerould)

Van Straten, Florence Wilhemina

Neurosciences

Brown, Barbara B.

Crosby, Elizabeth Caroline

Diamond, Marian Cleeves

Estrin, Thelma Austern

Fromkin, Victoria Alexandria (Landish)

Goldman-Rakic, Patricia

Graybiel, Ann Martin

Hockfield, Susan

Jameson, Dorothea A.

Jan, Lily

Kanwisher, Nancy

Leeman, Susan (Epstein)

Levi Montalcini, Rita

Micheli-Tzanakou, Evangelia

Pert, Candace Dorinda (Bebe)

Spelke, Elizabeth

Treisman, Anne

Wexler, Nancy Sabin

Nutrition & Home Economics

Aberle, Sophie Bledsoe

Brody, Jane Ellen

Brooks, Carolyn (Branch)

Calloway, Doris (Howes)

Carey, Susan E.

Cori, Gerty Theresa Radnitz

Edwards, Cecile Hoover

Emerson, Gladys Anderson

Guttman, Helene Augusta (Nathan)

Leverton, Ruth Mandeville

MacLeod, Grace

Macy-Hoobler, Icie Gertrude

Marlatt, Abby Lillian

Mitchell, Helen Swift

Morgan, Agnes Fay

Pennington, Mary Engle

Roberts, Lydia Jane

Rose, Flora

Rose, Mary Davies Swartz

Stadtman, Thressa Campbell

Stanley, Louise

Stearns, Genevieve

Stern, Frances

Stiebeling, Hazel Katherine

Van Rensselaer, Martha

Ocean Sciences

Avery, Susan K.

Bunce, Elizabeth Thompson

Clark, Eugenie

Colwell, Rita (Rossi)

Crane, Kathleen

Earle, Sylvia Alice

Harvey, Ethel Browne

Haschemeyer, Audrey E. V.

Hibbard, Hope

La Monte, Francesca Raimond

Lubchenco, Jane

McCammon, Helen Mary (Choman)

McNutt, Marcia Kemper

McWhinnie, Mary Alice

Owens, Joan Murrell

Ray, (Marguerite) Dixy Lee

Romanowicz, Barbara

Sullivan, Kathryn D.

Sweeney, (Eleanor) Beatrice Marcy

Tharp, Marie

Tilden, Josephine Elizabeth

Young, Roger Arliner

Paleontology

Davis, Margaret Bryan

Edinger, Tilly

Gardner, Julia Anna

Goldring, Winifred

Leopold, Estella Bergere

Lochman Balk, Christina

Loeblich, Helen Nina Tappan

Maury, Carlotta Joaquina

Michel, Helen (Vaughn)

Owens, Joan Murrell

Palmer, Katherine Hilton Van Winkle

Shipman, Pat

Talbot, Mignon

Physics

Ajzenberg-Selove, Fay

Ancker-Johnson, Betsy

Anslow, Gladys Amelia

Blodgett, Katharine Burr

Chasman, Renate (Wiener)

Conwell, Esther Marly

De Planque, E. Gail

Dewitt-Morette, Cecile Andrée Paule

Dreschhoff, Gisela Auguste-Marie

Dresselhaus, Mildred (Spiewak)

Edwards, Helen Thom

Gaillard, Mary Katharine (Ralph)

Garmire, Elsa (Meints)

Goeppert-Mayer, Maria

Goldhaber, Gertrude Scharff

Greene, Laura

Herzenberg, Caroline Stuart (Littlejohn)

Jackson, Shirley Ann

Jan, Lily

Karle, Isabella Helen Lugoski

Keller, Evelyn Fox

Kuhlmann-Wilsdorf, Doris

Laird, Elizabeth Rebecca

Levelt-Sengers, Johanna Maria Henrica

Libby, Leona Woods Marshall

Lubkin, Gloria (Becker)

Maltby, Margaret Eliza

Micheli-Tzanakou, Evangelia

Mielczarek, Eugenie Vorburger

Mitchell, Joan L.

Nickerson, Dorothy

Phillips, Melba Newell

Prichard, Diana (Garcia)

Quimby, Edith Hinkley

Ride, Sally Kristen

Sarachik, Myriam Paula (Morgenstein)

Spaeth, Mary Louise

Stoll, Alice Mary

Thornton, Kathryn (Cordell)

Warga, Mary Elizabeth

Way, Katharine

Weertman, Julia (Randall)

Wu, Chien-Shiung

Yalow, Rosalyn Sussman

Primatology

Altmann, Jeanne

Fossey, Dian

Hrdy, Sarah C. (Blaffer)

Small, Meredith F.

Psychiatry & Psychology

Attneave, Carolyn (Lewis)

Bartoshuk, Linda

Brothers, Joyce Diane (Bauer)

Carey, Susan E.

Chesler, Phyllis

Cowings, Patricia Suzanne

Delgado, Jane L.

Densen-Gerber, Judianne

Downey, June Etta

Fromkin, Victoria Alexandria (Landish)

Gibson, Eleanor Jack

Gilbreth, Lillian E. Moller

Gleitman, Lila R.

Goldman-Rakic, Patricia

Goodenough, Florence Laura

Gordon (Moore), Kate

Graham, Frances (Keesler)

Graham, Norma

Graybiel, Ann Martin

Harrison-Ross, Phyllis Ann

Hatfield, Elaine Catherine

Hollingworth, Leta Anna Stetter

Horner, Matina (Souretis)

Howard (Beckham), Ruth Winifred

Howes, Ethel Puffer

Jameson, Dorothea A.

Johnson (Masters), Virginia (Eshelman)

Kanwisher, Nancy

Kübler-Ross, Elisabeth

Ladd Franklin, Christine

Maccoby, Eleanor (Emmons)

Mitchell, Mildred Bessie

Nice, Margaret Morse

Partee, Barbara (Hall)

Payton, Carolyn (Robertson)

Rand, (Marie) Gertrude

Reinisch, June Machover

Scarr, Sandra (Wood)

Spelke, Elizabeth

Spurlock, Jeanne

Treisman, Anne

Turkle, Sherry

Washburn, Margaret Floy

Weisstein, Naomi

Westheimer, (Karola) Ruth (Siegel)

Wexler, Nancy Sabin

Woolley, Helen Bradford Thompson

Zoology

Adams, (Amy) Elizabeth

Bailey, Florence Augusta Merriam

Berenbaum, May Roberta

Boring, Alice Middleton

Braun, Annette Frances

Brooks, Matilda Moldenhauer

Cady, Bertha Louise Chapman

Carothers, (Estrella) Eleanor

Carson, Rachel Louise

Clark, Eugenie

Guthrie, Mary Jane

Hamerstrom, Frances (Flint)

Harvey, Ethel Browne

Hibbard, Hope

Hoy, Marjorie Ann (Wolf)

Hughes Schrader, Sally (Peris)

Hyman, Libbie Henrietta

La Monte, Francesca Raimond

Lewis, Margaret Adaline Reed

McCracken, (Mary) Isabel

Morgan, Ann Haven

Moss, Cynthia Jane

Nice, Margaret Morse

Patch, Edith Marion

Payne, Nellie Maria de Cottrell

Peckham, Elizabeth Gifford

Peebles, Florence

Pitelka, Dorothy Riggs

Poole, Joyce

Ray, (Marguerite) Dixy Lee

Rudnick, Dorothea

Russell, Elizabeth Shull

Scharrer, Berta Vogel

Stickel, Lucille Farrier

Waelsch, Salome Gluecksohn

West-Eberhard, Mary Jane

Witkin, Evelyn Maisel

Young, Roger Arliner

Other

Angier, Natalie

Bunting (Smith), Mary Ingraham

Baca Zinn, Maxine

Jackson, Jacquelyne Mary (Johnson)

Kanter, Rosabeth (Moss)

Menken, Jane Ava (Golubitsky)

Nelkin, Dorothy (Wolfers)

Reskin, Barbara F.

Riley, Matilda (White)

Chronology

1902   **Florence Sabin** appointed the first female faculty member at Johns Hopkins Medical School

1903   Marie Curie shares Nobel Prize in Physics with her husband, Pierre Curie, and Antoine Henri Becquerel

1911   Marie Curie awarded Nobel Prize in Chemistry

1919   Toxicologist **Alice Hamilton** appointed to faculty of Harvard Medical School, the first female faculty member in any Harvard department

1920   American women gain the right to vote with passage of the Nineteenth Amendment to the U.S. Constitution

1921   Margaret Sanger founds American Birth Control League

1923   Chemist **Louise Stanley** becomes head of the Bureau of Home Economics, the first woman to lead a division at the U.S. Department of Agriculture (USDA)

1925   **Florence Sabin** is the first woman elected to the National Academy of Sciences

1928   Anthropologist **Margaret Mead** publishes *Coming of Age in Samoa*

1935   Irène Joliot-Curie, daughter of Pierre and Marie Curie, shares Nobel Prize in Chemistry with her husband, Frederic Joliot

1937   Mount Holyoke chemistry professor **Emma Carr** is the first recipient of the Garvan Medal of the American Chemical Society, awarded to a woman chemist each year

1940   **Elsie Clews Parsons** named the first female president of the American Anthropological Association

1942 The U.S. government begins secret project known as the Manhattan Project to develop nuclear weapons, employing many female scientists, engineers, and researchers

1943 Committee of the U.S. Food and Nutrition Board (FNB), chaired by nutritionist **Lydia Roberts** and including several other female researchers, publishes new Recommended Dietary Allowances (RDA) guidelines for nutrients and vitamins

1947 Biochemist **Gerty Cori** shares Nobel Prize in Physiology or Medicine with her husband, Carl F. Cori

1948 Electrical engineer and mathematician **Edith Clarke** is first woman elected a fellow of the American Institute of Electrical Engineers

1950 Physician, nutritionist, and anthropologist **Sophie Aberle** is first female member of the National Science Board

   Rachel Brown and **Elizabeth Hazen** develop the antibiotic nystatin

   Beatrice Hicks helps found the Society of Women Engineers (SWE) and serves as first president

1953 Physician **Virginia Apgar** publishes her Apgar scale, which becomes standard test for assessing responses and health of babies at birth

1955 Chemist **Patsy Sherman** is co-inventor of Scotchgard Fabric Protector for 3M

1958 U.S. government creates National Aeronautics and Space Administration (NASA)

1960 The U.S. Food and Drug Administration (FDA) approves a combined hormone oral contraceptive ("the pill")

1962 Environmental biologist **Rachel Carson** publishes the book *Silent Spring*

   Nobel Prize in Physiology or Medicine is awarded to Francis Crick, James Watson, and Maurice Wilkins for the discovery of DNA, research to which British crystallographer Rosalind Franklin also contributed

1963 Russian cosmonaut Valentina Tershkova is the first woman in space

   Maria Goeppert-Mayer is co-recipient of Nobel Prize in Physics

1964 British chemist Dorothy Crowfoot Hodgkin receives Nobel Prize in Chemistry

U.S. Congress passes Civil Rights Act, which includes legislation against sex and race discrimination in employment and federal programs

1965 Chemist **Stephanie Kwolek** develops Kevlar synthetic material for DuPont

Engineer and industrial psychologist **Lillian Gilbreth** is first woman elected to National Academy of Engineering

Endocrinologist **Helen Taussig** named first woman president of the American Heart Association

1966 National Organization for Women (NOW) is founded

1968 Biological and environmental scientists **Anne Ehrlich** and Paul Ehrlich publish the controversial book *The Population Bomb*

1970 Economist **Marina v. N. Whitman** is first woman named to the President's Council of Economic Advisors

1971 **Mina Rees** named first female president of the American Association for the Advancement of Science (AAAS)

1972 Title IX of the Educational Amendments to the Civil Rights Act prohibits discrimination on the basis of sex in federally funded educational programs

1973 The book *Our Bodies, Ourselves* is published by Boston Women's Health Book Collective

1976 **Margaret Burbidge** is named first woman president of the American Astronomical Society

1977 Economist **Juanita Kreps** is named first woman secretary of the U.S. Department of Commerce

Rosalyn Yalow is co-recipient of Nobel Prize in Physiology or Medicine

Psychologist **Carolyn Payton** is named first woman director of the Peace Corps

1978 NASA opens astronaut program to first group of six women

Organic chemist **Anna Harrison** elected the first woman president of the American Chemical Society

1981   The U.S. Centers for Disease Control first identifies the HIV virus that causes AIDS

        NASA launches first space shuttle

1983   **Sally Ride** is first American woman in space

        Julia Robinson is first woman elected president of the American Mathematical Society

        Geneticist **Barbara McClintock** receives Nobel Prize in Physiology or Medicine

1986   Neurologist **Rita Levi-Montalcini** is co-recipient of Nobel Prize in Physiology or Medicine

        Nancy Fitzroy is first female president of American Society of Mechanical Engineers

1987   Anthropologist **Johnnetta Cole** is first black woman president of Spelman College, the United States' oldest historically black college for women

1988   **Gertrude Elion** is co-recipient of Nobel Prize in Physiology or Medicine

1990   Pediatrician **Antonia Novello** is named first female (and first Hispanic) U.S. Surgeon General

        Hubble Space Telescope is launched

1991   Cardiologist **Bernadine Healy** is first woman to head the National Institutes of Health

1992   American Association of University Women (AAUW) publishes report on *How Schools Shortchange Girls*

1993   Economist **Alice Rivlin** named first director of the new Congressional Budget Office

        Aeronautics engineer **Sheila Widnall** named Secretary of the U.S. Air Force, the first woman to lead a branch of the military

        Pediatrician **Joycelyn Elders** is second woman (and first African American) to be named U.S. Surgeon General

        Ms. Foundation begins "Take Our Daughters to Work Day"

1995   German biologist Christiane Nüsslein-Volhard is co-recipient of Nobel Prize in Physiology or Medicine

Theoretical physicist **Shirley Ann Jackson** is first woman to serve as chair of the U.S. Nuclear Regulatory Commission (NRC)

1998 Marine scientist **Rita Colwell** named first female director of the National Science Foundation (NSF)

Jane Henney named first female Commissioner of the U.S. Food and Drug Administration (FDA)

1999 Physicist **Shirley Ann Jackson** becomes first female president of Rensselaer Polytechnic Institute

2000 **Rodica Baranescu** elected first woman president of the Society of Automotive Engineers (SAE)

2001 U.S. National Institutes of Health (NIH) establishes Office of Women to focus medical studies and research specific to women

Biologist **Shirley M. Tilghman** is named first female president of Princeton University

Physicist **Shirley Ann Jackson** is first African American woman to be elected to the National Academy of Engineering

2002 **Peggy Whitson** is first woman commander of the International Space Station

2003 A draft of the full Human Genome Project is completed

2004 Neurobiologist **Susan Hockfield** is named first woman president of the Massachusetts Institute of Technology (MIT)

Biologist **Linda Buck** is co-recipient of the Nobel Prize in Physiology or Medicine

2005 Harvard University President Lawrence Summers delivers controversial remarks at conference on "Diversifying the Science & Engineering Workforce"

2006 Chemical engineer **Alice P. Gast** named first female president of Lehigh University

Human papillomavirus (HPV) vaccine is made available for prevention of cervical cancer

2007 Astrophysicist **France Cordova** is named first female president of Purdue University

2008   French virologist Françoise Barré-Sinoussi is co-recipient of Nobel Prize in Physiology or Medicine

Karen LuJean Nyberg is the fiftieth American woman in space

Oceanographer **Susan Avery** is named first female director of the Woods Hole Oceanographic Institution

2009   Marine ecologist **Jane Lubchenco** is named head of the National Oceanic and Atmospheric Administration (NOAA)

Geophysicist **Marcia McNutt** is named head of the U.S. Geological Survey (USGS)

Biologists **Carol Greider** and **Elizabeth Blackburn** are co-recipients of Nobel Prize in Physiology or Medicine

Elinor Ostrom is co-recipient of Nobel Prize in Economics, the first woman Nobel Laureate in that category

Israeli scientist Ada Yonath is co-recipient of Nobel Prize in Chemistry

Index

About the Author

TIFFANY K. WAYNE, Ph.D., is an independent scholar who resides in Santa Cruz, California. A specialist in U.S. history and women's history, she is a former Affiliated Scholar with the Institute for Research on Women and Gender at Stanford University. Dr. Wayne's previous books include *Woman Thinking: Feminism and Transcendentalism in Nineteenth-Century America*, *Encyclopedia of Transcendentalism*, and *Women's Roles in Nineteenth-Century America*.